God's Philosophers

God's Philosophers

HOW THE MEDIEVAL WORLD
LAID THE FOUNDATIONS
OF MODERN SCIENCE

JAMES HANNAM

ICON BOOKS

Published in the UK in 2009 by
Icon Books Ltd, Omnibus Business Centre,
39–41 North Road, London N7 9DP
email: info@iconbooks.co.uk
www.iconbooks.co.uk

Sold in the UK, Europe, South Africa and Asia
by Faber & Faber Ltd, Bloomsbury House,
74–77 Great Russell Street, London WC1B 3DA

Distributed in the UK, Europe, South Africa and Asia
by TBS Ltd, TBS Distribution Centre, Colchester Road
Frating Green, Colchester CO7 7DW

This edition published in Australia in 2009
by Allen & Unwin Pty Ltd, PO Box 8500,
83 Alexander Street, Crows Nest, NSW 2065

Distributed in Canada by
Penguin Books Canada,
90 Eglinton Avenue East, Suite 700,
Toronto, Ontario M4P 2YE

ISBN: 978-184831-070-4

Typeset in Caslon 540 by Marie Doherty

Printed and bound in the UK by
CPI Mackays, Chatham, ME5 8TD

To Vanessa

CONTENTS

LIST OF ILLUSTRATIONS

❧

Map of medieval Europe

The Truth about Science in the Middle Ages

The most famous remark made by Sir Isaac Newton (1642–1727) was: 'If I have seen a little further then it is by standing on the shoulders of giants.'[1] Most people assume that he meant his scientific achievements were built on the discoveries of his predecessors. In the same letter, he alludes to René Descartes (1596–1650), the French philosopher and mathematician, so presumably he was one of Newton's giants. Few people realise, however, that Newton's aphorism was first coined in the twelfth century by the theologian Bernard of Chartres (who died around 1130).[2] Even fewer are aware that Newton's science also has its roots embedded firmly in the Middle Ages. This book will show just how much of the science and technology that we now take for granted has medieval origins.

The achievements of medieval science are so little known today that it might seem natural to assume that there was no scientific progress at all during the Middle Ages. The period has had a bad press for a long time. Writers use the adjective 'medieval' as a synonym for brutality and uncivilised behaviour. Recently, the word has affixed itself to the Taliban of Afghanistan whom commentators routinely describe as throwbacks to the Middle Ages, if not the Dark Ages. Even historians, who should know better, still seem addicted to the idea that nothing of any consequence occurred between the fall of the Roman Empire and the Renaissance. In 1988, Daniel Boorstin's history of science *The Discoverers* referred to the Middle Ages as 'the great interruption' to mankind's progress. William Manchester, in his 1993 book *A World Lit Only by Fire*, described the period as 'a mélange of incessant warfare, corruption, lawlessness, obsession with strange myths and an almost impenetrable mindlessness'. Charles Freeman wrote in *The Closing of the Western Mind* (2002) that this was a period of 'intellectual stagnation'. He continued, 'It is hard to see how mathematics, science, or their associated disciplines could have made any progress in this atmosphere.'[3]

Closely coupled to the myth that there was no science worth mentioning in the Middle Ages is the belief that the Church held back what meagre advances were made. The idea that there is an inevitable conflict between faith and reason owes much of its force to the work of nineteenth-century propagandists such as the Englishman Thomas Huxley (1825–95) and the American John William Draper (1811–82). Huxley famously declared: 'Extinguished theologians lie about the cradle of every science, as the strangled snakes beside that of Hercules.'[4] Draper was a participant in the notorious debate on evolution between Huxley and the bishop of Oxford, Samuel Wilberforce (1805–73), in 1860, when the question arose of whether Huxley was descended from an ape on his mother's or father's side. Draper wrote the massively influential *History of the Conflict between Religion and Science*, which cemented the conflict hypothesis into the public imagination.

More recently, we have seen a real-life conflict between evolution and creationism. Conservative Christians and Muslims have launched an all-out assault on Darwinism. As this phenomenon shows, it is certainly true that particular religious doctrines can be in conflict with scientific theories. However, it does not follow that such hostility is inevitable. During the Middle Ages, the Catholic Church actively supported a great deal of science, but it also decided that philosophical speculation should not impinge on theology. Ironically, by keeping philosophers focused on nature instead of metaphysics, the limitations set by the Church may even have benefited science in the long term. Furthermore and contrary to popular belief, the Church never

supported the idea that the earth is flat, never banned human dissection, never banned zero and certainly never burnt anyone at the
stake for scientific ideas. The most famous clash between science and
religion was the trial of Galileo Galilei (1564–1642) in 1633. Academic
historians are now convinced that this had as much to do with politics
and the Pope's self-esteem as it did with science. The trial is fully
explained in the last chapter of this book, in which we will also see
how much Galileo himself owed to his medieval predecessors.

The denigration of the Middle Ages began as long ago as the
sixteenth century, when humanists, the intellectual trendsetters of
the time, started to champion classical Greek and Roman literature.
They cast aside medieval scholarship on the grounds that it was convoluted and written in 'barbaric' Latin. So people stopped reading
and studying it. The cudgels were subsequently taken up by English
writers such as Francis Bacon (1561–1626), Thomas Hobbes (1588–
1679) and John Locke (1632–1704). The waters were muddied further by the desire of these Protestant writers not to give an ounce of
credit to Catholics. It suited them to maintain that nothing of value
had been taught at universities before the Reformation. Galileo, who
thanks to his trial before the Inquisition was counted as an honorary Protestant, was about the only Catholic natural philosopher to be
accorded a place in English-language histories of science.

In the eighteenth century, French writers like Voltaire (1694–1778)
joined in the attack. They had their own issues with the Catholic
Church in France, which they derided as reactionary and in cahoots
with the absolutist monarchy. Voltaire and his fellow *philosophes*
lauded progress in reason and science. They needed a narrative to
show that mankind was moving forward, and the story they produced
was intended to show the Church in a bad light. 'Medieval philosophy, bastard daughter of Aristotle's philosophy badly translated and
understood', wrote Voltaire, had 'caused more error for reason and
good education than the Huns and the Vandals.'[5] His contemporary Jean le Rond d'Alembert (1717–83) edited an immense encyclopaedia that became the epitome of the *philosophes'* achievement.
D'Alembert's influential *Preliminary Discourse* to this magnum opus

set out the now traditional story of how scientific progress had been held back by the Church during the Middle Ages. He blamed 'the condition of slavery into which almost all of Europe was plunged and the ravages of superstition which is born of ignorance and spawns it in turn.'[6] But now, D'Alembert said, in his own time rational men could throw off the yoke of religion.

John William Draper and Thomas Huxley introduced this thesis to English readers in the nineteenth century. It was given intellectual respectability through the support of Andrew Dickson White (1832–1918), president of Cornell University. The hordes of footnotes that mill around at the bottom of each page of his book *A History of the Warfare of Science with Theology* give the illusion of meticulous scholarship.[7] But anyone who checks his references will wonder how he could have maintained his opinions if he had read as much as he claimed to have done.

The great weight of the assault on the Middle Ages carried on into the twentieth century. Popular historians based their work on previous popular histories and perpetuated the myth that the period was an interruption to mankind's progress. Television shows by Carl Sagan, James Burke and Jacob Bronowski handed the thesis on to a new generation. Even when someone discovered evidence of reason or progress in the fourteenth and fifteenth centuries, it could easily be labelled 'early-Renaissance' so as to preserve the negative connotations of the adjective 'medieval'.

The fight back began 100 years ago with the work of a French physicist and historian called Pierre Duhem (1861–1916). While researching an unrelated matter, he came across a vast body of unread medieval manuscripts. What Duhem found in these dusty tomes amazed him. He quickly realised that science in the Middle Ages had been sophisticated, highly regarded and essential to later developments. His work was carried forward by the American Lynn Thorndike (1882–1965) and German Anneliese Maier (1905–71), who refined and expanded it. Today, the doyens of medieval science are Edward Grant and David Lindberg. They have now retired, but their students already occupy exalted places in the universities of

North America. As scholars explore more and more manuscripts, they reveal achievements of the natural philosophers of the Middle Ages that are ever more remarkable.

Popular opinion, journalistic cliché and misinformed historians notwithstanding, recent research has shown that the Middle Ages was a period of enormous advances in science, technology and culture. The compass, paper, printing, stirrups and gunpowder all appeared in western Europe between AD500 and AD1500. True, these inventions originated in the Far East, but Europeans developed them to a far higher degree than had been the case elsewhere. The Italian doctor, mathematician and astrologer Jerome Cardan (1501–76) wrote that next to the compass, printed book and cannon, 'the whole of the ancient world has nothing to compare.'[8] A compass allowed Christopher Columbus (1451–1506) to navigate his way across the Atlantic Ocean, sailing far from the sight of land to discover the New World in 1492. The development of printing and paper meant that an incredible 20 million books were produced in the first 50 years after Johann Gutenberg (c.1398–1468) had published his printed Bible in 1455.[9] This dwarfed the literary output of antiquity. Printing probably had an even greater effect than gunpowder which, like the stirrup before it, revolutionised warfare and allowed Europeans to dominate the rest of the world.

Meanwhile, the people of medieval Europe invented spectacles, the mechanical clock, the windmill and the blast furnace by themselves. Lenses and cameras, almost all kinds of machinery and the industrial revolution itself all owe their origins to the forgotten inventors of the Middle Ages. Just because we don't know their names, this does not mean that we should not recognise their achievements.

Most significantly, the Middle Ages laid the foundation for the greatest achievement of western civilisation, modern science. It is simply untrue to say that there was no science before the 'Renaissance'. Once medieval scholars got their hands on the work of the classical Greeks, they developed systems of thought that allowed science to travel far further than it had in the ancient world. Universities, where academic freedom was guarded from royal interference, were first founded in the twelfth century. These institutions have always provided scientific research with a safe home. Even Christian theology turned out to be uniquely suited to encouraging the study of the natural world, because this was believed to be God's creation.

Today, when we talk about 'science', we have in mind a clear and specific meaning. We picture a laboratory where researchers are carrying out experiments. But the word 'science' once had a much broader definition than it does now. The word comes from *scientia*, which means 'knowledge' in Latin. Science encompassed all intellectual disciplines, including politics, theology and philosophy. Theology was, famously, the queen of them all. The study of nature as a separate subject was called 'natural philosophy', and it is this term that will be used throughout this book. One of the essential lessons of history is that if we use our own categories to describe the past we will seriously misjudge it. Instead, it is important to understand where people in the Middle Ages were coming from and to understand them on their own terms. Part of doing this involves looking at subjects that we would consider unscientific today. To medieval people magic, astrology and alchemy were all considered to be 'sciences'. More surprisingly, these arcane disciplines contributed directly to modern science by providing alternative ways of comprehending and manipulating the natural world.

The distinction between medieval natural philosophy and modern science is a subject of some debate among scholars today. However, one difference is immediately clear; modern science is naturalistic with no room for the supernatural. From the beginning of the nineteenth century, science has excluded God from the laboratory. In contrast, for the medieval natural philosopher, God was invariably central to any considerations about nature.

Modern science is a very specific kind of knowledge that blends empirical experimentation with rational analysis. Today we take it for granted and trust it to provide us with accurate information about nature. It is hard to believe that a few centuries ago, this scientific way of thinking hardly existed. Before the edifice of modern science could be built it required the strong foundations that were laid for it in the Middle Ages. The cornerstone was a widespread acceptance of reason as a valid tool for discovering the truth about our world. Clearly, this could not happen without the approval of the Church, which at the time was the guardian of almost all intellectual endeavours. This meant that the development of reason and its relationship with faith are both important parts of our story. So prevalent did rational argument become among philosophers during the Middle Ages that the period deserves to the thought of as the beginning of the 'Age of Reason'.[10]

Some historians of science have had a habit of lauding individuals who seem to echo our own prejudices or appear more 'modern' than their contemporaries. When we hear about someone from the past who anticipated our own beliefs, we tend to label them 'ahead of their time'. In fact, no one is ahead of his or her time. On closer examination, we always find that people are rooted firmly into their own cultural milieu. The best example of this is probably Leonardo da Vinci (1452–1519). A recent biographer, Michael White, even called him 'the first scientist'.[11] But surprisingly, despite being a genius, Leonardo had no impact on the development of western science at all. His influence was entirely in the arts. His lack of focus and constant experimentation prevented him from having as much success even in that field as he could have had. The reason no one followed Leonardo's scientific ideas is that he didn't tell anyone about them. His reputation today as a man of science is based on his famous notebooks, but these did not become known until centuries after his death. His secrecy was nothing to do with fear of prosecution or a belief that the Church would try to curtail his work. It was simply a character flaw that made him refuse to share his insights.[12] He even disguised his notes by using mirror writing to make them illegible unless they were seen reflected in a mirror. Consequently, and despite his enormous reputation, we will hear no more about him in these pages.[13]

Another common mistake is to divide up history into discrete periods and then give them names containing clear value judgements. This can be extremely misleading. For example, we are commonly taught that there was a Renaissance, which was 'a good thing', the Dark Ages, which were 'bad' and the Enlightenment, which was 'very good indeed'. How could anyone disapprove of being enlightened when the alternative, presumably, was to be benighted? Renaissance means 'rebirth', with the clear implication that previously civilisation had been well and truly dead. The term 'Dark Ages' was coined in the fourteenth century by the Italian writer Francisco Petrarch (1304–74). What he meant was that between the ancient world of Rome and his own time, nothing much happened. For 1,000 years, mankind had stood still. As we shall see, the advance of science provides one of the best examples of the injustice of these historical labels. The first appearance of the term 'the Middle Ages', a less pejorative label, was in the fifteenth century when it is used by various Italian humanists.[14]

One might think that the other names we give to historical periods also date back centuries, but in fact they nearly all originated in nineteenth-century France. French historians had a very clear idea that the past was the story of mankind's progress towards their own civilisation, which they regarded as the pinnacle of human progress. The English were just as bad. The Victorians invented a story about the triumph of civilisation through Protestantism, free markets and a benevolent British Empire. They even believed that this triumph had been made possible by frequent victories over the French. If we really are going to understand history, we will have to do away with prejudicial labels like 'the Dark Ages' and 'the Enlightenment', or at least learn to treat them with considerable scepticism.

On the other hand, some of the customary names and adjectives used for historical periods are just too convenient and so we will have to employ a few of them. The dates assigned to each period are, inevitably, rather arbitrary. According to this schema, the early Middle Ages (which used to be called the Dark Ages) extend from the fall of the western Roman Empire in AD476 up until 1066; the Middle Ages proper start at that point and end in 1500 when we enter the early modern period. All dates are AD unless otherwise stated and AD/BC designators are used whenever there might be some confusion.

There is a trend among historians today to replace the old system of AD and BC with CE (for Common Era) and BCE (for Before Common Era) as a non-Christian alternative. That seems right for a history of China or Mesoamerica, but for the European Middle Ages AD and BC remain entirely appropriate.

Briefly stated, the plan of this book is as follows. It tells the story of how natural philosophy in the Middle Ages led to the achievements of modern science. We begin with a review of the early Middle Ages up to AD1000. During this period, western Europe recovered from the collapse of the Roman Empire and began to rebuild itself with the help of several important new inventions. We will see how agriculture improved and how much a well-educated person at the time could expect to know about natural philosophy.

In the third and fourth chapters, we will learn how the West recovered the heritage of ancient Greek learning. This had been lost to Europe when the Roman Empire collapsed, but was regained from Arab and Byzantine sources. This wave of new knowledge inevitably caused concern to the authorities. Chapter 5 tells of how the Christian Church became increasingly concerned about heresy in the twelfth century. However, it eventually came to terms with Greek philosophy. And as we will see in chapter 6, a great deal of debate and argument was resolved by the titanic figure of Saint Thomas Aquinas (1225–74), the greatest scholar of the Middle Ages.

Chapter 7 looks at why, if you fell ill in the Middle Ages, you would be better off praying at a holy shrine than visiting a doctor. Chapter 8 examines two subjects that the Church treated with suspicion but which nonetheless enjoyed great popularity – alchemy and astrology. In chapter 9 we meet Roger Bacon, a dedicated alchemist, who devoted his life to the study of nature because he thought it would be a useful tool for converting Muslims to Christianity before the imminent end of the world. In chapter 10 we meet another less celebrated but no less fascinating Oxford scholar – Richard of Wallingford (1292–1336). Besides his achievements in astronomy, he built one of the finest and most complicated clocks of the Middle Ages, despite suffering from the dreadful affliction of leprosy.

Once Thomas Aquinas had Christianised Greek philosophy, medieval scholars could build on it. Chapters 11 and 12 demonstrate the advances in scientific thought that were made at the universities of Oxford and Paris in the fourteenth century. Two areas saw particular progress – the implications of the earth's rotation and the motion of accelerating objects.

In chapter 13, we will see how new inventions in the late Middle Ages had a profound effect on European society and, thanks to the voyages of Columbus and others, the rest of the world as well. Ascertaining whether or not the earth is flat was the last thing on Columbus's mind.

Chapter 14 examines the impact of humanism and the Protestant Reformation on science and technology. Humanists recovered important ancient Greek mathematical texts but also rejected the advances made in philosophy during the Middle Ages. The Reformation broke the power of the Catholic Church to control science but also made it less tolerant of new ideas.

Although 'the Renaissance' is often associated with the beginning of modernity, it also saw a surge in magical belief that especially affected those who were at the cutting edge of science. Chapter 15 looks at these links. In chapter 16, we will see how human dissection arose in Europe and helped us to understand the machinery of our own bodies.

Chapter 17 relates the story of how Nicolaus Copernicus (1473–1543) decided that the earth orbited the sun, and not the other way around as everybody else thought. He was no isolated genius, though, and owed a great deal to his medieval and Islamic forebears. Chapter 18 shows how Copernicus's radical idea was adapted and proved by Johann Kepler (1571–1630).

The last three chapters look at Galileo and his contemporaries. He too took ideas from earlier thinkers and used them to construct his own theories about matter and motion. Galileo pulled together many of the strands of medieval thought to form the basis of modern science. It is with him that our story concludes.

CHAPTER 1

~∞~

After the Fall of Rome: Progress in the Early Middle Ages

To understand why historians no longer feel comfortable with the term the 'Dark Ages', you only have to visit the British Museum in London to admire the treasure found at Sutton Hoo in Suffolk. Archaeologists discovered the grave of an Anglo-Saxon king there in 1939. It was filled with the most marvellous objects, lying in the rotted hulk of an entire ship that was buried under a mound. The craftsmen of seventh-century East Anglia who produced these stunning artefacts in gold, glass and precious stones were certainly no savages. They used materials from all over Europe to fashion buckles and accoutrements fit for a king. Even with a magnifying glass, it is difficult to see all of the exquisite detail on the jewelled purse lid and shoulder clasps. The silver drinking cups were manufactured in the eastern Mediterranean (although admittedly they are not of the highest quality)[1] and travelled to England along trade routes that probably took English wool and slaves in the opposite direction.

Although invented much earlier, the term 'Dark Ages' became popular in the nineteenth century. It is clear that dismissing half a millennium as being filled with gloom was not intended to flatter the people who lived during it. Some historians explained that by 'Dark', they only meant that relatively few written sources survive for the period compared to those immediately before and afterwards. What they actually meant is that very little of interest happened. Today we have come to realise that we cannot so easily write off the period. Roger Collins, in his standard work *Early Medieval Europe 300–1000*, states: 'The centuries covered by this book constitute a period of the greatest significance for the future development, not only of Europe, but in the longer term, of much else of the world.'[2] So little credence does Dr Collins give to the term 'Dark Ages' that he does not even bother to mention that he refuses to use it.

Ploughs, Horseshoes and Stirrups: New Technology in the Early Middle Ages

The royal treasure buried at Sutton Hoo tells us something about the luxuries available to the Anglo-Saxon elite; however, to learn about the lives of common men and women we have to look elsewhere. Luckily, we have a very good idea about everyday life at the end of the early Middle Ages because of a great administrative project set in motion by William the Conqueror (1028–87). After he had subjugated England following the Battle of Hastings, William wanted to know exactly what resources the country possessed. The resulting census, the *Domesday Book*, gives us a fantastic opportunity to step back in time and see the world through medieval eyes. Near the start of the *Domesday Book* is a short entry for Otham in Kent. This village, where much of the present book was written, lies on the southern bank of the River Len, just to the east of the county town of Maidstone. It contains more than its fair share of grand medieval manors, because the local ragstone quarry provided employment for stonemasons who could afford big timber-framed houses. The stone had been extracted since Roman times and is rumoured to have been used to build the ancient walls of London itself. However, by the time William the Conqueror's agents arrived to compile the *Domesday Book*, the quarry was silent and they don't mention it.

Anglo-Saxons preferred to build with wood, largely because trees were so plentiful. Many more ancient stone buildings do survive around the Mediterranean than in England, but this is more to do with the comparative lack of wood suitable for large-scale construction. In Japan, stone was rarely used right up until the nineteenth century.[3] Anglo-Saxons reserved stone for when they wanted to make a big impression, which was usually when they were erecting a cathedral. Otham had no stone buildings. Even the surviving houses of the stonemasons are largely made of wood.

The *Domesday Book* entry for Otham, expanded from the rather terse original, reads:

> Geoffrey of Rots holds Otham from the bishop of Bayeux. It contains three hides of land. There is land enough to provide work for two and half ploughs of which the Lord holds land for one. Nine villagers and three smallholders share one plough and the Lord has another. There are a church, two slaves, a mill generating five shillings a year, meadow of three acres and woodland supporting eight pigs.[4]

The lord of the manor in 1086, Geoffrey of Rots, was a knight from a small village near the city of Caen in Normandy. Nine villagers and their families lived in Otham together with three smallholding farmers. There was a church, which was certainly wooden too. The present church, set across the fields from the rest of the village, dates from the thirteenth century but may well occupy the same site as the previous Anglo-Saxon one. There were three acres of meadow where the villagers' cattle would graze, and enough woodland for eight foraging pigs.

So far, the scene doesn't sound all that different from how we might expect a Roman village to appear. But the entry in the *Domesday Book* contains two details that tell us we are dealing with a medieval settlement. The first of these is the mention of ploughs. From almost the dawn of agriculture, peasants had tilled their land with nothing more than a metal-tipped wooden spike, perhaps pulled by an ox, that gouged a furrow out of the ground. Then, in the tenth century AD, another method of tilling the soil arrived in England from the continent.[5] A team of eight oxen, yoked two abreast and pulling a heavy iron plough, now worked the fields of Otham. The new machine had a blade that cut into the earth; a ploughshare that dug in at right angles and a mouldboard behind that actually turned the soil over as it went. This had many beneficial effects. Turning over the soil buried any weeds growing in the field so that they died and improved the soil's fertility. It also increased the amount of water that the ground could hold. Finally, it was much more efficient to operate than the old scratch plough because it attacked a larger cross-section of soil.[6] However, the new plough was large and a single peasant could hardly afford one. In fact, the *Domesday Book* makes it clear that the peasant families of Otham had just one plough between them, with another belonging to the Lord of the Manor. Such was its effectiveness, however, that one plough was all the peasants needed to till their land.

Also on show in the fields of Otham were Geoffrey of Rots' horses. One of the factors behind the Norman victory at the Battle of Hastings had been the superior military technology at their command. The invention that most transformed warfare in the early Middle Ages was the humble stirrup. Ancient horsemen had clung to their mounts with their knees or had had the help of high pommels on their saddles to steady them. Without foot supports, a horseman was quite unable to bring his horse's weight to bear through the weaponry that he was carrying because he was always at risk of slipping off the side of his mount. The only strength that could go into the delivery of either a spear or a sword thrust was his own. And fighting with a sword was risky because if the rider should miss his opponent he would have the grip of his knees or saddle horns to prevent himself from becoming overbalanced and falling off. In battle, falling off was worth avoiding at all costs. As a result, ancient cavalry enjoyed the advantages of speed and manoeuvrability that made them good skirmishers, but were less useful as shock troops. The stirrup changed all that. Now, a horseman could sit firmly supported in his saddle both laterally and dorsally. He was able to move as one with his mount and bring its momentum into play. He could brace his spear against his side and transmit the full force of his charging horse into the enemy, transforming the cavalryman into the mounted knight.[7]

The Normans were among those who took full advantage of these developments. When Geoffrey of Rots went into battle, he rode a horse and carried a lance. The Anglo-Saxons ranged against him at Hastings fought on foot and with battleaxes. Saxon swords were actually marvels of metallurgy, but they were extremely expensive compared to an axe so only the richest nobles could afford them. That is not to say that the Saxon battleaxe, combined with their use of a wall of shields to defend themselves, was not a formidable weapon. But once the shield-wall was breached, as happened at Hastings when the Saxons pursued the retreating Normans, they were no match for the mounted knights.

Horses were just as valuable during peacetime as they were in a war. In eleventh-century Otham, oxen still made up the plough teams, but these were not the ideal draft animal. Even given their more expensive upkeep costs, horses were better at ploughing because they were able to pull faster. However, to be effective, the horse needed a harness that allowed it to use all of its strength. The new horse collar, developed at some point after AD700, was a huge improvement on previous harnesses that had tended to put pressure on the windpipe if the animal tried to pull with anything like its full might. Roman law had restricted the load to which a horse could be attached to 1,100 pounds, about half what they are capable of hauling, in all likelihood to protect the animal from exhaustion.[8] From the eleventh century, haulage became yet more efficient as the 'whippletree' began to make an appearance. This oddly named device was just a log chained horizontally in front of a plough or cart. The draft animals themselves were harnessed to the whippletree rather than directly to the load. Using the log equalised the force from the horses or oxen, so that turning was more efficient and animals of different strengths could be harnessed together.[9]

The iron horseshoe also added to the effectiveness of the beast. In wet terrain, unprotected hooves could be quick to rot and the shoe increased their durability. Taken together, their enhanced effectiveness as both a weapon of war and a draft animal made horses increasingly indispensable as the Middle Ages wore on.

The fields in which the villagers grew their crops would also have looked different from how they had in Roman times. Much of the change was due to the introduction of three-field crop rotation. Farmers had long been aware of the importance of rotating their crops – allowing some fields to lie fallow as pasture while varying the crops planted in others. This idea advanced further in the early Middle Ages when three-field rotation began to appear. Under this system, the fields of Otham were split into three groups. The first group lay fallow as pasture for the villagers' animals, especially the team of plough oxen whose manure added to the richness of the field. The villagers planted the second group in the autumn with grain as they had done since time immemorial. However, they also planted beans in the spring in the third group of fields, which further improved the soil and provided a broader diet.[10] Beans, we now know, take nitrogen out of the air and bacteria in their roots turn it into natural fertiliser. Today in Otham, beans are still planted purely to improve the productivity of the soil.

Increased volumes of agricultural produce drove the need for new technology to process it all. This is the relevance of the second revealing detail in Otham's entry in the *Domesday Book* – the presence of a mill. Wheat and barley were no longer ground by animal or manpower, but with the aid of a watermill. The River Len in Otham provides the perfect location for one of these. It is too narrow to be navigable, but powerful enough to turn the twelve waterwheels that once lined its banks. The *Domesday Book* states that the Otham mill generated an income of five shillings a year, which made it medium-sized. Watermills had existed in the ancient world, but the Romans did not adopt them in large numbers until the end of the Empire. In the early Middle Ages, they became increasingly common and the *Domesday Book* lists 5,624.[11] Tidal mills were adopted on suitable estuaries, where a dam harnessed the high tide and released it through a channel containing a watermill. Finally, the first recorded European windmills appeared in Normandy and East Anglia during the twelfth century and they quickly spread all over those parts of northern Europe where rivers suitable for watermills were not available.[12]

Taken together, these improvements in agriculture led to a population explosion because better farming techniques meant that the same acreage could yield more food and support more people than before. Estimates for the population of France and the Low Countries rise from 3 million in AD650 to 19 million just before the arrival of the Black Death in AD1347. For the British Isles, the equivalent figures are 500,000 people and 5 million. In Europe as a whole, the population increased from less than 20 million to almost 75 million. These figures are of course estimates, if not guesstimates, but the upward trend is clear. For comparison, at the height of the Roman Empire about 33 million people lived in Europe. Well before AD1000, the population far exceeded what it was when the continent had been ruled by Rome, and remained above that level even after the Black Death had killed a third of the inhabitants of Europe in the fourteenth century.[13]

In his study of early-medieval technology, the great American historian Lynn White Junior (1907–87) concluded that the period 'marks a steady and uninterrupted advance over the Roman Empire'.[14] The popular impression that the early Middle Ages represented a hiatus in progress is the opposite of the truth. Even so, the fall of Rome and the replacement of the imperial administration with a patchwork of barbarian kingdoms meant that this was a very unsettled period to live through. In order to shed some light on these times, it will be helpful to summarise events in Europe between the fifth and eleventh centuries.

The Fall of Rome and the Rise of Islam

The Roman Empire had ruled much of Europe until the beginning of the fifth century AD. Beyond its frontiers, in modern Germany, barbarian tribes gathered and looked on the Empire with resentful eyes. When the Rhine froze over in AD406, they poured across the river and spread throughout the vast territory of the Empire. In 410, a barbarian tribe called the Goths sacked Rome, the first time it had fallen to a foreign army in seven centuries. This event caused deep shock as the news reverberated around the Empire. Although the state religion was Christianity, there were still plenty of pagans, especially among the noble families of Rome. They blamed the abandonment of the old religion for provoking the gods into inflicting this unprecedented disaster.[15] It would not be the last time Rome was to fall. After narrowly avoiding the attentions of Attila the Hun (406–453) the city was sacked again in 455, this time by the Vandal tribe. The Goths had at least respected the sanctuary of the city's churches, but the Vandals showed no such restraint and caused even greater devastation. This is the reason that the Vandals have given their name to anyone causing needless damage. By the late fifth century, the Roman Empire in the West was no more. The traditional date for the final fall is 476 when the last emperor abdicated.

Despite this disaster, it was by no means the end of the Empire. As well as western Europe, it had straddled a huge sweep of land from Egypt, around the Levant, through Asia Minor and thence to the Balkans. These provinces remained firmly in Roman hands. The Emperor ruled from the city of Constantinople, the capital of the Byzantine Empire, which was so called after 'Byzantium', the old Greek name for Constantinople. The Emperor Constantine (c.AD272–AD337) had re-founded the city in 330 to be his capital instead of Rome. This was a remarkably prescient act given the fate of Rome a century later, and it meant that the Empire's centre of gravity swung to the East. Constantine's other great claim to fame is that he made Christianity the official religion of the Empire, although he did not altogether outlaw paganism. His successors were less tolerant of the old religion, and by 400 most forms of pagan practice were illegal even if actually being a pagan was not.[16]

One of the most distinctive features of the Byzantine Empire, as opposed to the old Roman Empire, was that the Byzantines were predominantly Greeks. Although they continued to call themselves Romans, the use of Latin died out and Greek became the dominant language. In the ruins of the western half of the Empire, knowledge of Greek practically ceased to exist when the barbarian tribes took over or destroyed the Roman infrastructure. Previously, the best-educated Romans would have been fluent in Greek as well as their native Latin. On the other hand, the barbarians all spoke their own tongues and although Latin did survive as the language of the Church, a linguistic divide opened up between East and West. This was doubly unfortunate because, in the ancient world, it was Greek and not Latin that was the language of scholarship and philosophy. Suddenly, the West lost access to this tradition.

Western Europe was also cut off from imperial influence, as the Byzantine Emperors' rule did not extend into the new kingdoms of the Saxons, Goths and Vandals. In this power vacuum, the local rulers took temporal control while the Church exercised supranational spiritual authority. The bishop of Rome had long maintained that he was at least first among equals with regard to the other Christian patriarchs, but this was largely irrelevant when real power, both secular and ecclesiastical, was in the hands of the Emperor. Then things changed. The retreat of the Emperors' power to the eastern Mediterranean gave the bishop of Rome a free rein in the West. Only now did he become the Pope in the sense that we understand today as unquestioned head of the Catholic Church. The Popes set about organising the evangelisation of the barbarian kingdoms, and slowly Catholic missionaries converted them all to Christianity.

As the conversion of the barbarians gathered pace, the eastern half of the Empire faced a threat that sought to conquer it in the name of a new religion. Arab invaders from the East would certainly not convert to Christianity if they succeeded in occupying the territory of the Empire, because they possessed a vigorous faith of their own – Islam.

Islam owes its origin to the Prophet Mohammed (c.571–632), a native of the Arabian city of Mecca, who claimed that the angel Gabriel had visited him and dictated the text of the Holy Koran. At this time, neither of the great empires to the north, Byzantium and Persia (modern Iran), had annexed the deserts of the Arabian peninsula. They remained in the hands of nomadic tribes who made their living by trade and banditry. At the time that Islam was founded, Mecca had an important marketplace where traders could meet under truce and do deals before retiring back to the desert. The tribes were a nuisance to the settled people, especially when they raided caravans, but conquering the vast wilderness was out of the question. Most of the tribes still worshipped their traditional gods and Mecca also served as a place of pilgrimage for them. It was home to a holy rock called the Black Stone of the Kaaba. This had been held as sacred since before recorded history and it is still an object of veneration by Muslims after Mohammed co-opted it for Islam.

The Prophet began preaching the new religion in Mecca but his fellow countrymen rejected his message, forcing him to flee north to Medina, another desert trading centre. This journey in 622 marks the beginning of the Muslim calendar. In Medina, Mohammed met with much greater success and was eventually able to return to his home town of Mecca as a conqueror. Then, he turned his attention to the rest of the desert people, united them under the banner of Islam and formed an army capable of conquering much of the known world.

Beginning in the mid-seventh century, Muslim armies marched out of Arabia and rapidly conquered Persia and a large part of the Byzantine Empire. It was the perfect time for the Arabs to launch an attack. Persia and Byzantium had been deadly rivals for centuries but neither could ever defeat the other. This changed in the early seventh century when the Byzantine Empire finally smashed the power of Persia in a series of wars ending in 628. This was a great victory but after fighting each other to a standstill, neither of the empires was in a fit state to resist the Muslims' advance. They annexed the entire Persian Empire and Byzantium fared little better, losing Palestine, Egypt and Syria in quick succession.

Islam's success was partly down to its simplicity. It eschewed the complicated legal codes of Judaism (although it would later develop a legal system) and the rarefied theology of Christianity. Instead, Mohammed proposed a basic five-point plan for getting to heaven known as the five pillars of Islam.

First, Muslims must reject all Gods except Allah and accept Mohammed as his final, definitive prophet. Christians and Jews, as the recipients of older and incomplete revelations, could opt out from this requirement but paganism was beyond the pale. We often hear of Islam's relative tolerance in that it accepted Christianity as a flawed but legal faith, whereas Christians considered Muslims to be infidels. In fact, this is more a matter of chronology than of tolerance. It is similar to Christians' ill-tempered acceptance of Judaism as a faith that pre-dated their own. Like Jews in medieval Europe, Christians living under early Islamic rule were very much second-class citizens.

The second pillar of Islam is prayer five times a day. At the call of the muezzin from the mosque's minarets, the faithful either assemble at the mosque or else unroll a prayer mat where they are. Muslims face Mecca while they pray. As Islam spread from the Atlantic to India through the eighth century, it became more difficult to determine in which direction Mecca lay. Scholars had to study the position of the stars to ensure that it was properly calculated and this helped to stimulate astronomy and trigonometry.

The third pillar is giving alms to the poor; the fourth, fasting during daylight hours in the holy month of Ramadan. Because the Muslim calendar follows the moon's orbital cycle rather than the sun's, its liturgical year is only 355 days. This means that Ramadan is a little earlier each year and moves through the seasons. The final pillar of Islam is the Hajj, which is a pilgrimage to Mecca, ideally undertaken once in the lifetime of every Muslim.

It would take another book to do justice to the great advances achieved by Arabs in the fields of mathematics, medicine and philosophy, let alone art and literature. However, it is essential to give a brief overview of this legacy because the inheritors of the Islamic tradition in science were western Christians. We saw earlier how a lack of knowledge of Greek cut off the West from much of its classical heritage. The Arabs did not have this problem because they had conquered a large number of Greek-speakers. They were also able to call upon the services of Syrian Christians who spoke a language called Syriac, which is related to Arabic. To take advantage of this, the Caliph, ruler of all Muslims and successor of Mohammed, founded a school in Baghdad called the 'House of Wisdom' where the cream of Greek science and philosophy was translated into Arabic. Scholars spread these works through the Islamic Empire, including Spain where western Christians first came across them.

However, it would be quite wrong to say that Muslims acted only as a conduit through which ancient learning could reach the West. The Byzantines independently preserved almost all of the most important surviving scientific texts in the original Greek, and few of them would have been lost without the Arab scribes.[17] Rather, the importance of Muslim science lies in the innovative works of philosophy, mathematics and medicine that the Islamic world produced. The Arabic origin of mathematical terms such as *algebra* and *algorithm* are further indications of how much we owe to the Islamic Empire.[18]

After conquering much of the Byzantine Empire, Muslim armies carried on westwards along the North African coast, taking Carthage in modern Tunisia on their way to Morocco. When they had reached the Atlantic Ocean, they turned north and crossed the Straits of Gibraltar into Spain in 710. Within two decades they had conquered the Christian kingdoms of the Iberian peninsula, leaving only a strip of land along the northern coast in Christian hands. Finally, the Muslim armies traversed the Pyrenees and invaded France. Here, at last, they met their match in Charles (688–741), chief adviser to the king of France, at the Battle of Poitiers in October 732. His army formed a shield wall 'holding together like a glacier' which the enemy could not penetrate. Abdurrahman, the Muslims' general, died in the battle and his army slipped away under cover of darkness to Spain, never to return in such numbers.[19] Although Muslims at the time saw the defeat as merely a temporary setback, they never again seriously threatened France and Christian Europe was secure. Edward Gibbon (1737–94), the irreligious English historian, reflecting on the military conquests made by the Arabs, considered what might have happened if Charles had lost the battle. 'Perhaps', mused Gibbon mischievously, 'the interpretation of the Koran would now be taught at the schools of Oxford and her pulpits might demonstrate to a circumcised people the sanctity and truth of the revelation of Mohammed.'[20]

The Foundation of the New Roman Empire

Charles was awarded the epithet of *Martel*, or 'Hammer', for defeating the Muslim invaders. Following his victory, he saw his power wax until he was effectively king of France rather than just the power behind the throne of a puppet monarch. Under him and his successors, France rapidly became the major power of western Europe. The most famous scion of the dynasty was his grandson Charlemagne (742–814), who expanded the territory under his control to include much of Italy and Germany. 'Charlemagne' simply means Charles the Great, and he is probably the only ruler actually to have greatness incorporated into his name. For Charlemagne, being king was not enough. He wanted more and at Rome on Christmas Day 800, the Pope crowned him Emperor. Allegedly, Charlemagne was unaware that this was about to happen, but we should take such pious anecdotes with a pinch of salt. Charlemagne also required a fine capital and he built his at Aix-la-Chapelle (or Aachen) in the Rhineland. His octagonal stone cathedral, still standing today, dominates the town and serves as the Emperor's mausoleum.

Charlemagne is significant partly because he was a strong ruler who was able to control enough resources to fund a cultural revival, usually called the 'Carolingian Renaissance'. Charlemagne himself was barely literate but he appointed the celebrated scholar Alcuin of York (c.735–804) to help foster learning in his capital and at other centres in his enormous Empire. Today we still have reason to be grateful for these efforts, as many works of classical Latin literature have come down to us because of them. Often the oldest manuscripts date from this period, recognisable from their distinctive *caroline miniscule* handwriting.[21]

Charlemagne also ordered that schools be set up at the cathedrals in his realm to ensure that there would be enough literate people to administer his Empire.[22] Many of those who attended these schools went on to become clerics, but this was by no means compulsory.

Merchants, lawyers and physicians could all expect to begin their education at the feet of a master appointed by the cathedral chapter.

Today Charlemagne is criticised for being so aggressive in spreading Christianity. His forced conversion of the newly conquered Saxons, his merciless treatment of prisoners and his acceptance of a role as the Popes' enforcer all strike us as unchristian behaviour. However, the conversion of disparate tribes to a single religion brought them all together into a single spiritual unit. As a result the Church could, to some extent, enforce its prohibition against fighting between Christians and insist that their martial energies were directed externally.

On the Emperor's death in 814, his sons divided his vast realm and then rapidly fell out with each other. The Empire had dissolved within two generations. The next dynasty to stand supreme in western Europe arose from the Saxons of Germany (not to be confused with the Saxons of England) whom Charlemagne had conquered and converted to Christianity. The historic divide between France and Germany, united under Charlemagne, dates from this period and specifically the refusal of the French monarchs to buckle under the Saxon yoke. The Holy Roman Empire, founded by the Saxon monarchs, included Germany, much of central Europe and Italy but never France. The French maintained their independence and prevented Europe ever again becoming a single political as well as religious unit. The first of the new Saxon Emperors was Otto I (912–973), crowned in 962. He was succeeded by his son Otto II (955–983) who was, in turn, followed by his son Otto III (980–1002). It is not surprising that historians call this period the Ottonian Age.[23]

By the time Otto III took the imperial throne, the barbarian invaders of the Roman Empire had coalesced into kingdoms and converted to Christianity. Agricultural production was being driven by improved technology and the population was expanding rapidly. western Europe was still a backward corner of the world, but it was well on the way to catching up.

CHAPTER 2

✎

The Mathematical Pope

In the year 999, on Palm Sunday, Gerbert of Aurillac (c.940–1003), the most learned man in Europe, was crowned Pope.[1] As the papal tiara was placed on his head, Gerbert's elevation to such a height from his humble beginnings in rural poverty must have seemed miraculous. During the Middle Ages, the princes of the Church tended to be related to the princes of the state, and there would not be another lower-class pope until the thirteenth century.[2] Gerbert's coronation took place in the Basilica of St John Lateran, the cathedral of Rome, which lies on the other side of the city from the Vatican. The building that stands on the site today is largely a seventeenth century creation, but we can have a good idea of what the old basilica looked like from studying ancient churches that have survived.[3] At the end of the first millennium, the interior would have been a long box-like space called the nave, with stone walls and a wooden ceiling. The walls would have been supported by a row of pillars which allowed access to the aisles that ran down each side. At the far end of the nave was the apse, a semi-circular alcove before which stood the altar. The roof of the apse was a half dome, providing an artistic space that could dominate the whole length of the basilica. Typically, a golden mosaic of Christ triumphant stared out and would have made it clear who was the true master of the building. Along the walls of the nave, frescos of the saints would make the building a riot of colour. Just below the roof, a line of windows would have let in light and ensured that the building was not too gloomy despite the clouds of incense.

Packed into this richly decorated hall on that coronation day would have been a crowd of people as finely arrayed as the church itself. The clergy would be distinguished by their tonsured heads – the ceremonial shaving that marked them out as men of God. Any monks present would be plainly attired but most priests would wear their most splendid vestments. Foremost among them were the cardinals who represented the parish churches of Rome. They did not yet have the power to elect the Pope. That was the preserve of the noble families of the city who fought among themselves for control of the papacy. At the coronation, the members of each family would have sought to surpass their rivals with the splendour of their jewels and robes. On that day, however, they would have had to stand aside for a still greater power because the Holy Roman Emperor was in town. In truth, Gerbert owed his elevation to his patron and pupil Emperor Otto III, who was still only in his teens. Otto had seized control of Rome and, when the previous Pope had died, ensured the appointment of Gerbert in his stead. The new Pope knew where he stood. He took the pontifical name Sylvester II because the first Pope Sylvester had been a councillor to the Emperor Constantine.[4] Thus, Gerbert's ascent of the throne of St Peter in St John Lateran was supposed to signal a new partnership between Church and state. It never happened. Within four years, both Emperor and Pope were dead.

The Career of Gerbert of Aurillac

We know a good deal about Gerbert's life because one of his pupils, a monk called Richer of Saint Remi (who died around 998), wrote a history of France giving plenty of attention to Gerbert's career.[5] We also have a collection of letters that Gerbert wrote, although most of these relate to church business and not to his personal life.[6] He was born near Aurillac in south central France and entered the local monastery at an early age. Initially, he would not have been one of the monks – his origins were too humble for that. Monasteries had two ranks of membership; the monks who spent their time praying came from well-to-do or even noble families. Below them were the lay brothers who carried out the day-to-day housework, farming and manual labour. However, Gerbert had great ability, which his superiors must have decided to nurture. The abbot allowed him to profess as a monk and devote himself to scholarship. Gerbert was always grateful for the opportunity that his monastic teacher afforded him and in later life wrote that to him 'I owe everything.'[7]

In around 967, a Spanish nobleman visited the monastery and was struck by Gerbert's talent. He persuaded the abbot to let him take the young monk back home with him to Barcelona, probably to act as his secretary. At this time, Barcelona was on the border between Christian and Muslim Spain. Living there presented Gerbert with an excellent opportunity to learn the latest mathematics and philosophy from the Arabs. He may even have travelled to Islamic Seville and been taught by Muslim scholars.

After two years in Spain, Gerbert and his noble patron travelled to Rome on a pilgrimage where the young scholar was introduced to the two most powerful men in western Europe, Pope John XIII (d.972) and Emperor Otto I. Otto had asked the Pope to keep an eye out for anyone skilled in mathematics because he was keen to encourage its study in his court.

Gerbert must have made a very favourable impression on the Emperor because, from then on, the imperial family promoted him incessantly. First, he became tutor to the Emperor's son. Then he was sent to complete his education at the cathedral school of Reims in northern France. This was one of the schools set up at the behest of Charlemagne to improve literacy. While there, Gerbert introduced some of the knowledge of the Arabs to a Christian audience. His surviving letters show that he considered arithmetic to be a useful skill; in a series of letters to a monk called Constantine of Fleury, he patiently explained the rudiments. Richer mentions how Gerbert used a musical instrument called a monochord to teach harmonics to his students. He also gave instructions on how to make astronomical apparatus and frequently mentioned his excitement at finding lost manuscripts. His greatest claim to fame is that he helped to introduce Arabic numerals into the West. He incorporated them into the abacus, which was used for almost all calculations in his day, to produce a more efficient instrument. His modified abacus used beads with numbers inscribed on them, rather than having each of the beads represent a single unit.[8]

We know that Gerbert was extremely knowledgeable about astronomy because he spent much of his time building models of the universe. Richer gives us details of several spheres that his master constructed, showing the locations of the stars and different sectors of the sky. These were assembled from horsehide stretched over a wooden frame that was then painted according to the requirements of the customer. Gerbert didn't take money for his work, but rather manuscripts. In 988, he wrote to one Remi of Trier offering him a celestial sphere in exchange for a copy of an epic poem by the Roman author Statius. Remi was clearly intrigued at the prospect of owning such an unusual object and duly wrote out the manuscript himself. Unfortunately, the version he copied from was incomplete, much to Gerbert's annoyance.[9]

Being interested in maths at this time could be a frustrating experience. Even those who knew enough to ask the important questions rarely had access to the right answers. Two near contemporaries of Gerbert, Ragimbold of Cologne and Radolf of Liège, liked to set each other geometrical problems and then circulate their discussion to anyone else they thought might be interested. What comes out of their correspondence is not only how little information these two had, but how passionate they were about what they did know. They were completely confused about the angles that make up a triangle, which as the ancient Greeks knew must always add up to 180°. Then, they could not make sense of the square root of two, which is a number that cannot be expressed as a fraction. Radolf got hold of a new astronomical instrument called an astrolabe, probably introduced to Europe from Arab Spain by Gerbert. An astrolabe should allow its user to tell the time from the positions of the stars or planets as well as carry out astronomical observations. Radolf was beside himself with excitement. He desperately wanted to show the instrument to Ragimbold but refused to part with it, so insisted that his friend came to visit him. Judging by the rest of their correspondence neither of them could have understood how an astrolabe works, but that failed to detract from their boyish enthusiasm for the new toy.[10] Thankfully, it did not take long for knowledge of the instrument to spread, at least in part due to an astrolabe instruction manual attributed to Gerbert himself. In 1092 a monk called Walcher, from Great Malvern in England, used his astrolabe to carry out an observation of a lunar eclipse that enabled him to reconstruct the entire lunar calendar.[11]

1. A later Islamic astrolabe of similar design to those which Gerbert would have seen in Spain

Meanwhile, as Gerbert's fame spread, so did speculation about where he had gained his great learning. He acquired an undeserved but unshakeable reputation as a sorcerer. Some held that Arabian alchemists had taught him forbidden arts while he lived in Spain. The English chronicler William of Malmesbury (who died around 1143) claimed that his knowledge came from a magical head made of brass. This powerful artefact supposedly spoke to Gerbert and revealed the secrets of nature to him.[12] His ability as a teacher and closeness to the imperial family also attracted jealousy from his colleagues. Richer wrote that the headmaster of the palace school in Magdeburg in Germany accused Gerbert of promoting mathematics at the expense of philosophy. Gerbert was not impressed. 'Do not let some half-educated sophist let you think that arithmetic is contrary to the liberal arts or philosophy', he wrote to a student.[13] However, the impudent headmaster would not relent and Otto II, who had by then succeeded to his father's throne, summoned both protagonists to Ravenna in Italy to debate the point. Gerbert, of course, obliterated the arguments of his opponent, and the emperor awarded him with the abbacy of the famous and venerable monastery of Bobbio in the Apennine Mountains. He also raised his former tutor to the rank of count.

Sadly, Gerbert's appointment was not a success. Bobbio had fallen on hard times under its previous abbots. The treasury was empty and the monks dissolute. He tried to turn things around by improving discipline but the monks defeated his every move. They were not going to take orders from a lowborn abbot imposed on them by the Emperor. Besides Gerbert was a scholar, not an administrator. Eventually, things got so bad that he fled back to Reims and a position as secretary to the archbishop there.

As the right arm of his master, Gerbert effectively ran much of the Reims diocese's business himself. Many of his letters were written on behalf of the archbishop. Such was Gerbert's reputation for both scholarship and piety that plenty of people wanted to see him become the archbishop himself when the incumbent died. Unfortunately, politics got in the way and in 995 Gerbert gave up the fight. Instead, he travelled back to Italy to join Otto III who had ascended the throne aged only thirteen. Gerbert travelled with the imperial court to take up the position of adviser and teacher to the young monarch. In return, Otto appointed him archbishop of Ravenna and then Pope.

He was not, it has to be said, one of the great popes. He was probably too intelligent and unworldly to be much of a politician, and he owed his position so plainly to the Emperor that he never had much of a power base in his own right. When the people of Rome rebelled, as they frequently did, he was forced to flee the city and only made it back shortly before his death in 1003. On the other hand, several of the popes who had reigned shortly before Gerbert had been so awful, sunk into pits of corruption and sexual depravity, that the entire period was later dubbed the 'pornocracy' – literally the 'rule of the prostitutes'.[14] Compared to such popes a scholar, whatever his imperial connections, was a vast improvement. The young Emperor Otto III had died in 1002 at the age of 22, his dreams of rebuilding the Roman Empire unfulfilled.

Gerbert's Knowledge of Ancient Philosophy

Although Gerbert knew more about science than any other Catholic in his day, he was still well behind the achievements of the ancient world. The split between the barbarian kingdoms of the West and Greek-speaking Byzantium in the East did not help either. However, before the final collapse of the western Roman Empire, a few Christian scholars had written important books that they passed on to their Latin-reading successors. Foremost among them was Saint Augustine (AD354–430), the bishop of Hippo whose magisterial *City of God* and autobiographical *Confessions* remain classics to this day. The philosophy of Plato (427–347BC) was the dominant school of thought when Augustine was writing and it did much to inform his own ideas. Although the most important neo-Platonists were pagans, it was easy for Augustine to adapt Platonism's mystical and supernatural tenets into something compatible with Christianity. The resulting synthesis became one of the major sources of Christian theology throughout the Middle Ages, while the original pagan sources remained unknown.

As well as the works of early Christians, Gerbert also had access to a few scraps of Greek philosophy that had been translated into Latin before all knowledge of the former language was lost. These included half of *Timaeus*, a dialogue by Plato about the creation of the world. The largest surviving fragments, however, were the complicated treatises on logic by Plato's greatest pupil, Aristotle (384–332BC).

The man who had carried out the task of putting Aristotle's complicated Greek philosophy into Latin was Anicius Manlius Severinus Boethius (AD480–525), a Christian and Roman aristocrat (hence the long name) who lived through the final collapse of classical civilisation in the West. He was orphaned before he was ten but raised by another upper-class family who ensured that he achieved an exceptional level of education. He could read Greek fluently at a time when such knowledge was already rare in the West and he was familiar with all the major works of ancient philosophy. Like a true Roman, Boethius combined his private scholarship with a career in public service. He served as consul in 510 and steered his sons in the same direction so that they too became consuls. In fact, both offspring achieved that honour in the same year, 522.

By this time, of course, Rome no longer ruled itself. It was under the sway of a barbarian king called Theodoric (454–526) whose warriors had conquered much of Italy in the 490s. Boethius saw it as his duty to serve this ruler in order to keep the old ways alive. For a while, the plan worked and Theodoric handed much of the administration of his kingdom over to the old Roman aristocracy. Boethius was the master of offices, effectively prime minister. Such accommodation with the new order was not universally popular and many Romans looked to Constantinople, where an Emperor still ruled, as their true capital city. Unfortunately for Boethius, one of his friends became embroiled in a plot to overthrow Theodoric and restore Roman rule. Being an honourable man, Boethius came to his friend's defence, but this only meant that he fell under suspicion himself. He was imprisoned in Pavia and sentenced to a brutal death.[15]

While in prison awaiting his fate, he wrote his famous treatise *The Consolation of Philosophy*. It is impossible to overstate the popularity and influence of this book. Originally written in Latin, even kings and queens translated it into their own languages – both Alfred the Great (849–899) and Elizabeth I (1533–1603) tried their hand at rendering it into English. The *Consolation* takes the form of a dialogue between Boethius, sulking in his cell after the bottom has fallen out of his world, and Lady Philosophy who tries to cheer him up. There is nothing explicitly Christian in the *Consolation* but plenty of stoicism of the sort popular with Roman aristocrats. Neither does it say anything incompatible with Christianity. Boethius asks the same big questions that have troubled thinkers throughout the ages: 'Why do bad things happen to good people?', 'Why bother being virtuous?' and 'How can I have free will if God knows what I am going to do?' Lady Philosophy assures Boethius that goodness is its own reward and that the evildoer is really only hurting himself. Once he accepts that wealth, power and status are meaningless, evil men can do him no harm. As for free will, although God knows the future, he does not make it happen. If God does not cause us to act, then his mere knowledge cannot impinge upon our freedom. These answers do not satisfy everyone, but they have provided succour and comfort to readers for 1,500 years.

Even before his fall from grace, Boethius recognised that the classical world was ending and that few of his descendants would have direct access to Greek scholarship. To remedy this, he set out to provide textbooks and translations into Latin that would form a core syllabus for future students. Initially he wrote short treatises based on Greek originals, covering arithmetic and music. These textbooks became the centrepiece of elementary education in their respective subjects until the sixteenth century.

For more advanced students, he set himself the task of translating the entire *oeuvre* of Aristotle, but managed to complete only some of the logical treatises before his arrest. At first these were not especially popular, largely due to their being excruciatingly difficult to understand. But they came into their own in the eleventh century when, as we shall see, there was a flowering of rational thought.

The Christian Vision of the Natural World and the Music of the Spheres

On the meagre rations of Boethius, Aristotle and a few other authors, Gerbert and his contemporaries fed their appetite for science. It is hardly surprising that when early medieval people looked out onto their world, they perceived it in a way that is quite alien to us. The concept of 'worldview' is important because it underlies other ideas, such as science, which would simply not make sense without it. The way we imagine our universe and ourselves is often deep-seated, almost unconscious, and we rarely think about it. We find it hard to imagine that the world could function in any other way and feel that the ideas we have learned to believe are, in some sense, self-evident. The modern secular western 'worldview' is naturalistic. We believe that nature blindly follows laws that we can describe using mathematics; and that we live in an impersonal universe that is unimaginably old and vast, on a planet orbiting an ordinary star. We also assume the laws of physics that apply to us also apply to the rest of the universe. Almost nobody asks why gravity actually works and we regard it as inconceivable that it should cease to do so. The only way that we can enjoy reliable knowledge is through science, which works because the laws of nature never change.

Medieval people also believed that the world worked in a completely reasonable way; it just wasn't the same way that we think it works. The central idea of the medieval worldview was that everything and everybody had a purpose. Nothing just happened. Nothing existed purely for its own sake. The ultimate governor of the universe was God and he had endowed everything with a reason for its existence.

For modern people who hold a naturalistic worldview, nothing ultimately has a purpose. The universe just is, and has no guiding hand. We do not need to look for a conscious reason for anything to know how it works. In fact, any such explanation involving a purpose is scientifically invalid. To a medieval mind, such a view would be completely irrational. They would say that rationality itself required a reason for everything. Take an example from the animal kingdom: today, when we want to know why a lion has sharp teeth and claws, we will look to the theory of evolution to explain it. To the medieval mind, the correct question to ask was what purpose the lion served. The answer would be that God designed it to catch its prey and it therefore had the attributes that enabled it to do that. Furthermore, God gave the world the lion to act as a symbolic reminder of his son, Jesus Christ, who is king of men just as the lion is king of the beasts.[16] This seems a million miles away from a mentality that could lead to a modern scientific outlook, but as we shall see, that is what happened.

In some ways, the medieval worldview was closer to ours than we sometimes imagine. For example, Gerbert and all his fellow men and women of any education in AD1000 were perfectly well aware that the earth was a sphere. They also knew that the universe was very large compared to earth. As Boethius wrote in his *Consolation of Philosophy*:

It is well known and you have seen it demonstrated by astronomers, that beside the extent of the heavens, the circumference of the earth has the size of a point; that is to say, compared to the magnitude of the celestial sphere, it may be thought of as having no extent at all.[17]

The myth that a flat earth was part of Christian doctrine in the Middle Ages appears to have originated with Sir Francis Bacon (1561–1626), who wrongly claimed that geographers had been put on trial for impiety after asserting the contrary.[18] There were a few authentic flat-earthers in late antiquity, but none among the scholars of the Middle Ages proper. One of the main reasons that some historians previously fell for the flat earth idea is because of the existence of *mappae mundi* (Latin for 'maps of the world') like the famous example at Hereford Cathedral. These follow the so-called T-O pattern. The O represents the encircling ocean which surrounds the entire inhabitable earth. The T comprises the Mediterranean Sea, the River Nile and either the River Volga or Don which split the O-shaped landmass into the three continents of Europe, Africa and Asia.[19] In this simplified plan of the world, east is towards the top and Jerusalem usually occupies a point very close to or at the centre. The Latin for east is *oriens*, which is why we now say we 'orientate' our maps, although with north at the top. It is understandable that, faced with such a map, modern scholars mistakenly believed that the people who drew it thought the earth was flat. What they did not realise was that it was only intended to map the quarter of the earth's surface that medieval people believed to be inhabited.

2. A T-O map from a 1472 printed edition of *Etymologies* by Isidore of Seville (d.636)

Gerbert and his contemporaries evidently also knew the approximate size of the earth, as a figure for its circumference of 29,000 miles appeared in Latin sources.[20] Like the ancient Greeks, Gerbert believed that the only inhabitable part of the planet was the part of the northern hemisphere that was sandwiched between the frozen poles and the burning equator. They were aware that it was a very long way from the far eastern tip of Asia over the ocean to the western coast of Europe. Thus, the inhabited region filled much less than one quarter of the earth's surface and could easily be mapped onto a flat projection. The *mappae mundi* may not be very accurate, but there is no reason to believe that their drafters did not intend them to give a reasonably literal view of their corner of a spherical world.

As for the southern hemisphere, medieval people believed the extreme temperatures at the equator to be so great that no one could cross it. This meant that the deadly heat of the 'Torrid Zone' would completely cut off the antipodes, if indeed any land existed in the south. In the eighth century, Virgil of Salzburg (c.700–784), an Irish monk who had travelled to mainland Europe, was accused of teaching that the antipodes were inhabited and, furthermore, that the inhabitants were not descended from Adam. The Pope condemned this doctrine but Virgil himself was cleared of any unorthodoxy. In the thirteenth century, he was even canonised.[21]

Gerbert's picture of the universe differed from ours in other ways too. He did believe that the earth was both stationary and the centre of the universe. According to him, while the earth stood still, the whole of the heavens revolved anti-clockwise once a day. Each planet was believed to be embedded in the rim of a great sphere which rotated clockwise, with periods ranging from one month for the moon to 30 years for Saturn. This meant that all the planets were orbiting the earth at a speed determined by the rotation of their respective celestial spheres. Of course, we now know that Gerbert was wrong, but his belief was perfectly rational. For a start, it is self-evident to us all that the earth is not moving. When you are on a merry-go-round, turning at high speed, you have to hang on tight. Gerbert knew how large the earth was, and he knew that if it were rotating every 24 hours then he would be travelling on its surface at close to 1,000 miles an hour. He should hardly be able to hang on! Or at the very least, he would feel the winds rushing past him as the atmosphere struggled to keep up with the spinning earth. Likewise, if he threw something into the air, he should note that it fell well behind where it started, as the earth would have moved on by the time it landed.[22] Anyway, this is what he learned at school. All his textbooks and his teachers agreed that the earth neither moved nor revolved. True enough, the Bible said this too but it was in agreement with all the pagan Greek sages. Besides, it was not always to be taken completely literally. The Bible also strongly implied that the earth was flat, for instance with reference to it having 'edges' in Job 38:13. Yet medieval people sided with the astronomers on this matter. Where the Bible seemed to conflict with good sense or reason, medieval thinkers were happy to interpret it figuratively rather than literally.

Before we criticise Gerbert and his compatriots for their foolish adherence to ancient Greek and Hebrew authority, consider this. If someone asked you today to demonstrate that the earth orbits the sun, you almost certainly could not do it. You could show them every book and ask every expert, but you could not provide them with direct evidence without a telescope, a lot of time and a lot of mathematics. Gerbert lacked the telescope and the maths, so we cannot blame him for believing his books when they so clearly echoed common sense. The idea that the earth moves was absurd, and it would take a great deal of careful thought before people realised that it was even possible.

Another modern misconception about the medieval Christian worldview is that people thought the central position of the earth meant that it was somehow exalted. In fact, to the medieval mind, the reverse was the case. The universe was a hierarchy and the further from the earth you travelled, the closer to Heaven you came. At the centre, underneath our feet, the Christian tradition placed Hell. Then, surpassed in wickedness only by the infernal pit, was our earth of change and decay. Above us, acting as a boundary between the earthly and the heavenly, was the sphere of the moon. This marked the dividing line between the perfect unchanging heavens and the transient sub-lunar region containing humanity, which was doomed to die. Next, there were the crystalline spheres of the seven planets – the moon, the sun, Mercury, Venus, Mars, Jupiter and Saturn – eternally orbiting with uniform, circular motion. The spheres were thought to consist of a transparent and imperishable fifth element called ether or quintessence. Above them were the fixed stars whose positions relative to each other never appeared to change. Beyond even them was the firmament and outside that was the realm of God. This hierarchical system gave people absolute directions of up and down, one towards the heavens and one down to earth at the bottom of the celestial ladder. To move the earth away from the centre of the universe was not to downgrade its importance but to raise it up towards the stars.

3. A diagram of the medieval universe from *Cosmographia* (1539) by Peter Apian (1495–1552)

The attraction of this model of the universe was its harmonious order. Everything had its correct place in the celestial hierarchy and it provided an exemplar for good governance on earth. Harmony was especially important to the theory. At this point we should recall that Gerbert had taught harmonics to his students at Reims. The relevance of this topic was that the crystalline spheres were believed to move in harmony, emitting as they did so the 'music of the spheres'.[23] This wonderful sound, the very resonance of the universe itself, was unfortunately inaudible to human ears. Instead, through proper training, students could learn to experience it second-hand through the mathematics of harmony. These ideas were not specifically Christian, but came from the pagan classical world with adjustments made to fit Christianity as necessary. The most important of these pagan philosophers in Gerbert's time was Plato because he had so strongly influenced Saint Augustine.

Alfred North Whitehead (1861–1947), the Cambridge and Harvard philosopher, once wrote, 'The safest general characterization of the European philosophical tradition is that it consists of a series of footnotes to Plato.'[24] Even if this is an overstatement, few men's ideas have had such great influence as his. He was born in Athens to an upper-class family and was a pupil of Socrates (469–399BC), a controversial philosopher who had a habit of exposing the hypocrisy of his compatriots by asking penetrating and leading questions. When a time of war and plague arrived, the authorities could no longer tolerate Socrates and they convicted him of blasphemy and corrupting the young. While these were undoubtedly trumped-up charges, Socrates did his level best to ensure that he suffered the highest penalty by refusing to compromise with his accusers or accept exile. In consequence, shortly after his trial, he was executed in prison by being forced to drink hemlock.

After his master's death, Plato founded a philosophical school in Athens called the Academy which paid its way by educating the aristocracy in rhetoric and culture. Exactly what he taught from day to day is hard to know because his surviving works are all highly polished literary dialogues between his dead teacher, Socrates, and various interlocutors. Apart from *Timaeus*, Gerbert could not have known any of Plato's dialogues at first hand because they remained in Greek until the fifteenth century when, newly rendered into Latin, they became extremely influential. What made Plato so conducive to Christianity was that he emphasised the intellectual and spiritual aspects of life over the material. For Plato, the ultimate reality was not the everyday physical world, but a transcendent realm of ideas. He called these ideas 'Forms' and claimed that they were the templates from which our world had been constructed by a divine creator from pre-existing matter. Thus, the material world was simply a shadow of the perfect reality of the Forms. A philosopher could gain access to this true reality through the act of contemplation. Plato believed that this ability to approach perfection gave the thinking man the advantage over the man of action.

Clearly, some of this would sound acceptable to a Christian. He would agree with the Platonist that God created the world, as the book of Genesis affirms at the start of the Bible. However, the Christian would deny that the material from which God made the world had already existed. Rather, God had created the world out of nothing, *ex nihilo*, which gave it an intrinsic value as the result of the direct will of God. Both pagan and Christian would agree that the world is not perfect but the Platonist would blame the material the creator had to use, while the Christian would believe that it was created perfect and only fell later due to human actions. Plato's support for a life spent in contemplation could easily be interpreted as praise for a life dedicated to Christian prayer and meditation rather than to secular deeds.

There is little doubt that Plato was the most influential of the ancient Greeks in Gerbert's time. This, however, would shortly change and Plato's pupil Aristotle would become known as 'the master of those who know'.[25]

CHAPTER 3

⧞

The Rise of Reason

Cold and hungry, a young man staggered through the snow of the Mount Celis pass in the western Alps. At its highest point, his journey towards the Rhône valley in France had taken him almost 7,000 feet above sea level. He had always loved the beauty of the mountains and in later years spoke about his dreams of God feeding him on white bread as he wandered in the peaks. Now, though, neither God nor anyone else was around to give him food and he was close to starvation. Leaving home without money or supplies no longer seemed a wise idea, but he was not about to return to his hostile father. However, his biographer Eadmer (who died around 1124) tells us, rummaging around in his pack, his servant found some more white bread that had miraculously appeared there.[1]

Saint Anselm and the Logical Proof that God Exists

Posterity would come to know the young man as St Anselm of Canterbury (1033–1109). He had been born in the town of Aosta, in the Italian foothills of the Mont Blanc range, where his family enjoyed considerable influence in the church. His beloved mother died when he was still young and his father sent him to be educated by an uncle. The experience was deeply unpleasant because his uncle, like many other schoolmasters of the time, was very quick to pick up the birch and beat his charges. As an adult and a teacher himself, Anselm would one day criticise this brutal regime and suggest that beating children only made them less enthusiastic about their lessons. 'You have nurtured beasts out of human beings', he chided a schoolmaster who was partial to the rod, 'and they have grown up perverted and vicious because they have never been raised in genuine affection for anyone.'[2] On returning home, Anselm's relationship with his father deteriorated and in 1056, after a blazing row, he left Aosta for the last time and set off for France.

In truth, he had little idea where he was going except that he was heading north. This was a providential choice because in the eleventh century, northern France was the most dynamic region of Europe. The cathedral cities of Paris, Chartres, Reims and Orleans were a magnet for scholars looking to make a name for themselves. The schools attached to these cathedrals specialised in teaching the logical works of Boethius and Aristotle. But to attend them, you needed money and Anselm had none.

He barely made it across the Alps and then, for three years, he wandered from place to place. Finally, while staying in Normandy, Anselm heard about another exile from Italy who had found a safe haven at the nearby abbey of Bec, teaching the novices. This was Lanfranc (1005–89), to whom Anselm presented himself in 1059. They hit it off at once and Anselm became Lanfranc's assistant at the monastery's school. Bec was a new foundation, a long way from the great schools of France, so its subsequent growing reputation was almost entirely down to these two men. Then, just as Anselm was settling in, news came from Aosta that his father was dead and he could return home to claim his inheritance. He had to make a decision either to renounce his worldly position to become a monk, or leave Bec and travel back to Italy. Lanfranc consulted the local archbishop who advised Anselm to stay at Bec. He followed this advice and formally professed as a novice monk. Continuing to help Lanfranc in the schoolroom, Anselm proved that he was a good teacher. When his master left for another monastery, he took over as head of the school. He taught his pupils Latin grammar and the rudiments of logic so that they would be intellectually equipped to tackle the Bible.

At last, Anselm had found peace. The Abbey of Bec provided him with a haven in which to write and meditate. The *Proslogion*, the theological work for which he is famous, dates from this period. This short treatise is important because it uses reason in the service of theology. In it, Anselm does something that must have seemed daring at the time. He tries to use pure logic to prove that God exists. His attempted proof, known as the 'ontological argument', is one of the classic conundrums of philosophy. It goes something like this:

> We have the conception of God in our minds as the being greater than any other thing of which we can conceive. However, in order for God to be truly the greatest thing that we can conceive, he would also have to exist, because if he did not then he would not be the greatest. A real greatest thing is certainly greater than an imaginary one. Thus, if God did not exist, he would not be the greatest thing we can conceive and hence he must exist.[3]

What the argument categorically does not attempt to do is to prove the existence of God to a sceptic. Instead, Anselm intended it to help someone who is already absolutely sure that there is a god to understand why such a god is necessary. His motto was 'faith in search of understanding'.[4] Note the order in which he places the words. For Anselm faith was prior to understanding and based on the mystical experience of God through meditation and prayer. After all, there was very little reason to deny personal experiences when they were consistent with the traditional teachings of the Church. On the other hand, for Christians there is no doubt that God has blessed man with rational faculties. It follows that we can also understand what we believe as well as accepting it from our religious experience. That is what Anselm's ontological argument attempts to do. In fact, most philosophers have found the argument unpersuasive, but actually refuting it turned out to be fiendishly difficult. Even Bertrand Russell (1872–1970) admitted that 'it is easier to feel convinced that it must be fallacious than it is to find out precisely where the fallacy lies.'[5]

Criticisms of the ontological argument by other monks started appearing immediately and Anselm tried to respond to them. He added the earliest such rebuttal to his manuscripts before appending his own rejoinder. Anselm's critics claimed that what we know by faith, we do not need also to demonstrate by reason. Indeed, using reason could open up a dangerous can of worms. It might not be long before people started to say that what we cannot prove by reason, we cannot know by any other means either. Anselm's critics were worried that once reason was allowed to have any sort of function in the religious sphere, it would eventually become the final arbiter of religious truth. In fact, there was nobody who actually claimed that reason had a monopoly on truth or that the Christian faith was wrong. Rather, the two sides drew the battle lines between those who objected to reason intruding on such matters at all, and those who felt that reason did have a role to play.

In 1078, Anselm was promoted to abbot of Bec. However, like Gerbert, he was a better scholar than he was an administrator and had trouble keeping the abbey's finances in the black. These worries meant that his later writings do not have the same freshness as the *Proslogion*. In 1093, he was promoted again to archbishop of Canterbury after the death of his mentor Lanfranc, who had previously held the see. Now, he had to throw himself into the stormy world of ecclesiastical politics and the concerns of kings. Anselm was a typical 'turbulent priest' who would not kowtow to royal authority. His poor relationship with William II (1056–1100) and Henry I (1068–1135), kings of England and the sons of William the Conqueror, meant that Anselm was twice exiled to the continent. If anything, he preferred this state of affairs to actually having to run his diocese and, while he was away, devoted himself to scholarship. His second exile ended only in 1107 and, two years later, he died.

The raw materials that eleventh-century theologians such as Anselm used to construct their rational arguments were the logical works of Aristotle. We have already seen how Boethius had translated and commented on these back in the sixth century. Unfortunately, the Aristotelian corpus is extremely difficult, even without the vagaries of Boethius's obscure language; because so many Greek logical terms had no equivalent in Latin, Boethius was reduced to inventing an entirely new vocabulary. These would eventually become the standard terminology for the study of logic, but they took five centuries to catch on.

Medieval logic was mainly concerned with how things are described and to what extent those descriptions are real. Here is a frivolous example; in the village of Otham today, you will often spy white fluffy animals gambolling in a field. We call them 'sheep'. But can we say that the concept of 'sheep' really exists, or should we just say that there are individual creatures that we collectively call by that name? In other words, is there such a thing as 'sheepness' or are there just a lot of sheep? This dispute ran on throughout the Middle Ages. Those thinkers who considered the word 'sheep' to be just a name that we have arbitrarily applied to white fluffy animals are called 'nominalists'. The other camp, who insisted that 'sheep' do in fact exist as a category independent of all the individual sheep, were 'realists'. We will come across these two schools of thought again.

Another key idea in logic, originating with Aristotle himself, is the distinction between substance and accident. To explain the meaning of these two terms, let us return to the white fluffy creatures mentioned above. We have already identified them as sheep, but one could easily ask what characteristics they have to have in order to be called sheep. The only properties we have said that they possess are whiteness, fluffiness and the ability to gambol. The trouble is that none of these characteristics actually captures the essence of what it means to be a sheep. Merely being white, fluffy and gambolling does not make an animal ovine. A crippled, freshly sheared black sheep would nonetheless be a sheep. On the other hand, it is impossible to be a sheep and not to be an animal. Thus, using the Aristotelian definition of substance and accident, we could say being white, fluffy and gambolling are *accidental* properties of a sheep. On the other hand, being an animal is a *substantial* property. It is possible to change the accidental properties of an object without it thereby losing its essential substance. On the other hand, any change in its substantial properties would alter its very essence and make it an entirely different kind of object.[6]

If all of this sounds far removed from everyday medieval concerns, in fact logic scored a notable success in the field of theology in the eleventh century by showing that reason could help defeat heresy. As the eleventh century wore on, the Church came to see unorthodox beliefs as an increasingly urgent problem. One of the earliest heretics who caused serious concern was Berenger (1000–c.1088), a French theologian who ran the cathedral school in Tours. His gripe was with the doctrine of the Eucharist and transubstantiation. Catholics believe that during the communion service or Mass, the bread and wine are miraculously transformed into the actual body and blood of Christ. Luckily for the congregation, though, they still look and taste like bread and wine rather than flesh and blood. Berenger had doubts about this and taught that the bread and wine merely came to represent the body and blood of Christ. If it looked like bread and wine, he said, then that was what it was. To a modern reader or a Protestant this sounds very reasonable, but it conflicted with mainstream Christian faith at the time. How could faith and reason be reconciled in this case? Anselm's teacher Lanfranc had the answer, and he found it by thinking carefully about Aristotelian logic. During the Mass, said Lanfranc, while the bread and wine maintained their accidental properties of looking and tasting like food and drink, their substance changed to the body and blood of Christ.[7] Thus, transubstantiation could have taken place even though the perceivable properties of the host stayed the same. As long as the bread and wine took on the substance of Christ's body, it did not matter what their accidental characteristics were. Most people agreed that this worked splendidly. Lanfranc had shown how reason could refute an argument even if it seemed to be based on common sense.

Between them, Lanfranc and Anselm pushed theology to centre stage. It became the subject that learned men aspired to master, but also one that involved high risks – not least possible accusations of heresy. Anselm, in particular, found that he had to be careful. In about 1090, a schoolmaster by the name of Roscelin (c.1050–c.1125) started circulating ideas about the Trinity. Worse, he claimed that his views had the support of Anselm and that the archbishop was a friend of his. Roscelin had already made quite a name for himself teaching logic to paying students in northern France, but in 1092 had been accused of heresy.

The doctrine of the Trinity has caused more trouble than any other. Christians believe that although God is one substance, he consists of three different persons – Father, Son and Holy Spirit. It doesn't take a master logician to see that one cannot equal three, so the Trinity seems to be inherently contradictory. In fact, it is hard to see how any rational treatment of the Trinity could be anything other than heretical. Most medieval people were comfortable with this because they were never so arrogant as to believe that human reason was able to understand everything. The Catholic Church is well justified in calling the doctrine a 'mystery'. But to thinkers like Roscelin, the Trinity was like a bad itch that they could not leave alone. He decided that logically there had to be three Gods rather than just one in order for the Trinity to make any kind of sense. Inevitably, his ideas brought him into conflict with the Church authorities, who chased him out of France.

Roscelin moved to England and appealed to Anselm for help. However, his reputation as a heretic had gone before him and the archbishop sent him straight back to the continent. Even being associated with such a man was excruciatingly embarrassing, and Anselm decided to go on the attack. He fired off a long letter to the Pope explaining where Roscelin had gone wrong.[8] The pontiff graciously accepted Anselm's explanation, which had the effect of giving all his ideas an official stamp of approval. The Pope was agreeing that it was permissible to use logic in theology, as long as it was conducted in Anselm's cautious way – putting faith before reason.

As it turned out, Roscelin does not appear to have been badly damaged by Anselm's refutation of his ideas and continued his career as a teacher. A short time later, a young man turned up at his classes who, as we shall see, became one of the most celebrated and controversial figures in the history of philosophy. His name was Peter Abelard (1079–1142).

The Doomed Lovers

Abelard was born in the town of Le Pallett on the southern edge of Brittany. His father was a knight who married a relative of the local lord. Unusually for the time, he arranged for all of his children to have some tuition in Latin. Peter, his eldest son, excelled in the classroom and abandoned any ideas of inheriting his father's estate. Instead, he set off eastwards to the centre of France and enrolled for Roscelin's lectures.

Abelard was a big man with an ego to match. He frequently made himself insufferable in classes by openly challenging his masters. After a few years, he fell out spectacularly with Roscelin. We do not know what the dispute was about, but neither ever forgave the other. Abelard moved to Paris and began to attend classes at the cathedral school of Notre Dame. The educational market at the time was unregulated and vicious. Each master had to attract students in order to earn the fees that gave him a living. Paris already had a reputation for being a place where the best masters were working, so students came from far and wide to study at their feet. There does not appear to have been any formal qualification required to teach at this time, and the masters set themselves up as freelance lecturers as soon as they thought they could attract enough students to support themselves. Reputation was everything. Without it, they could not hope to attract new students. Of course a teacher wanted people to know that he was not only very intelligent, but also an exciting and eloquent lecturer. An excellent form of advertising was to turn up at a rival master's lectures, engage him in argument and humiliate him in front of his students. Hopefully, they would be so impressed that they would abandon the rival as yesterday's man and all flock to hear the heckler.

Peter Abelard was unquestionably both clever and eloquent. We know a great deal more about his life than we do about those of most medieval intellectuals because he wrote a morose autobiography called *The Story of my Calamities*. He begins by telling us about how his success in debate caused resentment from his masters and fellow students.[9] He only lasted a couple of years in Paris before he had so effectively alienated his teachers that he had to flee. He moved 25 miles south to the city of Melun and set up his own school. There, he attracted some of the brightest and best students while he carried on his dispute with the masters of Notre Dame.

The subject at which Abelard excelled was logic. Like Anselm, he believed that he could harness the power of reason in the service of theology. However, he lacked Anselm's caution and put reason before faith when he claimed, according to his students, 'nothing can be believed unless it is first understood.'[10] Nor did he make any effort to avoid contentious topics. In one notorious discussion, he suggested that the Jews responsible for crucifying Jesus Christ did not sin because they thought they were doing the right thing. A sin, Abelard said, could only be a culpable act, punishable by God, if you knew it was wrong.[11] This, in itself, was not a new idea but Abelard picked the most outrageous example that he could think of. He was deliberately trying to generate controversy.

He was certainly a lively teacher. We have a description from one of his students who tells us: 'There [Abelard] was, detonating astonishing and unprecedented opinions for his audience, as if to make a mockery of the ranks of the sane and wise.'[12] Illustrations found in contemporary manuscripts show us what the scene would have looked like.

Abelard would have had a seat that was raised on a dais above his audience of students, who sat on the floor. He would also be equipped with a copy of the book from which he was lecturing. This gave him a natural advantage over any hecklers; the students would not have had their own copies of the book. Instead, they would make notes on sheets of wax held in a shallow wooden tray. Paper had not been invented yet and so books were made of parchment; that is, treated animal skins. This was far too expensive a material for just jotting down notes. To barrack the lecturer, a pupil could stand up but would still find himself at a lower level than his master. Abelard could also use his considerable height to face down opposition. Our source tells us how he tried to interrupt, but Abelard 'turned his savage eyes upon him and threatened him vehemently, "See that you

4. A fourteenth-century manuscript illumination of a master lecturing
his students

keep quiet and be careful not to disturb the course of my lecture."[13]
Of course, interrupting with difficult questions was exactly the tactic
that Abelard had used against his own masters. So, after failing to
intimidate the pupil, he had to let him say his piece or else risk looking weak to the other students.

After ten years of hard slog in the provinces, Abelard returned
to Paris in triumph in 1117 as a master in his own right. He was
appointed a canon of Notre Dame Cathedral, which gave him both
a steady income and plenty of prestige. The way was clear to a celebrated career as scholar and then, in all likelihood, promotion to
a bishopric. But this reckoned without Abelard's unerring ability to
completely wreck his chances of lasting success. His troubles began
because of his affair with a remarkable woman.

Although he is hardly known today, Abelard was once very famous,
not just as a philosopher but for his part in one of history's great tragic

love affairs. The object of his desire was Héloïse (d. 1164), the niece of another of the canons at Notre Dame Cathedral. Abelard tells us that 'in looks she did not rank lowest while in the extent of her learning she stood supreme.'[14] For a woman, she had acquired a remarkable education in the Latin classics. She could quote at will from Ovid, Virgil and Cicero as well as compose letters in the style of the ancients. She was a strong woman with opinions of her own who used her learning to win arguments with men with whom she disagreed. Her uncle, Canon Fulbert (d. 1127), who probably owed his position to family connections, expected that once Héloïse had completed her education, she would be appointed abbess of an important nunnery such as Argenteuil, just to the east of Paris. She had been brought up there before moving to Notre Dame to live with her uncle. Being an abbess was a position fit for an upper-class woman like Héloïse, and about the only role for a female that required any sort of education.

Abelard's own account of the affair explains that he was attracted to Héloïse because of her erudition. We do not know how old she was, but given her level of literary accomplishment, she is unlikely to have been younger than twenty. Abelard's account continues:

> All on fire with desire for this girl, I sought an opportunity of getting to know her through private daily meetings and so more easily win her over; and with this end in view, I came to an arrangement with her uncle, with the help of some of his friends, that he should take me into his house, which was very near to my school, for whatever sum he asked.[15]

Fulbert had a weakness for money and Abelard offered him a good rent for his lodging. Now that he was a canon of the cathedral, Abelard should have remained celibate. Although celibacy was not yet compulsory for all clerics, it was becoming impossible to get promoted to important positions like a canonry without abstaining from sex. As a celibate, Abelard was an ideal teacher for the headstrong Héloïse. His manifest suitability for the post led Fulbert to urgently request that he become her tutor. Abelard couldn't believe his luck and set about convincing Héloïse to be his lover. By all accounts, he didn't have to try very hard. His policy of seduction was an unqualified success and Héloïse fell head over heels in love with the dashing scholar. Before long, they were neglecting her literary education.

> Her studies allowed us to withdraw in private, as love desired, and then with our books open before us, more words of love than of our reading passed between us, and more kissing than teaching. My hands strayed oftener to her bosom than to the pages.[16]

The affair was intense to say the least. Abelard admits to beating Héloïse (as expected between a young student and her master), but he gives the impression that they both enjoyed such rough play. Even more disturbingly, he raped her when she tried to resist him. Years later, he wrote in a letter to her:

> Even when you were unwilling, resisted to the utmost of your power and trying to dissuade me, as yours was the weaker nature I often forced you to consent with threats and blows. Your weaker nature meant you could not prevent me.[17]

Even this did not daunt Héloïse and she was as besotted with Abelard as he was with her. We do not know how long they kept up the pretence, but before long their affair was the talk of the cloister. The only person who didn't know about it was Uncle Fulbert. Abelard's teaching career was suffering as well because his mind was no longer on the job.

Inevitably, it ended in tears. Within months, Fulbert discovered his niece and her tutor in a very compromising position. It was impossible to deny they were lovers. Abelard fled from Paris with Héloïse and returned to his native town of Le Pallet. Soon afterwards, Héloïse found she was pregnant and stayed with Abelard's family until she gave birth to a son. They called him Astrolabe, after the astronomical instrument that had so excited Radolf of Liège a century previously.

Leaving Héloïse at Le Pallet, Abelard returned to Paris to try to thrash out a compromise with Fulbert. They agreed that Héloïse would have to marry Abelard, but that the wedding should be kept secret so as not to undermine his career. Héloïse seemed to accept this and came back to Paris, but without her son whom one of Abelard's brothers adopted. Sadly the compromise could not hold. Despite the supposed secrecy of the marriage, word got out. Héloïse tried to save the situation by swearing under oath that she was unmarried but this merely enraged Fulbert. Abelard lost heart and transferred Héloïse to Argenteuil, the nunnery where she was brought up, although he did not make her take her final irrevocable vows. For Fulbert, Abelard shutting his niece away like this was the last straw and he planned a terrible revenge on the man who had brought dishonour on his family.

One night, Abelard was asleep in his house in Paris. Unknown to him, Fulbert had bribed his servant to let in a posse once his master was unconscious. The plan worked perfectly. The men burst into the bedroom and set about their gruesome work before Abelard had any chance to react. We can be sure that the operation was performed professionally. The victim was pinned down with his nightshirt lifted over his head. One of the gang, skilled in surgery, would have taken a ligature and tied it around the base of Abelard's scrotum. He pulled it tight to prevent bleeding and used a sharp surgeon's knife to cut off the testicles. Then the posse fled. The process was so quick that Abelard later claimed that he had hardly felt a thing.[18] The mental anguish the attack caused was another matter entirely. Abelard roused the watch and they captured one of the posse, as well as the treacherous servant. The guards castrated the prisoners and then blinded them as well for good measure. As for Abelard, he could no longer act as a husband to Héloïse. He sought shelter in the Abbey of St Denis outside Paris where he professed as a monk. Héloïse was forced to take her final vows and their desperate affair came to a dreadful end. As for Fulbert, he lost his canonry and property for his part in the crime. However, many people considered that his actions were justified and within a couple of years he was reinstated.

The Trials of Peter Abelard

Abelard's castration had been a very public humiliation but it did nothing to curtail his hunger for controversy. He quickly quarrelled with the other monks at St Denis, accusing them of being insufficiently holy. This was thought to be a bit rich coming from Abelard and they packed him off to teach students somewhere outside Paris. He also started to write his own book about the Trinity. While he accepted that it was impossible to prove this doctrine was true by reason, he thought that it was possible to use reason to rebut any argument that claimed to show the doctrine was false. Reason could not conflict with faith because both were gifts that came from God. If there appeared to be a conflict then it was because there was a mistake in the argument. So, if a heretic, like Berenger of Tours or Roscelin, used reason to formulate an argument which produced a conclusion in conflict with Christian doctrine, it must be possible to use reason to refute him. This was exactly what Abelard thought he was doing, but he also had a habit of making powerful enemies who would not hesitate to use accusations of heresy against him.

Roscelin was still alive and could see exactly where Abelard was aiming his rhetorical barbs. He was still smarting from how Abelard had treated him twenty years earlier and saw an opportunity to take revenge on his pupil. He accused Abelard's own book on the Trinity of being heretical and after a show trial in 1121, Abelard was forced to cast it into the flames. He returned to his abbey in disgrace. This would have been another good opportunity to keep a low profile but instead he upset the monks again by suggesting that their patron, St Denis, had never really existed. As St Denis was also the patron saint of France, Abelard also found himself accused of treason. This time, he fled to live alone in a hermitage. His attempt at the quiet life was also a failure; a clamouring horde of students who wanted to hear his teaching followed him to his wilderness retreat. He was soon writing and lecturing again.

Abelard's most famous work, intended primarily as a textbook for his students, is called *Yes and No*. It contains little more than a host of quotations from the Church fathers arranged in such a way as to make it clear where they were contradicting each other. The aim was not to prove that they actually were contradictory, but rather to aid his students in resolving these perceived problems. Abelard wrote in the preface of *Yes and No*: 'By doubting we come to inquiry and by inquiry we see the truth.'[19] He meant that we can discover the truth, but only through a sceptical method of asking difficult questions. The main criticism of the book made by his opponents was that it made no effort to resolve these contradictions, and thus gave the impression (which we might not think so misleading) that they were real.

Abelard's controversial career continued. At one point he was the abbot of a monastery in Brittany, but left when his monks tried to kill him. In 1129, Héloïse too was in trouble. She had risen to the rank of prioress at Argenteuil but she and her nuns were expelled for 'notorious immorality'. We do not know exactly what this means, but Héloïse was a passionate woman and it is likely that it involved men. This time, Abelard was able to ride to her rescue. He gave her his old hermitage so she could found a new nunnery, which was a great success. The Paraclete, as the nunnery was called, quickly acquired papal recognition and large grants of land. It survived until the French Revolution of 1789, when almost all religious houses in France were closed down.

As for Abelard, such was his ability as a teacher that he was always in demand. In 1133, he even returned to Paris as a master of logic. Many of his enemies were dead and it seemed that he could finally work in peace. He began to revise his book on the Trinity (presumably he had kept a copy, as he had burnt the original). Unfortunately, despite the demise of his old opponents, he had little trouble stirring up new ones. In 1140, as an elderly man of 60, he found himself facing the most formidable churchman of his generation – St Bernard of Clairvaux (1090–1153).

In traditional histories of the Middle Ages, St Bernard tends to get pigeon-holed as one of the bad guys. His preaching helped to incite the second crusade and he did not have a lot of time for religious tolerance. He was also ambivalent about the use of human reason over faith. None of this has endeared him to modern authors. However, St Bernard was by no means a religious conservative. He was a radical who devoted his life to fighting against the sins of the Church. Simony – that is, the sale of clerical positions – and concubinage, where priests kept lovers even if the Church forbade them to have wives, both provoked his ire. Likewise, he campaigned for a clergy which was uncorrupted by worldly wealth and which lived and worked among ordinary people. He reformed the monastic system to root out the laxity and luxury with which it had become associated.

At this time, Abelard was involved in a new dispute with a certain William of St Thierry (c.1075–1148), who was a friend and student of St Bernard. Out of his depth arguing with one of the great debaters of his generation, William wrote to his mentor to ask for help in combating Abelard's novel ideas. He enclosed Abelard's updated treatise on the Trinity, and St Bernard was able to find much in it that he did not like. His disagreement with Abelard was probably not over an obscure point about the nature of the relationship between Father, Son and Holy Spirit, but rather about the whole question of how reason should be used in the field of theology. St Bernard was of a mystical bent, a man who believed that the best way to gain access to God was directly through prayer and meditation. Reason, he thought, would just get in the way. Furthermore, Abelard's alleged mistakes showed how dangerous reason could be when it led men astray. It was better by far, he thought, just to stick to faith.

Everyone knew that in a public debate, Abelard would get the best of even the great St Bernard of Clairvaux, so Bernard wanted to avoid any such set-piece occasion. In the end, St Bernard prevailed by denying his opponent any sort of fair trial. Instead, he accused Abelard of heresy at a church council in Sens in 1140. Abelard could see the way the wind was blowing and refused to cooperate. Ignoring the council, he appealed directly to the Pope in Rome. This sent St Bernard into a panic, and he launched a flurry of letters to the Vatican demanding that Abelard be condemned forthwith. His eloquence worked. The Pope sentenced Abelard to perpetual silence and confinement to a monastery.

The story of Abelard gives the impression that the hierarchy of the Church was implacably opposed to reason and logic. He was put on trial for heresy in 1121 and 1140, losing the case both times. We have to be careful, though, not to be misled by Abelard's high-profile career. However sympathetic we might be to his plight, the fact remains that he brought most of his problems upon himself. His blatant hypocrisy and breathtaking arrogance ensured that he had a ready supply of enemies who were quite happy to use accusations of heresy to bring him down. Yet, despite his obvious character flaws, Abelard always had powerful supporters. He was twice a master of the prestigious cathedral school in Paris and abbot of a distinguished monastery (even if, like Gerbert and Anselm, he was too much of a scholar to be any good at the job). Even after his final condemnation in 1140, he found support from Peter the Venerable (1092–1156), abbot of the great monastery of Cluny and equal in stature to St Bernard himself. Peter ensured that Abelard's confinement took place within his own monastery where he was comfortable and well looked after.[20]

It was Peter who gave Abelard's body to Héloïse at the Paraclete for burial when he died of old age a few years later. By then the Pope who had condemned him was dead; the new Pope was a supporter of reason. He even owned the two books by Abelard that his predecessor had demanded be burnt.[21] Once Abelard was dead, his controversial and acerbic personality no longer obscured his ideas and they quickly came to dominate Christian scholarship. In the end, despite his enormous reputation and the silencing of Abelard, St Bernard had missed the boat.

CHAPTER 4

The Twelfth-Century Renaissance

I f any period deserves the label of 'renaissance' then it is the twelfth century.[1] Peter Abelard had championed the primacy of logic and argued that it must even be applied to the sacred mysteries. At about the same time, other scholars were beginning to think hard about the physical world as well. Two of the earliest medieval natural philosophers were William of Conches (1085–c.1154) and Adelard of Bath (1080–c.1160).

Most significant of all for the future development of science was the movement to translate into Latin an enormous body of newly discovered scientific and medical writing from the ancient Greek and Islamic worlds. This flood of new knowledge meant that western Europe could assimilate it and then progress from all that had gone before.

William of Conches and the Natural Philosophy of Plato

William of Conches received his education at the cathedral school at Chartres and probably taught at Paris. He may also have been the tutor of Henry of Anjou (1133–89), who later became King Henry II of England. Beyond that, we know little for certain about his life. We do know that William of St Thierry, the friend of St Bernard who alerted him to Peter Abelard, also accused William of Conches of heresy. But without the cachet of Abelard's dangerous reputation, nothing very much came of this affair. William of Conches shrugged off the accusations with a few amendments to one of his books.[2]

His most important achievement was an attempt to reconcile the natural philosophy of Plato, as found in *Timaeus*, with the creation accounts in the book of Genesis in the Bible. *Timaeus* was the only work of Greek natural philosophy that existed in western Europe throughout the early Middle Ages. This made it very influential. Unlike most of Plato's dialogues, *Timaeus* consists mainly of a single long speech by the eponymous speaker who describes the origin of the world. Even though Plato admits that his explanation of creation is merely a suggestion,[3] his reputation was such that even a mere hypothesis enjoyed great prestige when it came from his pen. Speaking through Timaeus, Plato tells us that a divine creator made the world out of some sort of pre-existing chaos. Because the chaos was difficult to shape, the universe that resulted was not as perfect as the original Forms on which the creator had based his design. The creator also gave the universe a soul, which is the active force that keeps the heavens moving. Various inferior gods govern the day-to-day workings of the universe and the creator delegated to them the job of fashioning animals, including man. Man thus finds himself in the middle of a great celestial hierarchy. The creator god is at the top with other gods below, while mankind is in the middle and brute animals lie beneath him.

William wrote a commentary on *Timaeus* to set out how it could be reconciled with Genesis. This might appear fairly straightforward in some respects. It is possible, though naïve, for instance, to equate Plato's creator with the Christian God. Plato's world-soul sounds similar to the Christian Holy Spirit. Plato also believed in the immortality of the soul and largely approved of the mode of life favoured by Christian monks. William of Conches realised that matters were not so simple; to give one example, Plato says that the creator constructed the world from existing material, but the Bible implies that God created it from nothing. Furthermore, Plato's creator was a much more passive deity than the Christian God who frequently meddled in human affairs. While wrestling with these concepts, William tried to interpret both *Timaeus* and the Bible in a figurative sense. He states explicitly that a literal understanding of parts of Genesis would be absurd. Rejecting such a literalistic interpretation of scripture, he wrote: 'The authors of Truth are silent on matters of natural philosophy, not because these matters are against the faith, but because they have little to do with the upholding of such faith, which is what those authors were concerned with.'[4]

Such a view was widespread, if still controversial, in William of Conches' time. We have already seen how, in practice, biblical passages that support a flat earth were not read as accurate descriptions of nature. In a similar vein, the Genesis creation account refers to both the sun and the moon as 'great lights' which implies that they both produce their own illumination.[5] However, no less a churchman than Pope Innocent III (c.1160–1216) was perfectly aware that the moon's light is reflected from the sun, and seemed to assume that this was widely known.[6]

The contrast between the Christian God who intervened in human affairs and a deity who just left the universe to run itself was an important one for William. He could see that for natural philosophy to be worthwhile, nature had to enjoy autonomy. To appreciate why this idea was so important, consider one of the standard arguments purporting to explain why science and religion might find themselves in conflict. It has been suggested that religious people who see God as intervening in the world cannot do science because they believe everything that happens is God's will. In other words, they are unable to distinguish between something that is a result of the laws of science and something that occurs due to the direct actions of God. A good example is the question of what a Christian should do if she becomes ill. Is sickness a divine punishment for some sin that has been committed? Alternatively, is it just a natural phenomenon best understood in terms of the physical spread of disease? In short, is it better to visit a priest or a doctor? Actually, most medieval people would tend to hedge their bets and follow both religious and medical advice. The former was especially popular during the plague and against other diseases for which there was no cure.

We do hear plenty of preachers blaming pestilence on God's wrath and the sins of the people, but this did not preclude the use of the medical profession as well.

William of Conches and his contemporaries managed to reconcile the two points of view in a way that was acceptable to all of the natural philosophers who came after him. He suggested that when we consider the reasons behind an event, we have to look at both the primary and secondary causes. According to William the primary cause, the ultimate reason why anything happens, is God. His omnipotence is such that he is the ultimate reality that underlies everything else. Nothing happens that is contrary to his will. However, the natural philosopher can also ask in what way God has attained his aims. This is the secondary cause of an event. Nature obeys the laws that God ordained when he created the world. These 'natural laws' are the secondary causes that God usually uses, although the precise terms 'natural law' and 'law of nature' were not used in this sense during the Middle Ages. Thus, by investigating natural causes, the philosopher does not threaten God's omnipotence in any way. There is nothing to prevent God from intervening directly and causing a miracle. However, we can only recognise a miracle as contrary to the normal course of nature if we already have some idea of what the normal course might be.

To illustrate this, imagine that you have just put the kettle on. Inside it, some water is gradually coming to the boil. The primary cause of the water boiling is that you fancy a cup of tea. However, a physicist in your kitchen could happily study the thermodynamics of boiling water in your kettle without ever having to worry about why you had set it going in the first place. He is interested only in the secondary causes, such as the way in which electricity heats the element and how this heat is transmitted to the water. One day, you turn on the kettle but switch it off again before the water has boiled because you have been called out unexpectedly. This one-off event has absolutely no effect on the physicist's analysis. Your direct intervention in the operation of the kettle, by turning it off, is irrelevant to his work on the secondary causes of boiling water.

In the same way, a natural philosopher can ignore the possibility of miracles in his research even while admitting that they occasionally happen. As one twelfth-century theologian explained:

> All things have been made by God as their author, but certain things are called God's work just as they are, namely those that he makes by himself ... Others are called works of nature and they are created by God after some natural resemblance, as a seed from other seeds, a horse from other horses and similar things from similar things.[7]

William of Conches also believed that God is loving and consistent rather than capricious and arbitrary. This meant that he could expect natural laws to remain the same forever. Now, he not only had a justification for investigating nature which did not infringe on the sovereignty of God; he also had a reason for believing that nature is regular enough in its workings to be worth exploring in detail.

One of William's colleagues at the school in Chartres summed up these ideas when he wrote in a *Commentary on Genesis*:

> Because the things in the world are mutable and corruptible, it is necessary that they should have an author. Because they are arranged in a rational way and in a very beautiful order, it is necessary that they should have been created in accordance with wisdom. But, because the Creator, rationally speaking, is in need of nothing, having perfection and sufficiency in himself, it is necessary that he should create what he does create only through benevolence and love.[8]

A common metaphor at the time was to imagine nature as a book written by God. Another twelfth-century theologian wrote: 'The whole of the sensible world is like a kind of book written by the finger of God.'[9] The Bible, of course, was another book written by God and so Christians could learn about him in two ways – by reading either the book of nature or the book of scripture.

Adelard of Bath

Adelard of Bath was a close contemporary of William of Conches but had a much more active life. He was a gentleman by birth who, like Peter Abelard, chose a clerical rather than a military career. As well as being a scholar, he was accomplished at the noble art of falconry and played the harp. He was probably a man of independent means who took on a few students, including his nephew, in order to give himself something to do. However, his true vocation was mathematics and natural philosophy. He realised that he was never going to become a master of these arts in Europe, so he left his students in France where he had been teaching and travelled to Sicily in 1112. His goal was to learn the secrets of the Arabs, who had ruled Sicily until 1091 when they lost the island to the Normans, the same people who annexed England under William the Conqueror. Even in Sicily, though, he could not quench his thirst for knowledge, so he journeyed on to the Middle East. This was the era of the crusades when much of the area was under Christian rule. Adelard reached Syria in 1114 and set about learning Arabic.[10]

At the time, there was little writing on advanced mathematics available in Latin. Boethius's short treatise on arithmetic was not a practical manual and his book on geometry was no longer extant. Adelard decided to translate the greatest achievement of ancient Greek mathematics into Latin so that his fellow Christians could benefit from it. The book universally recognised, then and now, as the finest work of Greek maths was the *Elements* by Euclid of Alexandria (325–265BC). This contains an exhaustive study of geometry arranged so that each step leads inevitably to the next. From just a few axioms, Euclid builds an edifice of geometrical conclusions, all of which he grounds firmly on solid foundations. As well as being useful in practice, it showed how logic could construct a perfectly functioning system from simple axioms that were indisputably true. Euclid's *Elements* provides an excellent example of just how far it is possible to go with the power of pure reason. No wonder some Christians decided that God must be a master of geometry. He was even illustrated holding a pair of compasses to measure the universe.

The *Elements* also contains chapters on number theory, similar to the *Arithmetic* of Boethius, and on solids. Although Euclid's masterwork was originally in Greek, Adelard came across Arabic translations during his travels. He rendered it into a Latin version that would remain the standard text until the sixteenth century.[11]

5. A manuscript illumination from a thirteenth-century French Bible of God designing the world

As well as translating the *Elements*, Adelard also wrote books of his own. These are very helpful to the modern reader because they give a good idea of what an educated person of the time knew. In his *Natural Questions*, Adelard is in conversation with his nephew. The nephew bombards his uncle with queries about the world of nature ranging from 'Why don't the oceans start to overflow due to the constant inflow of rivers?' to 'What do stars eat?'[12] Adelard's answers are often ingenious, but his lack of scientific knowledge constitutes a serious handicap. He suggests that water in the ocean is recycled back into rivers via underground streams that emerge as springs. He knows about evaporation but does not understand that it involves water turning into invisible vapour. Likewise, he believes that stars are alive because they move of their own accord. And being alive, they must eat something! Adelard suggests that they might feast on the air that rises off the earth.

There is a temptation to dismiss all this as nonsense. When Adelard's nephew asks 'Why are women more lustful than men and yet also colder?', he is simply repeating the nostrums of his age. Like the ancient Romans, medieval people believed that lust came from women. They also associated lust with physical heat. In his answer, Adelard explained that it is actually an excess of moisture in women that causes lust and not excess heat.[13] This resembles a garbled misunderstanding of Greek medical theory, coupled with some medieval misogyny. In this book, Adelard is struggling because he is attempting to make sense of incomplete information while also sticking to the rules of rational enquiry. He never tells his nephew that a subject is impious or forbidden. Nor does he invoke concepts that he would class as supernatural (even if the idea of stars having souls seems that way to us). Adelard's science was wrong, often spectacularly so, but not because he was irrational or superstitious.

In William of Conches and Adelard of Bath, we see two contrasting ways of expanding our knowledge of nature. William took as the basis for his studies materials already familiar in the early Middle Ages – Plato's *Timaeus* and the Bible – but thought deeply about what they actually meant. Adelard searched out new materials in order to bring the cream of ancient learning to a western audience. Nevertheless, their lack of access to the best of ancient Greek science seriously hampered both William and Adelard. The few old books preserved after the fall of Rome were satisfactory as textbooks of logic, but they were inadequate for mathematics and natural philosophy. Instead, they served to whet the appetite of readers who picked up many hints from them about other works that were no longer available. This made them desperate to search out the lost wisdom of ancient Greece. The resulting movement to rediscover and translate the classics was one of the most important events in western history.

The Translators

From the early twelfth century onwards, western scholars translated a vast corpus of Greek and Arabic learning into Latin. Once recovered, these works quickly came to dominate learning throughout Catholic Europe. The translation movement occurred because western Christians knew that they were missing out on a great deal of knowledge that was already available to Muslims and Byzantines. They spoke of the 'poverty of the Latins' because only the most basic material was available in their language. To remedy this they set out to acquire the riches that they knew existed in other cultures.[14] Many societies are slow to accept that they can learn something from their enemies. This was not the case with medieval Europeans, who borrowed many foreign ideas and inventions. Acquiring this knowledge and rendering it into Latin for the first time was the work of a relatively small number of scholars who had the necessary language skills. Sometimes it was difficult to find anyone who did, and so several steps were necessary. First, someone had to translate the desired work from Arabic into Spanish or Hebrew, and then someone else re-translated it into Latin.[15]

Of course, the Arabs did not simply hand over their libraries to the Christians. Most were taken by military force. By the eleventh century, Muslims had started to lose their grip on the Iberian peninsula, which they had conquered in the eighth century. The process whereby Christians slowly annexed all of Iberia, usually called the *reconquista*, took centuries. Islam would maintain a strong presence in southern Spain until Granada fell to Christian invaders in 1492. Initially, the whole of Arab Spain had the benefit of a single ruler. Dynastic squabbles and civil war put paid to this unity and the Caliphate split into various small principalities. The kings of the Christian territories entered into a shifting network of alliances with these Muslim rulers, which resulted in the slow absorption of the old Caliphate into Christendom. By cleverly exploiting disputes between Muslim states, the Christians rarely came up against their combined strength until it was far too late to turn back the tide. In 1085, the great city of Toledo fell to Alfonso VI of Castile (1040–1109). With it, the fruits of Arabic learning came into the hands of the Christian invaders. They captured the famous library of the city intact and word of the riches contained within soon spread. Contrary to popular belief, Christians were no more inclined to burn down libraries indiscriminately than anyone else was.

The most prolific of the translators was Gerard of Cremona (1114–87), who spent many years in Toledo working on the manuscripts in the library there. He learnt Arabic (as an Italian scholar, he was already proficient in Latin) and translated over 60 books on science, maths and philosophy. Probably his most impressive achievement was his work on the *Almagest*. *Almagest* means 'The Great' in Arabic, but the original Greek name for this book was the much less glamorous *Mathematical System*. It was written by Ptolemy of Alexandria (fl.AD140–170) and represents the high point of Greek astronomy. Ptolemy gathered together the best descriptions of the heavens that he could find and put them all on a firm mathematical basis. The result is a very difficult treatise that explains how to calculate the positions of the planets using complicated geometrical models. For anyone seeking to plot the stars, whether for the purposes of making a calendar or an astrological prediction, the *Almagest* was the last word in accuracy. In practice, its contents were so hard to master that most people instead used ready-prepared tables to trace the planetary movements. The Arabic name *Almagest* is testimony to the importance attached to it by Muslim scholars. It had been translated out of the original Greek into Arabic during the ninth century. Copies spread around the Mediterranean Sea to Spain where Gerard found them. This meant that the version he made in Latin was at two removes from the Greek original and mistakes had inevitably crept in. With such a complex and technical work, that was a serious drawback. Regardless of these difficulties, with the *Elements* of Euclid and *Almagest* of Ptolemy both now available in Latin, the cream of Greek mathematical science was open to western Christians. Gerard also translated commentaries and original treatises by Muslim scholars. The two most important writers in Arabic were Avicenna (980–1037), who specialised in medicine, and Averroës (1126–98), whose annotations on the works of Aristotle were so important that medieval scholars called him simply 'The Commentator'.

Gerard was not the only translator at work in Spain, and there were other sources of ancient manuscripts awaiting exploitation. In Sicily, where Adelard had started his search for mathematical learning, other Catholics searched the old libraries. There they found manuscripts in their original language – Greek. Sicily had been ruled by the Byzantines until the ninth century and still held on to much of its Hellenistic culture. At the same time, monks started digging around in other Italian libraries and found more than a few manuscripts of their own.

The Greek author that everyone wanted to read was Aristotle. Boethius, of course, had translated some of Aristotle's logical works back in the sixth century. Now, the bulk of his natural philosophy became available. This included *Physics*, *On the Soul*, *On the Heavens*, *Generation and Corruption* and *Metaphysics*. These books would form the basis for European natural science until the seventeenth century, and in the coming chapters we will be hearing more about them. However, all of Aristotle's writing is difficult, even to the modern reader, and none of his books can be called literature in the traditional sense. This is in contrast to Plato's dialogues, which are masterpieces of prose as well as philosophy. Reading Plato can be a pleasure, whereas reading Aristotle is usually nothing of the sort.

There is a good reason for the different characters of the two philosophers' works. Plato actively wanted his dialogues to be published, and so he polished them for public consumption. Aristotle's treatises are simply lecture notes that he pulled together for his own use in teaching his classes. He never intended them to be published in their own right. This is why they appear rough and unrefined. We do know that Aristotle also produced his own literary dialogues, but these were lost during antiquity.[16] Although Aristotle's surviving treatises are not always clear, they are packed with huge amounts of profound thought about almost every subject imaginable. Aristotle practically originated the disciplines of natural history with his treatises on animals, which are apparently based on careful observation and collecting specimens. The subject matter of one of his greatest books was so novel that no one even knew what to call it. *Metaphysics* got its name because it followed *Physics* in the standard list of Aristotle's works, and 'meta' is ancient Greek for 'after'. Up until Aristotle's time, this subject did not have a name because he had not yet invented it. His many works on logic have already been mentioned, but he was also concerned with ethics, rhetoric, psychology and even drama. It is no wonder that Aristotle was known simply as 'the Philosopher' and that the great medieval poet Dante Alighieri (1265–1321) called him 'the master of those who know'.[17]

Aristotle was born in 384BC at Stagaria (which is why he is sometimes called 'the Stagarite') in northern Greece where his father was the personal physician of King Philip II of Macedonia (382–336). Philip came to the throne in 359 and launched an ambitious plan to conquer the whole of Greece. By the time of his death, only militaristic Sparta was still holding out. Aristotle's father sent his son south to Athens, then the intellectual centre of Greece, for his education. He attended the academy run by Plato and remained there until the older man died. Even though the two of them disagreed about many aspects of philosophy, Aristotle always evinced a healthy respect for his teacher, even long after his death. Nonetheless, Aristotle was passed over for the job of succeeding Plato as the head of the Academy and so he decided to leave Athens. He travelled widely, got married and then returned home to take up the job of tutor to King Philip's son Alexander (356–323). Alexander would later ascend the throne of Macedonia and totally eclipse his father's achievements. In a remarkable thirteen-year reign, he succeeded in destroying the mighty Persian Empire and leading his armies as far east as India. It is no surprise that Alexander earned the epithet 'the Great' or that he is still an object of fear in the countries he conquered. Aristotle did not travel east with his royal student, but returned to Athens where he set up his own school of philosophy called the Lyceum. Here, he taught a new generation of students. According to legend, his favoured method of teaching was to stroll around talking while a gaggle of listeners followed him about. For this reason, later scholars called his philosophy 'peripatetic', from the Greek word for 'walking around'. Sadly this delightful image is based on a mistranslation. The word 'peripatetic' actually refers to the covered portico in which Aristotle and his students sheltered from the Athenian sun.[18]

In 323BC, Alexander the Great died in Persia. The Athenians lost no time in overthrowing the Macedonian government that had been installed in Athens. Aristotle, as Alexander's former teacher, was indelibly associated with the Macedonian regime and became *persona non grata*. He had to flee for his life to prevent, as he said, the Athenians 'sinning twice against philosophy'.[19] The first sin had been the execution of Socrates. Aristotle died shortly afterwards in exile. His works have allegedly reached us via a rather circuitous route. They appear to have lain, hidden and forgotten, in a cellar for a couple of centuries after Aristotle's death. Eventually a rich bibliophile rediscovered them and brought them back to Athens. Shortly after that, the Roman general Sulla (138–78BC) captured the city and carried the moth-eaten books to Rome.[20] The conquest of Greece had started a fad in Rome for Greek culture. The Romans greeted Aristotle's books as fine examples of Greek philosophy, so they copied them and thus ensured their survival.

The First Universities

At the same time as the medieval movement of translation into Latin was gathering speed, a new development in education was just getting under way. This was the invention of the university. Higher education had of course existed in the classical world in Alexandria, Athens and Constantinople. In ancient Athens, a single individual such as Plato or Aristotle ran each school, which he effectively owned and could pass on to his successor when he died.[21] The schools of Alexandria and Constantinople were both royal foundations that depended on the whim of the monarch for money and patronage.

In contrast, the new universities of western Europe enjoyed a special kind of status that made them effectively independent. Using legal precedents originally developed by the monasteries and found within Roman civil law, the new universities formed themselves into 'corporations'.[22] This meant they were not just collections of individuals, but companies that faced the world with a united front. Meanwhile, the individuals who made up each corporation could make all their internal decisions behind the scenes. Thus, the university was not dependent on any particular person and could survive even after all the original members had left or died. Nor were they beholden to princes and bishops. Indeed, universities became adept at playing secular rulers off against the Church to maximise their own freedom of manoeuvre. Using their independence, the universities were able to defend their position and hold out for even more privileges. Such was the economic clout of a big university that towns and cities would bend over backwards to keep its members happy. Moreover, if they did not get their own way, they would either go on strike or walk away from the area.[23] The development of the corporation was an extremely important advance in other ways too, as it could also be used by commercial concerns. This eventually gave rise to the limited company, which is still essential for the smooth running of capitalism.

The world's first true university was the law school in Bologna. The Holy Roman Emperor Frederick Barbarossa (1123–90) granted it an imperial privilege in 1158 when he allowed the students to govern themselves.[24] An exodus of these students, unhappy with their treatment by the town council of Bologna, led to the foundation of the university of Padua in 1222.[25] The university of Paris grew out of the cathedral schools where Peter Abelard had taught, gaining royal recognition from the king of France in 1200.[26] During the thirteenth century, the masters of Paris were sometimes on strike for years at a time, but such was the university's importance and prestige that they almost always got their own way in the end. Mystery surrounds the origins of Oxford, but it first appears in the records during the early thirteenth century.[27]

Bologna specialised in civil and canon law. It was unusual in that the students were in charge and they employed the masters to teach them; elsewhere, the masters held the reins. Some universities excelled in particular subjects. Medical schools arose at Montpellier and Padua with the aim of using the newly discovered Arab and Greek works on medicine as the basis for training professional physicians. For this, the basic textbook was *Articella*, a collection of short treatises by Avicenna, Galen (AD131–201) and Hippocrates (fl. fifth century BC) gathered into a single convenient package.[28] The most important subject of all, though, was theology, the queen of the sciences. Teaching this became the preserve of the university of Paris and, to a lesser extent, Oxford. The central text for theology was, of course, the Bible but the writings of the early Church fathers were also highly significant. The problem, as Peter Abelard had identified in his *Yes and No*, was that the fathers did not always agree. To rectify this difficulty, Peter Lombard (1095–1160) produced a synthesis called the *Sentences*, which completed Abelard's project in its attempt to make sense out of the Church fathers' disparate convictions.[29] The *Sentences* gets its name from the Latin word *sententiae*, which literally means 'opinions'. It was second only to the Bible in its importance for the study of theology in the Middle Ages.[30]

The twelfth-century renaissance represented the triumph of Peter Abelard over St Bernard. Logic was established as a vital tool for theology and the universities founded to provide a place for its study. The monasteries, guardians of Europe's literary heritage through the early Middle Ages, lost some of their significance. It was at the universities, primarily Paris and Oxford, that natural philosophy would now make its home.

CHAPTER 5

꩜

Heresy and Reason

On 20 November 1210, in the marketplace of Champeaux near Paris, ten members of a small sect called the Amalricians were burned at the stake. As the heretics were led out to their deaths, a storm was brewing. The cloaks of the clergy flapped in the wind and then pouring rain soaked the straw that was to act as tinder on the pyre. After tying the victims to the uprights, the king's executioner more than earned his pay as he got a roaring fire going in the damp conditions. The Amalricians had reason to be grateful for the weather; when wet wood burns it generates a great deal more smoke than usual and they would have suffocated before the flames even reached them. Among the people who gathered to watch, some muttered that the storm demonstrated that the souls of the heretics were damned and surely destined for hell. On this occasion, the crowd were treated to a particularly gruesome spectacle. Next to the ten live victims, the executioner threw onto the pyre the rotten remains of another individual who had died three years previously. This was the corpse of an academic, Amaury of Bène (who died around 1207), who had taught at the university of Paris. The condemned men, who included a teacher and two students from the university, were his disciples.[1]

Amaury had enjoyed a long career at the university of Paris, studying philosophy before joining the theology faculty. There, he ran into trouble by espousing beliefs that no orthodox Christian could ever accept. Although it is hard to know for certain what these beliefs were, Amaury was apparently a pantheist who believed that God and the universe were one and the same. He also claimed that there was no life after death and that the material world was all that existed. Today, these beliefs would be unexceptional but in the early thirteenth century they constituted scandalous blasphemy. That didn't stop them being attractive, however, and Amaury gained a following both among members of the university and the outside world. It was not the Church who first ordered Amaury to recant, but his colleagues in the theology faculty. Rather than give in to these demands, he travelled to Rome and appealed to Pope Innocent III. The Pope listened and then backed the Paris theologians. Now Amaury had no choice but to deny his previous beliefs. He returned chastised to Paris and died shortly thereafter.[2] However, not all of his followers were willing to renounce their alleged errors and a small sect developed around Paris teaching pantheism and materialism. It was all very well to discuss heretical ideas in the rarefied atmosphere of the university, but once they leaked out beyond the faculty walls trouble was inevitable.

Among the Amalricians, as they became known, was a goldsmith by the name of William of Arria. He claimed to be a prophet and spread the word to whoever would listen. One day in early 1210, he started to proselytise a certain Master Ralph, who listened politely but with mounting concern once he realised that William's doctrines were heretical. When William had departed, Ralph sought out the bishop of Paris. At the time, the bishop's most pressing concern was the building of the enormous new cathedral of Notre Dame. After 40 years of work, the nave, choir and transept already dominated the centre of Paris and rapid progress was being made on the western front. Heresy, however, demanded the bishop's immediate attention. The prelate ordered Ralph to join the Amalricians undercover and report back all he heard after three months. He did as he was commanded, and travelled around the countryside with the group as they spread their new religion among the peasant folk. When they returned to Paris, he told the bishop about their activities. Fourteen of the ringleaders were quickly rounded up for questioning.[3]

At first, the Church authorities used a softly-softly approach to try to coax the Amalricians back into the fold. Initially this method was successful and several of the prisoners confessed their errors. However, it appears that they subsequently backtracked and refused to renounce the heresy. They believed that Amaury had spoken the truth and that the Catholic Church itself was mistaken. Faced with such obduracy, the Church authorities gave up the fight and handed ten of the Amalricians over to the secular authorities. That could only mean one thing – they would suffer the ghastly fate of heretics under the severe civil law rather than the gentler ecclesiastical alternatives. The king's executioner burnt ten of the sectarians and four others, who had probably recanted at the last minute, were confined to a monastery for the rest of their lives.[4]

The University of Paris Bans Aristotle

The sad story of the Amalricians was to have long-lasting consequences for natural philosophy. The bishop of Paris was shocked to find a nest of heretics so close to home and set out to find the root of the problem. It seemed to him that the worst fears of St Bernard of Clairvaux had been realised. Amaury had used reason and logic not to defend Christianity, but to undermine it. The bishop formed a local synod of his colleagues who issued the formal excommunication of Amaury and his followers. The decree went on to ban the books assumed to have inspired the heresy. 'Neither the works of Aristotle on natural philosophy', it read, 'nor their commentaries are to be read at Paris in public or private. This we forbid under penalty of excommunication.'[5] Aristotle, the bishops thought, was the original source of many of Amaury's ideas. It is no use wondering what exactly the synod found objectionable in the books that they banned because, almost certainly, they had not read them. There is nothing like a moral panic to give rise to bad law.

Nevertheless, there is no doubt that Aristotle's natural philosophy contradicted Christianity on some central points. Christians believe that we all have souls that survive death; that God created the world at a definite point in the past; and, most fundamentally, that God is all-powerful and not subject to the laws of nature himself. Aristotle disagreed with all of this. He insisted the world was eternal, that it had existed forever and always would exist. There was no moment of creation and no creator. He was also highly ambivalent about personal immortality. Although he believed that humans have a soul, he also held that it dissolves at death. There is no last judgement, heaven or hell. This was bad enough, but Aristotle hardly believed in God either. The idea of a personal God, who answered prayers and intervened in the lives of men, was complete nonsense as far as the Greek philosopher was concerned. He did believe in a 'prime mover' who kept the universe turning, but this impersonal being had no interest in mankind and was nothing like the God of the Bible.

The Church could deal with these problems quite easily by officially stating that Aristotle was wrong where he contradicted a specific Christian doctrine. However, there was another issue that ran like a vein all the way through his natural philosophy and metaphysics. This was the question of necessity, which was of critical importance to the development of science.

Both Aristotle and Christian thinkers accepted that there are fixed natural laws, but the Greek philosopher went further. For Aristotle, the iron shackles of logical necessity determined what the laws of nature had to be.[6] They were not just the ones upon which God had deliberately decided, they were the only ones he could have used. Even if God had actually created the world, he would have had no choice about how it turned out.[7] It was as if he had to follow a pre-existing recipe and, unless he followed it precisely, the result would be a mess. There was no divine freedom and the Creator was not really responsible for the way of the world. That is not to say that Aristotle thought that the world was without purpose. But for him, the purposes of nature were not the same as the divine plan of God.

Obviously, this was dangerous stuff and once it became clear that it had influenced Amaury, the bishop of Paris's synod promptly banned it. Scholars at the university of Paris continued to study diligently the rest of Aristotle's works, such as his logic and ethics, so they probably flouted the ban as well. Besides, it only applied to Paris. At the other universities, such as Oxford and Cambridge, the synod's writ had no authority. In 1229, advertisements for a new university at Toulouse made much of the fact that students there could freely read those of Aristotle's works that were prohibited at Paris. The Pope reacted to this impudent attempt to poach students from Paris by extending the ban to Toulouse as well.[8]

The professors at Paris, which was, after all, the biggest and most prestigious university in Europe, were furious that the bishop was presuming to tell them what they could and could not read. Almost everyone agreed that the synod had overreacted in banning Aristotle's natural philosophy outright. However, getting a synod to admit they had made a mistake was not an easy thing to do. The only person who could overturn the decision was the Pope, and he would not want to seem soft on heresy. The solution was simply to wait a while until the fuss had died down and then try to get the ban rescinded.

By 1230, Notre Dame de Paris's western facade was complete, together with the rose window that still dominates the nave. The famous pair of towers that would complete the structure was rising over the city. The cathedral represented the church's renewed self-confidence, and this was reflected in the decision to revisit the question of the ban on Aristotle. In 1231, the university suggested to Pope Gregory IX that the original synod may have gone overboard in banning all the natural philosophy books. His Holiness agreed and sent two letters to Paris. The first pardoned everyone who had ignored the initial ban. The second read:

> We have learned that the books on nature prohibited in Paris by the provincial council are said to contain both useful and useless matter. Lest the useful be tainted by the useless, we command that ... examining the books as is convenient, subtly and with prudence, you entirely excise what you find there to be erroneous or likely to give scandal or offence. Thus, with the suspect material removed, the rest may be studied without delay and without fault.[9]

We do not know if a committee was ever formed to expunge Aristotle's books because, as it turned out, no one bothered to remove anything. Reading all of the previously banned books on nature quickly became a central component of the course of study. The university of Paris syllabus from 1255 included all of Aristotle's books in the curriculum for students taking their bachelor's or master's degrees.[10]

If Aristotle was so dangerous, why did the authorities make reading him a compulsory element of courses at all European universities, even going so far as to take the highly unusual step of reversing a ban? The answer is that, as the Pope said, Aristotle was incredibly useful. Ironically he was, when properly understood, one of the most formidable weapons against exactly what the Church feared most – heresy.

Heresy and the Inquisition

During the Middle Ages, heresy was an increasing problem. Especially prevalent in the south of France, the Rhineland and northern Italy, it was not as if heretics had only recently appeared. They had existed throughout the previous 800 years but the Church had hardly bothered them. When heretics did not represent a serious threat to the growing power of the papacy, the Church could safely ignore heterodox belief. Only high-profile cases like Peter Abelard's attracted papal attention, and then only because his prestige as a theologian was so great. In the twelfth century, this relaxed attitude changed.

One important feature of Christianity at this time was the increased level of lay piety and the pressure for ecclesiastical reform.[11] This heightened religious feeling of lay people, perhaps increased further by the start of the crusades in 1095, meant that extremism became more common. Pope Urban II (1042–99) had preached the first crusade from his pulpit with the intention of recapturing the Holy Land from its Muslim rulers. Not everyone wanted to travel all the way to the Middle East, however, and mobs began to look for infidels closer to home. One notorious aspect of this was that Jews became the targets of violence. Canon law allowed Jews to practise their religion and stated that they could not be subject to violence or be forced to have their children baptised.[12] But to the mob, these edicts were iniquitous and the Jews fair targets. A series of pogroms by domestic crusaders along the Rhine killed thousands in 1096 despite the efforts of the local bishops to protect the victims. The mobs were not above attacking the clergy from whom the Jews had sought sanctuary. The archbishop of Mainz had to flee for his life for this reason.[13]

Heretics in the midst of the faithful were believed to be a further manifestation of the enemy within. The common people feared that they would be damned themselves if they allowed deviants to live peaceably among them. This is why Master Ralph had so quickly reported his encounter with the Amalricians to the bishop of Paris. And if the authorities were insufficiently zealous in dealing with the heretics, then the mob would do it for them. At Cologne in 1144, the crowd seized two leaders of a heretical sect even while the archbishop of Cologne was interrogating them. Without any thought for due process, the mob dragged their prey outside and burnt them there and then.[14]

The Church also grew more concerned about heretics because they were growing more common. The largest heretical sect was the Cathars. They seem to have originated in eastern Europe and may have been a descendant of an early offshoot of Christianity called Manichaeism. Both the Cathars and the Manichaeans believed that there were two gods, one good and one evil. There was disagreement over whether the two gods were equally powerful, but the evil one created and ruled the material world. This meant that matter itself was evil and mankind should try to escape it by concentrating wholly on the spiritual.[15] The most committed Cathars, called the *perfecti*, were strict ascetics who abstained from eating meat and having sex. They hoped that by their refusal to countenance material things they might be able to free themselves from the shackles of the flesh. They were strongest in southern France around Toulouse, where they enjoyed considerable influence and the protection of the nobility.

With heretics growing in numbers and the common people taking matters into their own hands, something had to be done. While it accepted that unrepentant heretics deserved death, the Church was perturbed about not giving them a chance to return to the faith. It was clear that dealing with heresy required a new system. For this reason, a series of dynamic popes developed a legal process called inquisition.[16]

During the Middle Ages, there was no single monolithic institution that we can call 'The Inquisition'. Inquisitors were simply individual agents of the Pope who travelled to areas afflicted by heresy and used their special powers to deal with it. They worked in conjunction with the local secular and ecclesiastical authorities. Several papal bulls granted the inquisitors their powers, culminating in two from Pope Gregory IX that were issued in the same year that he lifted the ban on Aristotle. Authorising the activities of two inquisitors, he wrote:

> We seek, urge and exhort you ... that you be sent as judges into different districts to preach to the clergy and people gathered together where it seems useful to you. And using those discrete people that are known to you, seek out diligently those who are heretics or reputed to be heretics. We concede [to your lay helpers] the free faculty of wielding the sword against enemies of the faith.[17]

What made the inquisitors novel was that they used the latest legal techniques to investigate heresy. This was a consequence of the new interest in Roman law at the university of Bologna that we touched on at the end of the last chapter. Until the thirteenth century, most countries continued to use the old legal codes that they had followed for generations. In these codes, the legal process was started when a member of the public made a formal 'accusation'. When criminal accusations were made, the defendant had a number of ways in which he could demonstrate his innocence. One was to produce character witnesses who would demand an acquittal. Another was to undergo trial by ordeal. In neither case did real evidence have much relevance. Furthermore, the accuser was vulnerable to punishment for defamation if the defendant was acquitted. Someone with a bad reputation could never win a legal battle against someone who was generally thought of as honest. The Church frowned on trial by ordeal and banned the clergy from participating in it in 1215.[18]

With the new system of 'inquisition', the 'accusation' method of justice was eventually abandoned altogether. Instead, the authorities appointed a magistrate to investigate the crime, interview witnesses, examine the evidence and reach a verdict. In the case of heresy, the magistrate was an inquisitor and appointed by the Pope. The system was an obvious improvement over the old ways and slowly spread to secular justice too. In fact, it worked so well that it still forms the backbone of criminal investigation in continental Europe to this day.[19]

The inquisitors had to follow strict rules and reserved the most serious punishment only for heretics who were obstinate in their error or were repeat offenders. Everyone had a second chance. When an inquisitor arrived in an area, he began his mission by declaring that he would deal mercifully with all heretics who gave themselves up. He then followed through any leads that he had received, made arrests, carried out interrogations and declared who he thought was guilty. While inquisitors had a special dispensation from the Pope to use torture, this was rare.[20] The popular image of dank inquisitorial dungeons equipped with a variety of imaginative means of torment is a later myth, popularised after the Reformation by Protestant writers.[21] If the inquisitor found someone guilty of heresy, then they had an opportunity to recant and perform a penance. Most people took this option and the resulting penances were often quite lenient. However, those convicted of heresy were on notice that the inquisitors would deal with a second offence much more severely. In that case, as relapsed heretics, they could face life imprisonment or worse. In the same boat as repeat offenders were those whom the inquisitor convicted but who refused to admit the error of their ways. In the most serious cases, the inquisitor would hand over relapsed and obstinate heretics to the secular authorities. Officially, the Church could not execute a suspect but inquisitors knew perfectly well what the fate of those they handed over to the secular arm would be.

In 1017, King Robert II (972–1031) of France had instituted burning as the punishment for heresy and the penalty spread through most of Catholic Europe.[22] Nevertheless, it was not universal. In England, parliament did not enact heresy laws until the fifteenth century, while in Venice the penalty was death by drowning.[23] The authorities also deemed that Amaury himself had continued to espouse his heretical ideas even after he had repented in front of the Pope. For that reason, they exhumed and burnt his body – gruesome revenge on a man who had caused so much trouble even after he was dead.

The inquisitors of the Middle Ages have a deservedly poor reputation. There is no defence for subjecting people to an agonising death over religious disagreements. Executions were uncommon (occurring in about five per cent of cases in the surviving records)[24] because when it came to the crunch, few people wished to be martyrs. There was, however, another side of the struggle for doctrinal orthodoxy. To effectively combat heresy, the church realised that a two-pronged approach was necessary. It was no good convicting people for false beliefs if they had no idea what they were supposed to believe. The answer was to provide good preaching for the common people so that they had the chance to learn the rudiments of the orthodox Catholic faith.

The New Orders: Dominicans and Franciscans

Dominic Guzman (1170–1221), later canonised as St Dominic, was one of those who could see that the Church was just not doing its job of spreading the faith well enough. Priests were absent from their parishes, the bishops were too interested in secular power and the Church as a whole could be a pretty unedifying spectacle. No wonder, Guzman thought, that the common folk were seeking their spiritual nourishment from alternative sources. He also realised that the men sent out to preach the word of God had to be of unimpeachable virtue. It was no use telling peasants about Jesus if you were clearly not taking much notice of his message yourself. To help solve these problems, Guzman founded the Order of Preachers in 1216. Its job was to combat the spread of heresy among ordinary people by going out and explaining Christianity. His new order dressed in a simple black habit (and hence they were known as the Blackfriars), lived on charity and did their work among the poor and dispossessed, often in cities rather than the countryside.[25] The Blackfriars had to be learned enough to promulgate and defend the Christian faith, so they began to take advantage of the educational opportunities available at the new universities.[26]

Guzman's idea was a great success. Quite soon, his order took on the name of its founder and became the Dominicans, although to this day they still have the letters OP (for Order of Preachers) after their names. Their mixture of piety and fearsome intellect meant that the Dominicans were also the perfect people to act as the Pope's inquisitors. The zeal with which they preached and acted against heresy earned them the nickname 'hounds of God' (a pun on the Latin version of their name, *domini canes*), of which they were very proud. Frescoes in Dominican churches, like Santa Maria Novella in Florence, sometimes show St Dominic surrounded by his dogs sniffing out heretics.

At about the same time as Dominic was setting up his order of preachers, Francis of Assisi (1181–1226) launched his own evangelising mission in the form of the Order of the Friars Minor, or Franciscans. They have always enjoyed a rather less sinister reputation than the Dominicans, partly because St Francis was such a gentle soul himself. However, after a slightly slower start, the Franciscans also began to seek out the best education and produced several towering thinkers to match the best of the Dominicans. The Franciscans also worked as inquisitors, although they never took to it quite as well as the Dominicans. This was partly because the Franciscans spent the first century of their existence torn apart by doctrinal disputes about just how poor and beggarly they needed to be.[27] When many members of the order stood accused of heresy themselves, the Franciscans were not best placed to lecture anyone else on orthodoxy.

It may well have been the Dominicans who were behind the lifting of the ban on Aristotle's natural philosophy in 1231. They had quickly realised that they needed Aristotle when they started to dispute with Cathar heretics. Cathars, to recap, were dualists who believed that the material world was evil and created by a wicked god. It was no good pointing out to them that the first chapter of Genesis clearly states that the one God created the world and said that it was good. The Cathars did not revere the Old Testament and thought that the God it described was a completely different one from the God portrayed in the New Testament.[28] It is difficult to have an argument without any common ground. Luckily, Christian and Cathar alike respected the works of Aristotle. Using them, the Dominicans could debate and prove that the Cathars were wrong to think that the material world was wicked.[29]

For example, although Aristotle did not believe that the world had been created by God, he did believe that nature is essentially benign. His books on natural history provide case after case to show how nature provides animals with the niche in which they can best live their lives. Aristotle's world is beautiful and exquisitely crafted for the benefit of the creatures that inhabit it. There was no way, the Dominicans insisted against the Cathars, that this fecund world could be the product of an evil deity.

To use his works successfully, the Dominicans had to know their Aristotle back to front, so it was necessary that his works remained a central part of the university syllabus. But what should be done about the obvious dangers of his unorthodox opinions? Luckily, Christians had centuries of practice in dealing with pagan philosophy. There was a debate in the early church about whether it was permissible to read Plato. Some extremists advocated ignoring Greek philosophy altogether and insisted that everything that you needed to know was in the Bible. St Augustine of Hippo disagreed. He wrote:

> Any statements by those who are called philosophers, especially the Platonists, which happen to be true and consistent with our faith should not cause alarm, but be claimed for our own use, as it were, from owners who have no right to them.[30]

In other words, it was perfectly acceptable to pick and choose the most agreeable-elements of pagan thought. Even if some of Aristotle was dangerous, that did not justify abandoning his works altogether. Rather, Christians should co-opt the parts of his philosophy that could serve Christianity. Everyone agreed that theology was still the most important branch of knowledge but natural philosophy now had a role as the servant of theology, or the 'handmaiden' as they put it in the Middle Ages.[31] Being a handmaiden had some practical advantages, especially when serving the queen of the sciences. It meant that natural philosophy had a protected status in the universities because it was a prerequisite for the study of theology. In any case, that was the theory. As it turned out, despite the Pope's lifting of the ban in 1231, the controversy about Aristotle rumbled on through the thirteenth century. Even with the handmaiden analogy, philosophers and theologians still could not agree where the line between their subjects was, nor to what extent reason should criticise or amend sacred doctrine. What everyone now needed were clear boundaries between the disciplines. Inevitably, it would be hard to agree where these should be.

CHAPTER 6

⚭

How Pagan Science was Christianised

The Dominicans had realised that while Aristotle's philosophy could be dangerous, it was also very useful in the battle with heresy. They needed to combat the extremists on both sides so that they could use pagan ideas for the benefit of Christianity. It is no surprise to find that the two biggest guns in the battle for Aristotle were Dominicans. More than anything else, it was the work of this pair of scholar-saints, Albert the Great (c.1200–80) and his most celebrated pupil, Thomas Aquinas, that made Aristotle not only respectable but essential to Christian theology. As a contemporary wrote of them, 'Doubtless many others were famous during this same time both in life and thought. But these two transcended and deserve to be placed before all others.'[1]

The Universal and Angelic Doctors

Albert the Great was born in a region of south-western Germany called Swabia and joined the Dominicans as a young man. The order was attracting many brilliant men and Albert wanted to be part of the exciting new thinking that it had spawned. He spent his career at the university of Paris and at Cologne, where he was partly responsible for the foundation of a new university. His tomb can still be found in the crypt of Cologne's fine St Andreas church. Commemorated as one of the greatest of all German thinkers, Albert had to wait until

1931 before the Catholic Church finally declared him a saint. He acquired the honorific 'the Great' during his own lifetime by virtue of his enormous appetite for learning and the sheer size of his output. The standard edition of his collected works fills 38 capacious Latin volumes.[2] His marvellous mind scrutinised everything under the sun, and he questioned and analysed all that he read. His books covered the Bible, the early Church fathers and, above all, Aristotle's natural science and natural history. It is little wonder that Albert earned the title of 'Universal Doctor'.

All this knowledge came at a price. Almost inevitably, Albert gained a reputation as a magician among the common people. Rumours even claimed that he had inherited Gerbert's talking brass head and used it to gain illicit knowledge. In fact, Albert did write some suspect works and his infamy in this respect was not completely undeserved. The contemporary book *Mirror of Astronomy* takes a very liberal line on the permissibility of magic and may well have been written by Albert himself.[3]

Although he based his philosophical work primarily on the books of Aristotle, he was not an uncritical admirer. He warned, 'If Aristotle had been a god then we must think that he never made a mistake. As he is a man, he has certainly made mistakes just like the rest of us.'[4] Albert also restated some important principles about natural philosophy. 'Experience', he explained, 'is the only safe guide in such investigations.'[5] We need to be careful, however, not to over-emphasise the originality of Albert's remark. It did not mean that he performed experiments. 'Experience' meant passive observation rather than active interrogation. For him, investigating nature was a rational and literary activity more than an empirical and practical one. Often, it involved reading the works of many authorities and judging between them. That said, he did from time to time insist that his own experience contradicted Aristotle who must, therefore, have been mistaken. Even the Philosopher's authority was sometimes suspect. 'It is the task of natural science', Albert said, 'not simply to accept what we are told but to enquire into the causes of things.'[6] In other words, as William of Conches had realised, natural philosophy is about discovering natural secondary causes. Albert was clear that the existence of miracles did not prevent the natural philosopher from doing his work.

After noting all his achievements, we must accept that Albert's fame rested squarely on how much he knew and not what he did with all his knowledge. His learning was wider than it was deep and he could sometimes appear more as a collator of facts than as someone who was pushing back the boundaries of scholarship. There is no system of philosophy named after him and he is largely unknown outside his native Germany. On the other hand, Albert's treatises were systematic and he did lay foundations for the assimilation of the mass of newly translated Greek philosophy. Perhaps his greatest achievement was to recognise and nurture true genius when he saw it. His most gifted pupil ranks as one of the most famous philosophers who has ever lived, and today we still recognise him as a truly revolutionary thinker. That man was Thomas Aquinas.

Whereas Albert had to wait until the twentieth century to be canonised, Thomas Aquinas was made a saint less than 50 years after his death. He was a humble and devout man, as no one doubts, but he owes his canonisation to his phenomenal works of philosophy and theology. They have been one of the intellectual bulwarks of Catholicism ever since, to the extent that the Church has awarded him the title of 'Angelic Doctor'.

Thomas Aquinas had the best possible start in life. He was born in central Italy and was closely related to more than one royal family. A quiet and rotund boy, he was clearly unsuited to life as a soldier, so his family mapped out a career in the church for him. Bishops were as powerful and respected as secular rulers, so the clergy was an acceptable profession for such a well-born individual to enter. He could even be a monk without any impropriety. As a member of the Benedictine order, he would live in the luxury of a rich monastery and be a mover and shaker in society. In preparation for such a role, his parents sent him to the university of Naples for the requisite education.

Once there, however, Thomas did not act as his family wished. His piety was entirely genuine and he demonstrated it by joining the

Dominican order. This was an entirely different matter to becoming a Benedictine monk. Instead of being a prince of the Church, he had entered into a life where he could own nothing and even had to beg for his food. His family were not amused, and his mother ordered his older brothers to abduct their wayward sibling. They imprisoned him in the family castle. Even shut up in a darkened room to think things over, Thomas refused to change his mind. According to a story told by his earliest biographer (and also a noted inquisitor) Bernardo Gui (c.1261–1331),[7] the family went as far as to hire a prostitute to persuade him of the virtues of sin. Thomas threw her out of his chamber.[8] Eventually, the family relented and Thomas continued in his vocation. Wisely, the Dominicans moved him well away from Italy to be educated by Albert the Great in Cologne.

Thomas, the chubby and taciturn newcomer, did not impress his fellow pupils and they dubbed him a dumb ox. Albert was a better judge of academic potential. 'We call this lad a dumb ox', he said, 'but I tell you that the whole world is going to hear his bellowing.'[9] So he invited his quiet student to come with him to the intellectual capital of Europe – the university of Paris. Once there, Thomas rapidly made a name for himself as a thinker of serious note. He received his doctorate in theology from the Paris faculty in 1257 before embarking on a life of teaching, preaching and writing. Avoiding the honours with which the church attempted to shower him, most notably the archbishopric of Naples, he remained a simple friar until the end of his days.

As we might expect for a man so esteemed by the church, there are plenty of stories about Thomas that fill out the rather meagre official record of his life. A couple of these bear repeating even if not entirely true, because they reflect contemporary perceptions of the man. While still a student of Albert the Great, it is said that one of Thomas's peers took pity on him and assumed by his silence that he was having trouble following a lesson in logic. The kind student sat down with Thomas after class to run through the work so that the dullard might not fall too far behind. Thomas suffered the indignity with patience as his conscientious tutor laboriously explained matters that the future saint had mastered years earlier. Eventually, after the student had made an obvious error, Thomas politely intervened and stopped his would-be master in full flow. He proceeded to offer a far more lucid exposition on the point than even Albert the Great himself had achieved. Word quickly got around that Thomas Aquinas was not in need of remedial classes.[10]

Much later, while Thomas was a professor at the university of Paris with his reputation secure, King Louis IX (1215–70) invited him to a great banquet. Louis was later canonised himself, at least in part for his unsuccessful efforts as a crusader. Without much enthusiasm, Thomas politely attended the feast but took no part in the conversation. Soon he was lost in thought. Indeed, he quite forgot where he was and, upon receiving a sudden insight into an obscure problem that had vexed him, slammed his fist into the table and cried out aloud, 'That's it!' An uncomfortable silence greeted this outburst, and all eyes turned to the king to see what his reaction would be. Luckily, His Majesty was aware that scholars needed to be indulged and so merely ordered that quill and parchment be brought to the table so that Thomas's untimely thoughts should not be lost to prosterity.[11]

As soon as he became a professor of theology, Thomas was thrown into the latest wrangle about Aristotle that was raging in Paris. This time, the argument hinged on the work of 'the Commentator' on Aristotle – the Arab scholar Averröes. Christians had come across many of his works at the same time as they had been translating Greek philosophy into Latin. He was a Muslim who had lived in Islamic Spain during the twelfth century, at least when not in exile as a result of his radical ideas. He based his philosophy firmly on Aristotle and in some ways went even further. For instance, he suggested that the universe was completely deterministic. In other words, we have no free will and hence no moral responsibility. He also agreed with Aristotle that the universe had to be eternal and that there was no life after death.[12] These were exactly the sorts of ideas that had got Amaury into so much trouble. Merely discussing them was fine, but what you could not do was say that they were true. Averröes said they *had* to be true.

This emphasised Aristotle's idea that the laws of nature are necessary and even constrain God's freedom of action. Averröes was certainly not content for philosophy to be a handmaiden of theology. Reason, he implied, trumped faith every time.

Thomas Aquinas's Scholastic Method

Thomas Aquinas was the prime exponent of the extremely methodical and carefully organised system that medieval philosophers used to construct rational arguments. Today, we call this mode of argument the 'scholastic method' and hence the entire body of medieval thought is often described by the single word 'scholasticism'. We can see what this means in practice by way of an important example taken from the thought of Thomas himself: his attempt to prove that God exists.

A scholastic argument almost always begins with a question and so near the start of his greatest work *Summa Theologiae*, Thomas Aquinas asks 'Does God exist?'[13] Rather than launching himself immediately into explaining why he thinks God does exist, Thomas, in typical medieval fashion, first sets out a couple of arguments for the non-existence of God. Although he is very brief, the two arguments he chooses to use are strong ones. The first is the problem of evil, which states that if God existed there should be no such thing as evil. An all-powerful and good God should be able to ensure that no evil would exist, whereas clearly it does. The second of Aquinas's arguments for atheism is that the concept of God does not explain anything. We can use science to understand everything about the world and so we do not need to postulate a God.

The next stage is to state the contrary to the arguments for atheism. This is not a counter-argument, just a statement that Thomas intends to defend summarised by a quotation from the Bible or some other authority. In this case, it is that God says in Exodus 3:14 that he does exist.

Only now do we get to the meat of the argument. Thomas presents us with no less than five proofs for the existence of God, known to philosophers as the 'five ways'. Thinkers have been arguing about them ever since. While his five proofs, like the ontological argument of St Anselm, are not finally convincing (or there would be no such thing as an intelligent atheist), they did help demonstrate to Christians that reason was their friend and not something they needed to be afraid of. If they could use rational arguments to show what they believed by faith, then rational arguments were a course well worth pursuing. Here we will restrict ourselves to a combination of the first and second of Thomas's arguments because they give a good idea of how his mind worked.

Taking his cue from Aristotle, Thomas insisted that every effect in the world must have a cause. Things cannot happen for no reason at all. However, Thomas realised that this approach led to a problem because as soon as you have determined the cause of an effect, you have to ask what the cause of that cause is. Even if you can find that out, you are still faced with an infinite series of causes going back forever.

A modern example can illustrate the problem.[14] Imagine a car is stuck at the traffic lights at Trafalgar Square in London. These take an inordinately long time to change from red to green and London traffic being what it is, if you are at the front, a large queue often builds up there. Suddenly a van shunts into the back of the car and causes a dent. The driver of the car gets out in order to remonstrate with the van driver and obtain his insurance details, but finds that he is already remonstrating with the driver behind him. It turns out that the van shunted into the car because the vehicle behind the van had run into it. Likewise, the same thing had happened to the next car back in the queue and so on. Imagine that London traffic has now become so bad that the queue is infinitely long. As far back as it is possible to go, everyone has been shunted into the car in front of him or her by the car behind. What makes this doubly unfortunate is that although the repair bills and aggravation are also infinite, there is no one responsible for the initial shunt because the queue has no end. Clearly, Thomas would have said, this is impossible. Without an initial cause, there could never have been any effect. There has to be a first car colliding to begin the whole chain of events.

Thomas reasoned that this must always be the case: everything is caused by something else, and so there must be some first cause that set the ball rolling. Otherwise, nothing should happen at all. However, we have already stated that everything in the world has a cause. So we must conclude that the first cause, the first movement, came from outside the world. That supernatural cause is what Aristotle called the prime mover and Thomas Aquinas called God.

Thomas rounds off the question of God's existence by briefly responding to the objections that he started with. He replies to the problem of evil by claiming that God's greatness is such that he can draw good even from evil. Finally, he reiterates that the laws of nature require a 'first cause', that is God, to explain them. Thus, according to Thomas, science can never explain everything.

Averröes versus Aquinas

The most prominent Averröist at Paris while Thomas was teaching there was Siger of Brabant (who died around 1282), a radical philosophy professor who had gathered a group of like-minded students around him. Siger was a combative character who was fully involved in the rough-and-tumble of university politics. His name first appears in the records when he took part in an attempt to kidnap a rival. He also played a leading role in the conflicts between students of different nationalities that took place in the 1260s.

His philosophy was just as controversial. He could not resist the more radical interpretations of the work of Aristotle and Averröes, believing that philosophy proved that the earth must be eternal and that humans lack individual souls. We are, Siger said, all part of a single hive-intellect that animates the rational faculties of the entire human race. Of course, he could not safely say that these conclusions were actually correct, so had to content himself with claiming that reason demonstrated their truth. Contrarily, he continued to assert that we can know by faith that, in reality, God did create the world and that each man does possess an immortal soul. In other words, Siger claimed that reason said one thing and faith said another. This flew in the face of Christian orthodoxy, which stated that the two could never conflict because both came from God. Echoing St Augustine and St Anselm, the Church believed that reason illuminated faith and you could not have one without the other.

Worse was to follow. Some of Siger's disciples allegedly claimed that when reason and faith conflicted, they were both true! Thus, in the philosophical sphere, the world was eternal and in the theological sphere, God created it. This was called the doctrine of the two truths. The bishop of Paris roundly condemned the idea as complete nonsense.[15] If faith and reason seemed to conflict, he thought, it was down to human error. In 1270, he issued a decree outlawing some Averröist ideas. Siger then apparently recanted and returned to lecturing.

Thomas Aquinas's interventions in the great debate at Paris over Averröism were hugely significant. He was determined to show that he could make Aristotle into a servant of Christianity, and also to demonstrate that Siger was wrong to claim that reason led to conclusions in conflict with the faith. Thomas wrote direct rebuttals to some of Siger's work but concentrated on producing his own synthesis of Christian and pagan philosophy. This found its definitive expression in his *Summa Theologiae*. It is a massive work and, despite the fact he never quite finished it, remains the highest achievement of medieval scholarship. What he achieved was such a successful amalgamation of Aristotle's philosophy with Christian doctrine that some Catholics have since failed to distinguish between the two.

Thomas combined an enormous respect for non-Christian authorities with his reverence for the Church fathers. As well as Aristotle and Averröes, Thomas found inspiration in the writings of Moses Maimonides (1135–1204). Maimonides was the most renowned Jewish philosopher of the Middle Ages and spent all his life living under Islamic rule. He wrote his most important book, *The Guide for the Perplexed*, for a student who was having trouble reconciling what he had learnt about natural philosophy with the words of the Bible. As Maimonides explained in his introduction:

The object of this treatise is to enlighten a religious man who has been trained to believe in the truth of our Holy Scriptures, who conscientiously fulfils his moral and religious duties and, at the same time, has been successful in his study of philosophy.[16]

Thomas Aquinas thought that his problem was the same. He had to reconcile the Averröists, who believed philosophy was supreme, with their opponents, who thought it was dangerous. Like Maimonides, he united the two sides and showed the value of both. Thomas gave human reason a great deal of credit. While he did not hold that the existence of God was self-evident to the human mind, he did believe that unaided reason could rationally prove God existed.[17] He also thought that we could discover a great deal about the world around us. However, there are limits to reason. When it comes to properly understanding the mysteries of the faith, such as the Trinity, we also need supernatural illumination. Actually to know for certain that Christianity is true is, for Thomas, a gift from the Holy Spirit and not something we can ever achieve by ourselves.[18]

For natural philosophers, reason was the correct tool and Thomas showed how they could make best use of it. According to him, the world was real and it was subject to cause and effect. Things did not happen without any reason but behaved in predictable ways. Thus, it was worthwhile to study causes. Of course, Albert the Great's opinion had been that causes were the whole point of natural philosophy. In particular, Thomas stood up for the doctrine of secondary causes as a valid way for a Christian to investigate the world. He did not accept that it was impious to say that a plague was caused by a disease rather than attributing it directly to the will of God. Nor did he think the world was evil. After all, it had been created by God and he could not do anything wrong. Aristotle had said that the world must be eternal, but Thomas showed how the Greek's arguments were flawed. However, he admitted that he could not prove conclusively that God created the universe because of the limits of human reason. So, according to him, we must look to the book of Genesis to find the truth. Thomas also stood against determinism and in favour of human free will. Against the Averröists, he believed that humans each have their own souls.

Siger and Thomas aimed several of their tracts at each other before Thomas left Paris and returned to Italy in his last years. Shortly before he died, he had a mystical experience while at prayer that caused him to give up writing. He declared that all he had achieved was mere chaff, at least compared to the vision of God's glory which he had experienced. Few have agreed with Thomas's own assessment of his work.

The Condemnations of 1277

After Thomas Aquinas left Paris, Siger and his students continued to make trouble with their radical theories. In a new effort to end the dispute, the university instituted a reform that aimed to separate the warring factions. As we have noted, much of the trouble for the Church in accepting the work of Aristotle was down to the lack of a clear border between philosophy and theology.

Thomas Aquinas had gone some way towards providing that border, because he was always clear about what could be known through reason alone and what requires faith. Just to be on the safe side, the university of Paris required philosophers to agree that they would not meddle in matters of theology. From 1272, new graduates had to swear that they would never argue about sacred doctrines and, if they had to mention them, they would always come down on the side of orthodoxy.[19] From a medieval point of view, this was a sensible and understandable arrangement. Already in the thirteenth century, lawyers and doctors were taking action against amateurs who professed knowledge in their fields of expertise. Everyone agreed that theology was by far the most important and highly skilled profession of all. Only the most rigorously trained individuals could practise it. It took seven years to qualify as a theologian and that was after having spent at least four years working on a first degree. No wonder theologians were jealous of their prerogatives. If a philosopher did want to tackle matters of faith, then he was perfectly entitled to join the theology faculty and train as a theologian.

By 1277, it was clear that not even this reform had ended the trouble. The ideas of Averröes had continued to spread, perhaps because existing graduates did not have to take the new oath. Taverns and inns echoed with arguments between rival factions. The row threatened the reputation of the university as well as its unity.

Siger had decided that the situation was simply too dangerous for him to remain. The local inquisitor was pursuing him for questioning, although Siger was never convicted of heresy, even *in absentia*.[20] He travelled to Italy, possibly seeking to take the matter up with the Pope. It is unlikely that the Pope in question, John XXI (c.1215–77), would have been very sympathetic. He had already written to the bishop of Paris demanding that he do something about Averröism. It is worth noting that before he became Pope, John XXI (then called Peter of Spain) had been a student at Paris himself. In the course of his academic career, he wrote the textbook on logic that dominated the field until the close of the Middle Ages. He also trained as a physician and occupied the chair of medicine at the university of Sienna. A medical handbook attributed to him was enormously popular. Certainly, this Pope had no problem with the proper application of reason. On his election, he had a private study added to the papal palace in Viterbo near Rome so that he could get on with philosophical research in his spare time undisturbed. Sadly, after less than a year as Pope, John XXI was killed in a freak accident. He had been working in his new study when the roof collapsed on him.[21]

The bishop of Paris, even before he had received the Pope's letter, set about compiling a list of all the things that he found objectionable about Siger's and Averröes' ideas. It turned out to be quite a long list – 219 propositions were condemned in all. The document forbade anyone at the university from teaching or holding any of the condemned opinions on pain of excommunication. As for Siger, for reasons unknown he was murdered by his secretary. After his death, he enjoyed a reputation as an important thinker, cemented by Dante Alighieri, who placed him in paradise in the *Divine Comedy*.[22]

The 219 condemnations of 1277 hold a very important and controversial place in the history of science. The main point at issue was whether God was constrained by natural laws himself or whether he was above them. The condemnations themselves are fairly confusing to read because they consist of a list of things that people were *not* allowed to say, rather than setting out what people *should* believe. Several of the condemned statements deal with specific cases where God is not to be restricted. For instance, among the now-heretical statements were that God 'could not make several universes', that he could not 'move the universe in a straight line lest a vacuum result', nor 'make more than three dimensions exist simultaneously'. The bishop of Paris summed up the condemnations by prohibiting anyone from saying that 'God cannot do anything that is naturally impossible.'[23]

Nobody, as far as we can now tell, thought that God had actually created extra universes or that there really were more than three dimensions. That wasn't the point. The bishop of Paris was just telling everyone that they should never say never. Science could not make final pronouncements, especially those that limited the freedom of God. This meant that the bishop was not stopping people from investigating how the world worked, he was merely preventing them from saying that God was constrained in how he could organise the world. Rather than restricting the work of natural philosophers, the condemnations actually freed them up. They no longer had to doggedly follow Aristotle, but could invoke God's freedom to do things differently and develop theories outside the Aristotelian paradigm. We will see in the following chapters that they leapt at the chance to explore these possibilities.

Of the remaining banned propositions, several asserted that the world had not been created by God and that it was eternal. Others questioned human free will and moral responsibility. A few give us an idea of just how bitter the war of words between the Averröists and the theologians had become. Among the condemned statements were 'Theology is based on fables' and 'You can't learn anything from theology'.[24]

Reading through the 1277 condemnations, it is clear that they were written in a hurry. They are chaotically ordered and appear to be the product of some sort of ecclesiastical brainstorming session. No one sat down and thought carefully about what they were supposed to achieve. This makes their success all the more remarkable. They did put a stop to the Averröists and finally settled the row over how Christians should interpret Aristotle. But they did not prevent natural philosophers from pushing back the boundaries of their subject. Things could have gone so differently. If the Averröists had pushed the Church into a corner, then Aristotle might again have been banned outright. Wiser counsel prevailed and both the condemnations and their enforcement allowed space for natural philosophy to flourish.

Much of the credit for preventing the 1277 condemnations from being interpreted too widely must go to Thomas Aquinas, even though he had died two years previously. His frequent references to the books of Averröes ensured that they were not themselves prohibited. Furthermore, among the forbidden propositions, there were some that he had probably supported. By apparently seeking to correct the great scholar, the bishop of Paris had bitten off more than he could chew. The Dominicans lobbied hard for Thomas's canonisation and within half a century had attained their goal. When he was made a saint in 1323, a later bishop of Paris had to declare the condemnations amended so that they did not conflict with his thought.[25] After Thomas Aquinas was admitted to the ranks of the blessed, no one could claim that a combination of Aristotle and Christianity automatically led to heresy.

The condemnations and Thomas's *Summa Theologiae* had created a framework within which natural philosophers could safely pursue their studies. The framework first defined clear boundaries between natural philosophy and theology. This allowed the philosophers to get on with the study of nature without being tempted to indulge in illicit metaphysical speculation. Then the framework laid down the principle that God had decreed the laws of nature but was not bound by them. Finally, it stated that Aristotle was sometimes wrong. The world was not 'eternal according to reason' and 'finite according to faith'. It was not eternal, full stop. And if Aristotle could be wrong about something that he regarded as completely certain, that threw his whole philosophy into question. The way was clear for the natural philosophers of the Middle Ages to move decisively beyond the achievements of the Greeks.

It would be a mistake to see the restrictions that the 1277 condemnations placed on natural philosophers as evidence that the Church was anti-science. True, there was no such thing as completely free inquiry, but placing limits around a subject is not the same thing as being against it. The limits imposed on natural philosophy served a dual purpose. While they did prevent it from impinging on theology, they also protected natural philosophers from those who wanted to see their activities further curtailed. Like a country with secure borders, philosophy was safe to develop in peace and without fear.

The Gothic Cathedrals

With the dispute over natural philosophy resolved, the bishops of Paris could devote themselves to fitting out their new cathedral. The towers had been completed in 1250, but interior and exterior decoration continued until 1300.[26] Notre Dame de Paris is only the most famous of the Gothic cathedrals that sprouted throughout northern France during the twelfth century. If you ever have the chance to visit the region, it is worth heading for the equally beautiful churches in Orléans, Reims, Rouen or Chartres instead. There, you will be able to enjoy the architecture without the horde of fellow tourists that so blight a trip to Paris. Even the most one-eyed critic of medieval civilisation cannot deny that the Gothic cathedrals are among the wonders of the world. After more than seven centuries, they continue to hold visitors and pilgrims in awe. In recent years, restoration programs have left many of them looking better than ever. Soot and pollution have been laboriously cleaned off to reveal their shining veneer beneath. Now they glow golden in the sun.

Gothic architecture relied on three innovations – the pointed arch, rib vaulting and the flying buttress. The style was not called 'Gothic' at the time. The term was coined in the sixteenth century when it was intended as yet another barb aimed at the Middle Ages. The style was originally called 'French' because the earliest truly Gothic building was the new church of the Abbey of St Denis on the outskirts of Paris,[27] where Peter Abelard had once been a monk.

Pointed arches came from India and arrived in Europe some time in the eleventh century. All arches are intended to support a load, be it a roof or a bridge. Round arches send the entire weight of the load straight down the pillars on which they stand. A pointed arch does not. It has a natural tendency to splay outwards. There are two consequences of this. Firstly, it is not necessary to sit a pointed arch on top of enormous thick pillars, as they are not required to bear the full load. Secondly, some kind of exterior support is needed in order to stop the arches from pushing out the walls of the building. The support that medieval masons hit upon was the flying buttress. This holds in the outside of the wall and means that the interior of the building is spacious and uncluttered. The new idea caught on remarkably quickly. Some cathedral builders even changed their plans halfway through construction. At Bayeux, the bottom half of the nave is supported by round arches over massive stone pillars. Above them, fine pointed arches frame the windows with much more slender columns.

The final element of the Gothic style is the rib vault. The roof is held up by a series of arches but instead of these running perpendicular to the wall, they are sent across diagonally. This means that two ribs can cross over each other, mutually reinforcing and reducing the amount of space that has to be filled between them.

The triumph of Gothic architecture is that medieval masons moulded these three elements into a series of aesthetic masterpieces. They wanted to build upwards and create massive amounts of interior space. And because the flying buttresses were outside the building, they could fill the walls with glass rather than needing enormous trunks of stone. The results have stood the test of time.

Height was the ambition of the masons and every bishop wanted his cathedral to rise over that of his neighbours. The trend towards ever-taller buildings culminated with the magnificent choir at Beauvais. When you step through the doorway of Beauvais Cathedral, your eyes are drawn inexorably upwards. The vaults are an unprecedented and unrepeated 160 feet (48 metres) above the floor.[28] A thick screen of slender pillars in two tiers swings around the apse of the church to form the choir. Between them, stained-glass windows curve across the entire field of vision. The effect is like standing in an exquisite kaleidoscope of colour. Unfortunately, like the builders of the tower of Babel, the masons at Beauvais overreached themselves. Merely to set eyes on the cathedral is to see that something has gone very wrong. For although it towers over its environs, only half of it has been built. It has a choir and a transept but where the longest nave in Christendom should have been, just empty space. The far wall of the transept is simply boarded up.

Work on the choir was completed in 1272, but twelve years later its excessive height caused it to collapse. The dense forest of columns that holds it up today, although an aesthetic success, is the result of the masons desperately trying to stop it from falling down again. No more was done for two centuries but in 1500 work finally started on the transept. It took 50 years and, on completion, appeared to be holding up well. At that point, hubris overtook the builders again and they planted an enormous bell tower on top, complete with a spire and capped with a heavy iron cross. Briefly, Beauvais possessed the tallest building in Europe. All too briefly. On Ascension Day 1573, the congregation had just filed out of the church when the tower crashed through the roof. Although the vaults were quickly repaired, the cathedral's nave was never built.[29] Even today the building remains in a precarious state. A huge brace is strung across the transept in an effort to keep it standing while the opposite wall is buttressed by a giant wooden prop dug into the foundations. The cathedral's height makes it especially vulnerable to the high winds that sweep across the plains of Picardy. Despite standing for centuries, the days must be numbered for this monument to man's inability to know when to stop.

We must now temporarily leave behind the philosophers and cathedral builders to spend some time in less reputable company. For modern science also owes a debt of gratitude to those who dabbled in magic, alchemy and astrology.

CHAPTER 7

~~~

Bloody Failure: Magic and Medicine in the Middle Ages

Magic works. Sometimes.

That doesn't mean to say that the reason it works has anything to do with the supernatural. It might be luck, psychology or fraud. The medieval mind was alive to all these possibilities, but this did not mean that they automatically ruled out the uncanny. And even if magic had no effect whatsoever, that could still be a positive. One of the major factors behind the success of magical healers was that they rarely did any damage. Unlike the professional physicians of the Middle Ages, they would not bleed their patients or deliberately make them throw up. The medicines they prescribed were not usually poisonous and tended to be for exterior use only. Praying at a saint's shrine was the safest course of all and consequently, in all likelihood, the most effective. Scholarly medicine operated in competition with both magic and miracles, but its 'cures' were far more likely to be actively harmful. Physicians could make good money hastening their patients to the grave.[1]

Being Ill in a Medieval City
Place yourself in Paris in 1300. The narrow streets are full of students, craftsmen and beggars. Occasionally, moving through the throng, you might catch sight of a priest or one of the brightly attired administrators attached to the royal court. You hardly care because you are sick.

You have had a headache that has grown steadily worse. It keeps you awake all night and now your vision is becoming blurred. What to do?

Essentially, you have two options; three if you have money. There is the church, the local healer or a qualified doctor. All three will expect a fee, but the first two are far more likely to keep their charges down to a level that the poor can afford. The church can offer spiritual nourishment, of course, but it also runs Europe's hospitals. Next door to the great cathedral of Notre Dame de Paris was the Hôtel Dieu, an early hospital where medical assistance is given to those who can't stretch to a doctor.[2] But they are full and probably won't admit you. The establishment is intended for those who are too ill to look after themselves and have no family to help out.

You should, of course, visit the cathedral itself and pray for God to heal you. You might think that asking a priest for a blessing or praying to a saint for a cure would be no different from asking the local witch. However, as far as people in the Middle Ages were concerned, there was a fundamental difference between magical and miraculous cures. With magic, the magician was supposed to have the power to effect a cure of his own accord, as long as he does his job right. Properly carried out, a magic spell should always work. A religious cure is different. The supplicant realises that God may or may not cure him depending on a number of unknown factors. He has no right to expect a cure and no way to compel God to provide one. Whereas the magician might claim that he can bind demons to do his bidding, there is no obligation on God to do anything.[3] So, it was wise to try alternatives and most people would patronise more than one source of healing.

The Church had no objections to physicians or even village healers who abstained from diabolical magic. According to Christians, God was ultimately responsible for illness but had also provided the means in nature to deal with it. Canon law did forbid priests from moonlighting as physicians, partly because the secular physicians did not like the competition and partly because it was not the priest's job.[4] Only with great scourges like the Black Death of 1347–50, before which all the physician's arts were powerless, did people come to rely exclusively on religious help.

So if you have a chronic headache in the Middle Ages, you had better visit the local healer. She lives in a small, unremarkable house. Little differentiates her from her neighbours except that she is treated with extra respect and perhaps a tinge of fear. Her fees are very reasonable and as you are a local, you will already be known to her. Her consulting room is just the ground floor room of her house. It is full of the sort of paraphernalia that you would expect a wise woman to have, including clay pots of herbs and a small iron cauldron. In one corner are a couple of small books. These are not just for show; the healer can read and write. She was taught by her mistress who also handed down many of the recipes that she uses for her ointments.[5] Sitting down, you explain your symptoms to her. The close proximity of the university medical faculty means that the healer has pretensions towards the methods of the academic physicians. She takes your pulse and then produces a small glass flask, asking for a urine sample. Without embarrassment, you produce one and hand it to her. She examines the colour carefully and shakes the flask to determine if it contains any foreign matter. Finally, she has a good sniff and sips a little.

No matter what is wrong with your head, there is nothing therapeutic that the wise woman or any other physician in 1300 can do to cure it. You will either heal naturally or grow worse until you die. What she can do is offer palliative support that may both lessen your pain and strengthen the body's ability to get well by itself. She gives you two treatments. The first is an amulet to wear attached to your head. It consists of a small piece of parchment, made from the skin of a ram, upon which the healer has written a line of scripture. She rolls up the scrap tightly and ties it together with a piece of wool. The second treatment is a herbal tincture made from walnuts. As a mere patient, you are unlikely to know why these remedies are supposed to work, but if you believe in them anyway, then they might do some good. Nonetheless, there is a logic behind the wise woman's methods which depends on the theory of magic current at the time.

Magic in the Middle Ages

Medieval magic depended on two great dichotomies: microcosm versus macrocosm and sympathy versus antipathy. Through his understanding of these, it was believed that a magician could manipulate the hidden powers of the universe and harness them for his use. The macrocosm is the whole universe but most especially applies to the stars, the signs of the zodiac and the planets. The microcosm means our own domain of the surface of the earth, or alternatively the human body itself. A fourteenth-century alchemist, Bernard of Trevisan, laid out the essence of the microcosm/macrocosm dichotomy. 'When that which is above is made by that which is below', he explains, 'and that which is below is like that above, then miraculous occurrences will from thence appear.'[6] Everything on earth, the healer believed, had some sort of correspondence with something in the heavens. If you could figure out what these correspondences were, then you could use them for healing and magic.

The ancients recognised seven pure metals – gold, silver, mercury, copper, iron, tin and lead – which they associated with the seven planets, respectively the sun, moon, Mercury, Venus, Mars, Jupiter and Saturn. It is easy to see how gold corresponded to the sun or reddish tin to Mars. Some of the other associations seem rather more contrived. Copper was linked to Venus only because the goddess of that name was supposedly born on the island of Cyprus, which also hosted the ancient world's major copper mines.[7] Consequently, copper was the appropriate metal with which to make a love charm.

The planets were also linked to parts of the body and animals. For example *Picatrix*, a medieval tome of magic, explains:

Mercury is the origin of the power of understanding. Among the external parts of the human body, it has a special relationship with the tongue and among the internal parts, with the brain ... among minerals with quicksilver ... among animals with mankind, small camels, monkeys and wolves.[8]

In a similar way, the signs of the zodiac also related to different parts of the body. Many beautiful manuscripts from the Middle Ages illustrate these correspondences between specific organs and the constellations. The sign for the head was Aries, the Ram. This is why a medieval healer in Paris would have made an amulet out of the skin of a sheep (preferably a male one) and tied it together with wool. The passage of scripture would be an attempt to borrow the power of the Christian God for healing. It would be best that the Church did not know about this intrusion into its prerogatives.

Figuring out all the correspondences was an arduous task and there was very little agreement between authorities beyond the basics. People imagined that nature was like a giant medicine cupboard filled with pills and tablets of all shapes and sizes. But none of them had an unambiguous label revealing what they did. Instead, there was the equivalent of a cryptic picture on each packet of tablets to help in identifying its purpose, and people had to guess from the picture what a particular medicine was for. In reality, it was the appearance of individual plants that provided the visual clues as to their purpose, but the principle is the same.

The reason why the correspondences were believed to exist in the first place was tied up with the magical worldview. For the magician, different parts of the universe were connected together so that the various things that corresponded to each other showed some outward sign of this fact. This is what was meant by 'sympathy'. Sometimes, this was obvious and is even reflected in the names we still give to plants. For instance, the leaves of liverwort (*Hepatica*) were said to resemble the liver. This resemblance was called the plant's signature and enabled a skilled observer to identify the best herb to use for each condition. Of course, it was rarely so easy. Deciding how the shapes and forms of particular herbs demonstrated their uses required great skill. Or, alternatively, some inspired guesswork.

The reason our Paris healer would have offered a potion made from walnuts for a head ailment becomes obvious the moment you look at a walnut, even today. No inspired guesswork is required. The nut has a hard outer shell which, when cracked open, reveals a soft craggy interior divided into hemispheres. The similarity between the edible part of a walnut and the brain must have struck herbalists as undeniable. This signature was nature's way, or God's way for a Christian healer, of showing man which plant to use to treat a head complaint. Sadly, nature was being disingenuous. Neither walnuts nor liverwort really cure the ailments they were believed to. Healers had more success prescribing willow bark for pain relief. We now know that this plant contains salicylic acid, the active ingredient in aspirin.[9]

Medieval medicine

As a commoner, the local wise woman would have been your only option. However, a rich lawyer working in the royal courts would have had no need to visit a local crone when he became ill. He would have taken to his bed and the university-trained doctors would have come to him. The medical faculty at Paris was well regarded but comparatively small compared to the great schools of the south at Montpellier and Padua. This meant that Paris attracted physicians from abroad who were keen to cater for the large market of rich councillors and clergy. The university tried to ensure that only those who were properly qualified could practise as doctors, and especially wished to keep such undesirables as women and Jews away from patients' bedsides. The doctors that would have appeared at a lawyer's house – and a rich individual would usually insist on consulting more than one – used similar means of diagnosis to the wise woman, in other words taking both the pulse and a urine sample. The treatment, however, was likely to be much more involved, expensive and painful.

Doctors were trained to treat each patient as an individual. In order to offer effective care, they would take down their patient's entire medical history, details of his diet and as much information as possible about the illness. Feeling for the pulse and examining the urine sample were taken to much greater lengths by the professional physician than by the wise woman (who was probably only doing it for show). They noted the strength, timing, depth and rhythm of the pulse. Doctors classified the shape of a patient's arteries alone in over two dozen ways. As for urine, it was such a common medium of diagnosis that the standard way to recognise a physician in a medieval illustration is that he will be holding a flask of it up to the light.[10]

6. Thirteenth-century manuscript illumination of a doctor
with a flask of urine

Learned doctors tended to prescribe not just medicine, but a lifestyle or 'regimen'. Bed rest, a special diet and relaxation were most often recommended. This was, at least, unlikely to do much harm and might help the body heal itself. Unguents, potions and other herbal treatments would also be prescribed. The doctor himself would not prepare these, but instead would employ the services of a specialist apothecary to mix the ingredients. Despite the enormous array of herbs that doctors could use, the main function of any medicine that the patient imbibed was to make him vomit.

The most notorious form of medieval treatment, which remained common until the nineteenth century, was bleeding.[11] Again, the doctor would not do this himself because such manual work was beneath a professional person. Instead, a barber-surgeon would be sent around to drain blood from the patient at specified times. Bleeding, as we now know, would have been actively harmful but many patients insisted on it. Peter the Venerable, the abbot of Cluny who gave Peter Abelard shelter in the last years of his life, once complained that his respiratory infection was worsened because he had postponed his regular bleeding session.[12] In the Middle Ages, some physicians were browbeaten by their clients into bleeding them half to death. Even so, bleeding was not irrational; it was consistent with the theory of disease that all physicians respected. To understand its justification, we need to take a look at the intellectual foundation of academic medicine in the Middle Ages.

The Origin of Learned Medicine

Before about 1100, educated people did not consider medicine to be a professional field. Because doctors used their hands to treat patients, their skills were equivalent to those of a craftsman rather than a scholar.[13] All this changed with the appearance of Arabic and Greek medical texts in the twelfth century. It was clear that an impressive array of theory lay behind ancient medical practice. Laden with this cerebral ballast, medicine claimed its own faculty at many of the new universities.

Taddeo Alderotti (c.1223–95) of Bologna championed the new approach and made a fortune in the process. His prestige as a scholar helped to enhance his reputation as a healer so that even the Pope paid him a substantial salary.[14] After Alderotti, the medical faculty at the university of Bologna enjoyed almost as much influence as the law school.

An important reason for the acceptance of academic medicine was that it could trace its origins back to Greek scholars every bit as revered as Aristotle. For example, Galen of Pergamum, the source of much medieval medical knowledge, was born in about AD129. He trained in medicine in Alexandria before returning home to practise as a doctor. His first job was patching up gladiators who had been injured in the local amphitheatre. It was not the most glamorous of roles and he quickly realised that he was not going to find fame and fortune in a provincial city like Pergamum. In AD161, he departed for the imperial capital of Rome where he established a successful medical practice whose clientele reached the heights of society. He even ended up treating the emperor himself.

Galen appears to have been an accomplished showman who gave public lectures that featured the vivisection of animals. Torturing pigs was his favourite spectacle. He showed how, by severing the spinal cord of a live sow, he could paralyse first its hindquarters and then its front legs.[15] He was certainly a prolific writer and two million words of his medical works survive. That is without even his philosophical writing, destroyed by fire during his lifetime when the library of the Temple of Peace in Rome burnt down.[16] He died early in the third century, by which time his vast literary output meant that his legacy was secure. He became the standard authority on medicine in both the Roman and Muslim worlds. However, western Europeans in the Middle Ages did not have access to very many of his original works because he wrote in Greek. They had to make do with Arabic commentaries and medical encyclopaedias that had been translated into Latin during the twelfth century.

The essence of Galen's medical theory, which he inherited from even earlier Greek writers, was that the body contains four 'humours': blood, phlegm, yellow bile and black bile. These four substances were never found in their pure forms. Even real blood contained a mixture of all four humours. Yellow bile was the major component of pus, phlegm belonged mainly in the lungs and black bile was responsible for putrefaction. But all four humours were present throughout the body and they were all essential, in their way, to its proper functioning.

The key to good health was to ensure that the proportions of the humours remained in the correct balance. Different people thrived with a different balance so that one of the four normally prevailed over the others. Someone with a natural excess of blood was sanguine (from the Latin for blood). Alternatively, they might be phlegmatic, bilious or jaundiced (the latter derived from the Latin for yellow) if a different humour predominated. These adjectives have now slipped their medical moorings and become part of general English.

The best way to restore balance to the humours, once they got out of kilter, was to drain the one that was present in excess from the body. First of all, the physician had to determine which of the humours was causing the problem. The theory assigned the qualities of being 'hot' or 'cold' and of being 'wet' or 'dry' to each of the humours. For instance, blood was 'wet' and 'hot' while phlegm was 'wet' and 'cold'. If someone was suffering from too much of one of the humours, this would manifest itself in symptoms that reflected the qualities of the excess humour. For instance, a patient who was chilled and clammy to the touch was suffering from excess phlegm, which was cold and wet. If he was running a temperature and sweating from a fever, the doctor would recognise that the symptoms included being hot (from having a high temperature) and wet (from sweating). From these symptoms he could diagnose that the patient's illness resulted from an excess of hot and wet blood. The appropriate treatment was bleeding to siphon off the surplus blood. The properties of being hot and cold, wet and dry were assigned to herbs as well. To deal with a fever, plants with cold and dry properties, such as quince or rose, would be used to counteract the heat and moisture of the. fever. A doctor might also use a purgative to make the patient throw up. Vomit was held to contain bile and so being sick could reduce a glut of this humour.

The twin treatments of bleeding and purging must have been very unpleasant. Neither could have done the slightest bit of good and were much more likely to weaken patients just when they needed all their strength to fight off the illness. Any success that doctors had was due to the 'placebo effect' whereby patients' beliefs that they are being cured help to reinforce the body's natural defences. The situation was much the same for the healers who claimed to use magic, only they were less likely to harm their customers because their treatments were less invasive. The most effective medical treatment available in the Middle Ages was probably to visit a saint's shrine. Miracles aside, 'the placebo effect' can be just as easily activated by religious faith as by a doctor's bedside manner. And praying, at least, is unlikely to lead to any ill effects. Analysed in this light, it is surprising that the medical profession survived at all. The prestige of an ancient Greek pedigree, coupled with a university degree, must have made up for the doctors' complete inability to cure their clients.

The one genuine skill that medieval doctors did possess was the ability to identify diseases. This meant that they could often predict

the outcome of the patient's illness even if they could not do anything about it. Still, doctors were taught that it usually paid to be pes-
simistic about the prognosis. One instruction manual advises, 'When
you have left the patient, say a few words to the members of the
household. Tell them that he is very sick. For, if he recovers, you will
be praised more for your skill. Should he die, his friends will testify
that you had given him up.'[17] Taddeo Alderotti once forgot to follow
this advice and declared a nobleman he was examining to be on the
mend. The next day he returned to find his patient had declined
almost to the point of death. The family remonstrated with him and
it looked like he might lose his fee. Casting around for an excuse,
Taddeo noticed that a window had been opened in his absence. This,
he declared, was the reason for the near-fatal turn the nobleman had
taken. The physician could not be blamed.[18]

The Doctor of Physic in Geoffrey Chaucer's *The Canterbury Tales*
used a combination of natural magic, probably derived from traditional herb lore, and Greek medical theory based on Galen. Chaucer
writes, when he introduces the good doctor in his *General Prologue*:

> For, being grounded in astronomy,
> He watched his patient's favourable star
> And by his natural magic knew what are
> The lucky hours and planetary degrees
> For making charms and magical effigies.
> The cause of every malady you'd got
> He knew, and whether dry, cold, moist or hot;
> He knew their seat, their humour and condition.[19]

In the first line, Chaucer tells us something else about medieval doctors that we might not expect. They relied a great deal on astrology in the course of their work. The stars would determine when
bleeding or other treatment should take place. The famous physician Bernard Gordon (c.1258–c.1320), who is actually mentioned
by Chaucer, admitted that he once had got the time of a bleeding
wrong without any ill effect to his patient.[20] But even he still advised

sticking to the received wisdom. A horoscope would also form part of a patient's medical history because the disposition of the heavens at the moment of their birth helped determine their bodily chemistry. But astrology went much further than setting the time for medical treatments. As we will see, it was, like alchemy, a learned discipline in its own right.

CHAPTER 8

~~~

The Secret Arts of Alchemy and Astrology

Astrology was the most widespread and respected of the magical disciplines. It was a great deal more complicated than just dividing humanity up into twelve groups according to their sun signs. The astrologer needed to be a skilled mathematician so that he could calculate the positions of all the stars, major constellations and other celestial features. In ancient Rome, the word for an astrologer was *mathematicus* and they were a ubiquitous feature of life. Several emperors, both pagan and Christian, tried to keep the astrologers on a short leash, frequently expelling them from the city. The Emperor Tiberius (42BC–AD37) used to invite them to his clifftop villa in Rhodes for a private consultation. As they returned along the precipitous path that was the only way to reach the house, one of Tiberius's slaves would throw them into the sea. This kept the details of the future emperor's horoscope completely confidential. One astrologer, called Thrasyllus (d. AD36), foresaw his own imminent demise when Tiberius asked him to predict his future. This impressive piece of prognostication saved Thrasyllus's life and he joined Tiberius's household as court astrologer.[1] Later emperors were just as touchy about keeping their horoscopes under wraps, and it was a capital offence to try to read the imperial stars.

The Influence of the Stars

Christians always found it difficult to come to terms with astrology. Theoretically, the Bible forbade all forms of divination and there is a celebrated episode in the Acts of the Apostles when new converts burn their valuable books of magic.[2] Some Christians denied that astrology worked at all. In the fourth century, St Augustine pointed out that twins, who must be born under the same stars, could often have very different fates. He concluded from this that astrology was bunk, but had to concede that some astrologers did enjoy notable success. However, he said, 'when astrologers do give very many wonderful answers, this is to be attributed to the hidden prompting of spirits far from good ... The astrologers' success is not due to the art of observing and studying horoscopes, for there is no such art.'[3]

The complicated calculations required for a horoscope were probably beyond most people in western Europe during the early Middle Ages. However, the new learning from Arab and ancient Greek sources recovered in the twelfth century showed that even the most sagacious ancient authors, including the likes of Ptolemy himself, believed in astrology. Ptolemy had even produced his own astrological manual, known as the *Tetrabiblios*, and thus lent his enormous prestige to divining with the stars.[4] Working from this and other classical sources, Arab writers had produced guides to astrology, which Christian scholars translated along with all their other works. Despite the opposition of Augustine, these Greek and Arab authorities demanded respect and, when they arrived in the West, they brought about a rebirth of the astrological arts.

Even today, no one denies that heavenly bodies do affect events here on earth. The sun is the source of heat and light and even during the early Middle Ages some people thought, correctly, that the moon gives rise to the tides. In the twelfth century, Adelard of Bath discussed and dismissed this possibility in his *Natural Questions*.[5] From there, it is not much of a leap to ask what other effects the planets might have.

There are important differences between the practice of astrology in the Middle Ages and the horoscopes that are found in today's tabloid newspapers. A medieval astrologer did not place so much emphasis on the twelve signs of the zodiac. These were just a way to chart the movement of the planets. What was much more important was the position of each of the seven planets at the time that the reading related to. The planets, including the sun and moon, move across the sky while the fixed stars provide a background. The sky is divided up into twelve segments (technically called 'houses'), each one named for the sign of the zodiac prominent within it. Thus, at any time, the position of each of the planets can be defined by which house it is in.

When someone today talks about their star sign, what they mean is the house the sun was in at the time of their birth. Of course, you can't see the stars when the sun is out, but they are there invisible nonetheless. Everybody knows what their sun sign is, be it Libra, Leo or any of the others. The sun travels through all twelve houses during the course of a year. The moon, on the other hand, completes its orbit of the earth in just under a month and passes through any given sign in two-and-a-half days. For someone to find out their moon sign, therefore, they might need to know the exact time of their birth. We also have a Venus, Mars, Jupiter and Saturn sign depending on where each was in the sky at the time and in the place of our birth.

With all the details of exactly where and when someone was born, there are internet sites that will do the calculations and produce a full list of all the planets and other astrological facts about their birth. Armed with this information, an astrologer claims to be able to tell a great deal about a subject's personality and the stellar influences on their life. According to astrologers, the sun governs our outward personality. Venus, of course, rules our sex lives and Mars our active lives. In addition, for each of us, one particular planet predominates over the others and determines our personalities. If it is Jupiter we will be jovial, if it is Mercury we will be mercurial (or crafty). We still use these adjectives, along with venereal, martial and saturnine, although they have lost their attachment to astrology. The word 'lunatic' originally meant someone governed by the moon.

It is hard to have much time today for people who check their horoscope in the paper in anticipation of it telling them something useful. However, the astrologers of the Middle Ages, even though their art was ultimately redundant, do deserve some respect. They had good reason to believe that the stars influenced the lives of people on earth. The authority of the ancients backed them up on this and their methods required a fine grasp of mathematics. They needed plenty of skill to trace the paths of the planets, even using the astronomical tables that most astrologers relied on. They had to master the astrolabe to tell the exact time, they needed to know their precise location and they had to read an awful lot of difficult material. Of course, many astrologers were just frauds, but we know from surviving books and manuscripts that others worked incredibly hard to decipher messages from the stars.

The most successful astrologers were never short of patronage. One of the earliest practitioners we know about was Guido Bonatti (c.1210–c.1290) who worked under the protection of the Lord of Forli in north-eastern Italy. Reputedly, Guido's boss would do nothing without first consulting him on the position of the stars. In 1282, troops loyal to the Pope attacked the town of Forli. Guido made an astrological forecast so that he could advise the town's citizens on the best strategy to adopt against the superior numbers ranged against them. He suggested that a feigned withdrawal would cause the papal army to drop its guard and make them vulnerable to an ambush. The men of Forli did as he suggested and won a famous victory. Guido himself was wounded in the battle as he treated his fallen comrades. He assured his worried compatriots that he had foreseen his injury and would make a full recovery.[6]

Although Guido Bonatti did not suffer for his astrological interests while he was alive, they did his subsequent reputation no good at all. When Dante Alighieri wrote his *Comedy* in the early fourteenth century, he condemned Guido to hell for divining with the aid of demons. He was cursed to spend eternity with his head twisted around backwards as punishment for seeking too earnestly to look forward into the future.[7] Nonetheless, despite its author's confinement to hell,

Guido's guide to astrology remained a standard work until the seventeenth century.

Monarchs were also keen to have the stars on their side. In 1235, the Holy Roman Emperor Frederick II (1194–1250) acquired a young English bride. However, with admirable self-control, he resisted consummating the marriage until his astrologers told him it was a propitious moment to conceive a son. The stars did not lie and nine months later, an heir to the imperial throne was born.[8] Frederick was renowned for his patronage of the sciences. Among other achievements, he legislated to regulate the medical profession, set prices for drugs and wrote a remarkable study of birds.[9] His learning was so broad that he was called 'the wonder of the world', although the papacy, with which he was in constant conflict, did not agree.

Initially, the Church tended to follow the lead of St Augustine and disapproved of astrology because of its diabolical connotations. The astrologers defended themselves by insisting that, on the contrary, the planetary influences were simply natural forces that it was entirely legitimate to investigate. That being the case, there was not very much for the Church to object to. As astrology became more widespread, clerical opinion grew increasingly lenient. Thomas Aquinas considered the matter carefully. His views represent something approaching the medieval consensus and are worth quoting:

> If anyone attempts from the stars to foretell future contingent or chance events, or to know with certitude future activities of men, he is acting under a false and groundless presumption, and opening himself to the intrusion of diabolic powers. Consequently, this kind of fortune telling is superstitious and wrong. But if someone uses astronomic observation to forecast future events which are actually determined by physical laws, for instance drought and rainfall, and so forth, then this is neither superstitious nor sinful.[10]

So, as far as the Church was concerned, the science of astrology was acceptable when it was restricted to studying the natural effects that the stars have on earth. In that case, the astrologer could not see the future but could make forecasts based on the predictable motions of the heavens. However, true divination was not possible using purely natural means and hence was out of bounds to good Christians.

The Church also drew other lines that no astrologer should cross. A highly controversial question in the Middle Ages was whether the stars could fix people's fate or just exert an influence. Aquinas was sure that it was the latter:

> The stars and planets and their courses cannot directly cause the choices of man's freewill. Still, they can dispose and incline man to do this rather than that, in as much as they make an impression on the human body.[11]

This was important because if the facts of our birth determine the course of our lives, any idea of free will or moral responsibility becomes untenable. The Paris condemnations of 1277 made it very clear that as far as the Church was concerned, fate was not fixed and the stars could only effect predispositions.[12] The whole debate over astrological determinism closely resembles the argument about nature versus nurture that is raging today. Are our personalities determined by our genes or do the decisions we make in life affect the kind of person that we are? Substitute the stars fixing your fate at birth with your parents' genes doing so at conception, and the whole question of determinism becomes an urgent twenty-first-century concern. Like the Church in the Middle Ages, most of us are acutely uncomfortable with the idea that free will might be an illusion.

The medieval Church repeatedly attacked determinism and defended the concept of free will.[13] No one could be in any doubt that it was not merely wrong to say that we cannot escape our destiny, it was heretical. Inevitably, some astrologers would not toe the line and one or two galloped imperiously across it. Cecco D'Ascoli (c.1269–1327), a lecturer on astronomy at the university of Bologna, was one of these.

The Terrible Fate of Cecco D'Ascoli

Bologna in the fourteenth century was a city of lofty towers. When a Bolognese family wanted to keep up with the neighbours, they had to ensure that their tower was taller than the one next door. Few of these architectural monstrosities survive today but the Torre degli Asinelli, one of those that does, is an impressive 300 feet high. In Cecco D'Ascoli's day, there were hundreds of these turrets dotted around the city as ostentatious displays of the leading families' wealth. The university was thriving as well. It was already over two centuries old and had added a medical faculty to the original law school. Students who wanted to study medicine were expected to master Aristotle's natural philosophy and basic mathematics before they started the medical course proper. Although Cecco was a lecturer on astronomy, for students training to be physicians, in fact astrology was the rationale for learning about the stars. It is no surprise, then, that it was as an astrologer that Cecco built his reputation.

The text that he used for his classes was *The Sphere* by Englishman John Sacrobosco (who died around 1256). About Sacrobosco himself we know next to nothing except that he spent most of his career teaching at Paris. His book is a short and simple introduction to the knowledge of the heavens that was current in the thirteenth century, specifically intended for students at the new universities. The title refers to the fact that medieval people, like the ancient Greeks, thought that the universe was perfectly round, with the earth at its centre. Because *The Sphere* was so brief, lecturers tended to use it as a jumping-off point rather than treating it as an exhaustive survey. Its brevity also meant that Sacrobosco's book was extremely adaptable, and university lecturers were still using it in the early seventeenth century.[14] Not until it became widely accepted that the earth was not at the centre of the universe did *The Sphere* finally face obsolescence.

Cecco D'Ascoli was among the lecturers who went well beyond the basics laid out by Sacrobosco. He put his own astrological spin on the material and appended many other unrelated ideas. Sacrobosco himself does not mention astrology at all. Among the additional material that Cecco added to spice up his lectures was speculation about astral spirits and other even more risqué aspects of magical lore. When he actually gave the lectures, he may have gone even further than he was willing to admit in his published writings.

The professors of Bologna were respected and reasonably well paid, but Cecco had a lucrative sideline that allowed him to live very comfortably indeed. He traded on his reputation as an astronomer to moonlight as an astrologer, providing readings for clients who wanted to know their future. Cecco was in no doubt that astrology worked and that the stars were a sure guide to what lay ahead. He crossed the threshold of acceptable opinion, holding, contrary to the Church's instructions, that events on earth were absolutely determined by the position of the stars. His confidence in astrology caused him to make statements that even today seem shockingly foolish in their bravado. In an incredibly unwise move, he went as far as to calculate the horoscope of Jesus Christ. The result of his calculations, he declared, showed that the reason for Jesus's lowly birth and violent death was his being born under malevolent stars.[15] This was explosive. Under the doctrine of the Trinity, Jesus was God incarnate. By claiming that the Messiah's life was determined by the arrangement of the heavens at his birth, Cecco was subjecting the very Deity to the stars.[16]

Rumours of what Cecco was up to leaked out. We do not know if his students or his clients betrayed him, but someone found his pronouncements beyond the pale and reported them. The man who was most interested in such deviancy was the local inquisitor for Bologna, one Lambert de Cingulum (fl.1316–24). Like many inquisitors, he was a member of the Blackfriars, the Dominicans of Thomas Aquinas. This meant that he was no ignorant cleric frightened by things he could not understand. Lambert was an educated scholar, an expert on the ethical teaching of Aristotle and a professor in his own right.[17] He probably knew Cecco personally.

Like the secret police in a communist state, inquisitors depended on a network of informers and agents to keep them abreast of heresy in their district. The system was wide open to abuse by those who denounced their enemies for personal or venal reasons. An inquisitor was trained to be aware of this possibility and punish those who made false accusations. We cannot tell if Cecco was the victim of a vendetta or jealousy but nevertheless Lambert had little difficulty in finding him guilty of making heretical statements. In 1324, he stripped Cecco of his lectureship, fined him £70 and imposed a penance that he should listen to a large number of sermons over the following year. The size of the fine indicates that the crime was a considerable one and that Cecco was wealthy enough to pay it. Above all, Lambert forbade Cecco to continue either studying or practising astrology.[18] We should note that even for such a heinous offence as subjecting God to the zodiac, the inquisitor did not imprison, let alone execute, Cecco. This tells us the inquisitor was satisfied that the defendant had come clean and admitted to all his crimes. When summoned before an inquisitor, any strategy other than complete candour would be a serious and probably fatal mistake.

If only Cecco had stuck to mainstream astronomy from then on, he could have continued his career without much of a stain on his character. In time he would have expected to get his job back and been reintroduced to polite society. Unfortunately, he would not stop reading the stars. He left Bologna and travelled to Florence where he set up shop as a freelance astrologer. There are plenty of later and unreliable stories about what he got up to and how he made enemies of important people. None of them is necessary to explain what happened next. Inquisitors communicated with each other and it only took a couple of years for Cecco's past to catch up with him.

The inquisitor of Florence was a Franciscan friar called Accursius. He locked up Cecco in 1326 and then sent a messenger to Bologna asking for details of the previous trial and a copy of Cecco's book of lecture notes. Accursius was nothing if not meticulous and spent a year carefully investigating the case. This thoroughness was necessary for a very simple reason – Cecco's crimes were now of a capital nature. We have seen how the inquisitors would not hand a heretic over to the secular arm of government for their first offence but were utterly merciless to repeat offenders. By continuing to practise astrology, Cecco had deliberately flouted a direct order from an inquisitor.

If found guilty there would be no clemency. Eventually, Accursius satisfied himself that the evidence against Cecco was watertight. A court session in the Franciscan Church of Santa Croce in Florence found that he was a recalcitrant heretic and ordered that he be handed over to the secular authorities. As the prisoner was led out, he would have passed close to the Barbi chapel where the paint of Giotto's sublime fresco cycle on the life of St Francis of Assisi was barely dry. The same society that could create such tenderness and beauty on wet plaster could also send a man to an agonising death for a crime we hardly recognise. The great piazza in front of the church lay just outside the city walls. There Cecco was burnt on 16 September 1327.

Cecco and his peers had good reason to believe that they lived in a world where their art might have a physical basis. Even so, astrologers certainly had an interest in promoting the reliability of what they did. The influence of the stars may have been pure moonshine but the money to be made was quite substantial. Importantly, astrology also provided astronomy with an application beyond mere curiosity and a vital market for astronomical tables.[19] This led to demands for improved accuracy and encouraged the study of Greek works newly translated from Arabic. Only because astrology meant that it could be lucrative to do so did anyone bother to master Ptolemy's fiendishly difficult *Almagest*. Unless astronomy was a skill that brought with it the ability to earn hard cash, there was little point in learning it. The disinterested pursuit of useless knowledge was not something to which many people were rich enough to aspire.

The Philosopher's Stone and the Elixir of Life

Astrology's twin erstwhile science was alchemy. Everybody knows that alchemists were trying to transmute base metals into gold but there was a great deal more to it than that. To perform transmutation, they needed first to generate a special substance to act as a catalyst. This was called various names such as the 'Elixir' and the 'Philosopher's Stone'. Attempting to produce it was the principal aim of the alchemists' art and they attributed all sorts of wonderful properties to the elusive substance.

Alchemical texts tend to be confusing, contradictory and deliberately obscure. However, on one thing they all agreed – the subject should always be approached with a pure heart. The quest for the philosopher's stone was, like the search for the Holy Grail that was developing in contemporary romantic literature, only for those of the highest moral integrity. Those who were just out to make money would surely fail. This was sage advice, as alchemists were notorious for losing fortunes in their research. Many medieval writers satirised the fact that alchemists were far more adept at spending gold than creating it. Why, Georg Agricola (1494–1555) asked in his treatise on mining, is there no such thing as a rich alchemist?[20] Geoffrey Chaucer teased them in the *Canterbury Tales*. His Canon's Yeoman despairs of ever finding the philosopher's stone:

> We seek and seek, and were it once discovered
> We should be safe enough, expenses covered.
> But there is no way; whatever paths we trod
> The search was useless and I swear to God
> For all our cunning, when all's tried and done
> That stone won't yield itself to anyone.[21]

We can learn something of the theory of alchemy from a reasonably clear guide, supposedly written by Albert the Great, that lays out the process by which transmutation was supposed to take place. Mercury, pseudo-Albert writes, is 'the source and origin of all metals'. Other substances, including sulphur and arsenic, act as a sort of tincture to the quicksilver. The alchemist, through his experiments, is trying to manipulate the composition of a base metal to produce gold or silver. Pseudo-Albert provides further precepts that the fledgling alchemist should follow. He advises secrecy, a special laboratory and the use of glass apparatus. Finally, he reiterates the need to have a good supply of funds.[22]

Alchemy's poor reputation was mainly caused by concern about fraudulent activity in which crooks passed off fake gold as the real thing. In 1317 Pope John XXII (1249–1334), who ironically has an alchemical treatise ascribed to him, issued a decree against those who claimed that the gold they pretended to have created was genuine. While the decree does not outlaw alchemy in itself, it certainly denies the possibility of actual transformation. Alchemists are unable 'by the very nature of things', said the Pope, 'to produce real gold or silver.'[23] More sympathetically, Thomas Aquinas analysed the question of whether an alchemist could honestly sell the gold he created. He concluded that 'if genuine gold could be chemically produced, it would not be illicit to sell it as true gold, for there is no reason why science should not exploit natural causes to produce natural and true effects.'[24]

Philosophers also doubted the effectiveness of alchemy. The Muslim writer Avicenna had been certain that alchemy would never work because it was logically impossible for mankind to improve on nature. His attack was not just on alchemists, but on all kinds of technology and machinery.[25] Secular rulers, on the other hand, were prepared to hope there was something to it. Henry VI (1421–71) of England granted licences to people who said they could make gold at will, seemingly unaware of the inflationary problems this would engender for his economy.[26]

No one has yet been able to transmute base metals into gold. But medieval alchemists did make some quite impressive breakthroughs in the course of their research. Foremost among these was the discovery of acid. There are three main acids, hydrochloric, sulphuric and nitric. Together, these are known as the mineral acids because they are relatively easy to produce from the right sort of ore. Nitric acid was particularly exciting because it could be used to dissolve gold – a feat ascribed to Moses in the Bible.[27]

The acids, as well as alcohol, appear to have first been isolated by Christian alchemists in the thirteenth century using a technique called condensation,[28] but for many years it was assumed that the Arabs had produced them much earlier. We now know that this misconception was caused by Christians attributing their texts to Arabic writers. We've already seen the tendency of esoteric manuscripts to be ascribed to a famous author to increase their credibility. Alchemists were particularly prone to this and their favourite pseudonym was the Arab savant Geber, who was active in the ninth century. It is far from clear that any of the works ascribed to him are genuine, but the accretion of titles to his name has led to him being credited with all sorts of innovations, such as the distillation of acids, which he did not actually make.[29]

Gold was beyond them, but late-medieval alchemists did extend the range of metals beyond the seven known to the classical Greeks. By the sixteenth century they had isolated the metallic elements of zinc, bismuth and antimony as well as others that the extant documentation makes hard to identify.[30]

Besides the production of new materials, the major achievement of alchemy was in the field of techniques. Their experiments led to the perfection of distillation and calcinations, not to mention the development of the glass apparatus required to carry out these operations.[31] The methods were handed down through the generations until the eighteenth century when the experimental method began to make itself felt in chemistry. In order to make serious progress, chemists had to learn the discipline of careful quantitative measurement, especially the weighing of materials before and after chemical reactions. During the Middle Ages, no one seems to have had the patience or skill to carry out such exacting work.

Occult Forces

The common ground shared by the arts we have discussed in this and the previous chapter – magic and medicine, astrology and alchemy – was their reliance on occult forces. Magic favoured a holistic view of the world which supposed that invisible threads linked the heavens and earth, macrocosm and microcosm, into a tight web of influences. The stars emitted rays, the astrologers claimed, by which they affected the lives of men. The properties of metals were the result of hidden forces that worked within them. Control the forces, said the alchemist, and you could transform the materials. Even the learned doctor studied the active qualities within herbs that could heal an afflicted body.

Nowadays, the word 'occult' specifically means 'magical' or something connected to spiritualism. But it used to have a much wider sense, connoting any force or property that was hidden. Put bluntly, if you cannot see it, it could be classed as occult. Aristotle had little time for the concept and argued that all effects must be material. One thing, he said, can only affect another by touch. Modern science rejects this absolutism and recognises all sorts of actions at a distance, from gravity to magnetism. Aristotle had his own explanation for gravity, which we will come to in chapter 9, and he ignored magnetism. For a long time, that didn't matter too much because magnets were rare lumps of rock and hardly a reason to overthrow the laws of physics. Then, in the thirteenth century, they became rather more important. As we will see in the next chapter, the Arabs heard about a new navigational instrument from the East and before long it arrived in western Europe. This was, of course, the magnetic compass.

Perhaps the medieval attitude of suspicion rather than outright hostility towards astrology and alchemy, typified by the Church, struck the right balance. This prevented the magical worldview from dominating the alternatives while allowing the practical aspects of astrology and alchemy to feed through into modern scientific thinking. However, as we will now see, it was in the fields of natural philosophy and technology that progress was most marked in the later Middle Ages.

CHAPTER 9

〜〜

Roger Bacon and the Science of Light

In 1277, the bishop of Paris had launched his condemnation of 219 philosophical opinions in an effort to resolve the dispute over Aristotle's impact on theology. The repercussions were felt well beyond the bounds of his diocese. Just a few weeks later, the archbishop of Canterbury, Robert Kilwardby (d. 1279), issued a much shorter list of heretical statements that he banned from the university of Oxford. Like Pope John XXI, who had ordered the bishop of Paris to settle the row over Aristotle at Paris, Kilwardby was a formidable philosopher in his own right.[1] His brief set of condemnations looks more like a settling of scores with his academic rivals than an attempt to crack down on debate at Oxford. In any case, his actions do not seem to have attracted much notice. His successor but one as archbishop, John Peckham (d. 1292), another scholar of considerable note, had to write to the university to ask if they had a copy of Kilwardby's condemnations as he couldn't lay hands on one himself. He eventually had to ask the bishop of Lincoln to lend him a copy.[2]

Although it was not as celebrated as Paris, Oxford's university had already produced several thinkers of the first rank by 1277. Many of its brightest stars migrated across the channel to lecture at Paris. The shared religion of western Europe, as well as widespread knowledge of Latin, meant that medieval scholars formed a single international intelligentsia that was more closely knit than it has ever been since. Unfortunately, despite a common language among the elite, Europeans still spent plenty of time fighting each other – and new technology could help them do it.

The Physics of War

Medieval western Europe benefitted from the advantages that are enjoyed by large-scale empires without the disadvantages of unified secular control. The combination of political fragmentation with religious and cultural unity meant that scholars within Christendom could exercise a great deal of freedom. No secular ruler could control them and they enjoyed the overarching protection of the Church. Likewise, a theologian like William of Ockham (c.1287–1347), who fell out with the Pope, could join the Emperor's party for protection. Other scholars could travel around to wherever the Church or their careers took them. Thomas Aquinas was brought up in Italy but made his name in Paris, while his longstanding opponent Siger of Brabant came from the Netherlands. Pope John XXI was Spanish and Albert the Great German.

Catholics saw themselves as a unit that was collectively called Christendom. Despite the frequent conflicts between the nobility and royalty, everybody accepted that it was, in fact, unacceptable to make war on fellow Christians. This was one of the reasons why no single ruler was ever able to dominate in the way Charlemagne had done. There was still a Holy Roman Emperor, but he had enough trouble holding together the fissiparous Germans and Italians to contemplate conquering France, England or Spain. The popes had no desire for a single political ruler to rival their own spiritual power and made it their priority to prevent the Holy Roman Emperor from ever achieving such supremacy. This was not difficult because all the other crowned heads of Europe were squarely behind limiting the Empire's influence.

An effective way to control the martial ardour of the European nobility was to send them off on a crusade. The crusades were church-sanctioned wars but otherwise leadership was firmly in the hands of secular lords. Today, we tend to remember the ill-fated expeditions to the Middle East. These crusaders failed to conquer Islamic Palestine permanently, and they probably caused more damage to the Christian Byzantine Empire when they sacked its capital Constantinople in 1204. Less well known and more successful were the northern crusades launched against the remaining pagans of the Baltic region, the last of whom converted to Catholicism only in 1386.[3]

Secular rulers found the clergy extremely useful. After all, they were literate, urbane and well-travelled. Kings employed them as councillors and ambassadors, while anyone who could afford a tutor for their children would hire a cleric. As we have seen, the church even had to pass legislation to prevent monks and priests from practising as doctors. More surprisingly, siege engineers were sometimes in holy orders.[4] While the clergy were not supposed to shed blood, this law was very loosely applied. Dropping large boulders on people from afar appears to have been acceptable behaviour. As priests were more likely than most to have the necessary mathematical and engineering knowledge required to accurately direct the fire of siege machines, those willing to perform this function were in great demand. If the enemy were infidels or heretics, it is unlikely that the priest in question would have had many scruples about carrying out this task.

The weapon of choice for knocking down walls was the trebuchet. This enormous machine, first reported in western Europe around 1200, consisted of a long wooden beam pivoted close to one end.[5] This meant that one end of the beam was much further away from the pivot than the other. From the short side, a box of rocks or some other heavy counterweight was attached. The far end of the long side was extended even further by having a sling attached to it. Into this sling was inserted the stone that the trebuchet would fire. For a large trebuchet, the shot could weigh 300 pounds and be propelled into a target 500 feet away.[6] To fire the weapon, the operator winched the counterweight into the air by pulling down the long side of the beam. When the counterweight was released, it fell due to gravity and the beam shot up to a vertical position. The sling attached to the end of the beam unleashed its load towards the target. No one knows who invented the trebuchet but whoever they were, they had a good understanding of how a lever works. The differential lengths of the beam on each side of the pivot meant that the sling was travelling at the highest possible speed at the moment when it released the stone. Likewise, the lever made it possible to haul the heavy counterweight back into the air for the next firing.

7. A large trebuchet

The lever is the sort of practical mathematics that the trebuchet's inventor probably had only an instinctive feel for, even though the correct numerical formula to calculate its effects had been known by the ancient Greeks. Unfortunately, the Greeks' ideas about the dynamics of a stone as it flew through the air, or about the flight of an arrow, were much wider of the mark. This is worth a brief explanation because advances in understanding of the motion of projectiles were among the great achievements of medieval science.

In his *Physics*, Aristotle explained that there are two kinds of motion, which he called 'natural' and 'violent'.[7] Natural motion means falling by the action of gravity. Aristotle explained that this was 'natural' because falling objects were inherently striving to reach their proper place. The proper place for an object depended on its weight. Rocks are heavy, so they want to be as low down as possible. Water is less heavy and so is content to rest between the rocks below and the air above. Fire is lightest of all, so it rises even through the air. The Greeks speculated that an invisible region of fire formed a boundary between the atmosphere and outer space, rather like the popular conception of the ozone layer today. Thus, according to Aristotle no *force* of gravity was necessary. It was the natural desire of objects to occupy their proper place that caused them to fall or rise.

Violent motion was any kind of movement apart from a gravitational one. Thus, when you lift a rock, the motion is violent, but when you let go the subsequent fall is natural. One surprising implication of Aristotle's ideas was that the two kinds of motion were completely incongruent. That is, it was thought to be impossible for an object to move naturally at the same time that it was being forced violently. This doctrine gives rise to a very strange result for anything that is thrown, for example a trebuchet's boulder. Because violent and natural motion could not co-exist, the boulder must in theory keep going in the direction in which it has been fired until it slows down and finally stops. At that point, it drops out of the sky. This means that a projectile will move in straight lines rather than the curved path that modern science predicts. Opinion in the Middle Ages was divided, but there is no doubt that Avicenna and those who followed his thinking took this idea completely literally.[8] The effect of this would be much like the fate of a cartoon character who has inadvertently run over the edge of a cliff. As long as the poor creature's momentum keeps him moving forward, he does not fall, but when he comes to a halt he drops like a stone.

Historians have long been puzzled about how anyone could believe that a projectile could travel in a straight line and then drop out of the sky. After all, experience should have taught otherwise. But experience can be misleading. Bowmen were well aware that they could shoot straight at a target for maximum accuracy or fire into the air for maximum range. Those under a hail of arrows would have noted that they came from above and, under the circumstances, no one would have bothered to measure the exact angle of incidence. The trebuchet also propelled its rock into the air and, by the time this landed, it had lost a good deal of its forward momentum to air resistance. It would have appeared to those under attack that the projectiles were coming from above.

By the mid-thirteenth century, Aristotle's *Physics* was a central part of the undergraduate curriculum at Paris, Oxford and most other universities. We can be sure that Aristotelian ideas about motion had an equally wide currency. And the fact that clerics were capable of being employed as siege engineers provides us with a link between the academic realm of the universities and the military camp. Directing the siege train is probably what a certain Peter the Pilgrim (fl.1269) was doing with the army of Charles of Anjou (1225–85) while it was besieging the city of Lucera in southern Italy during 1269.[9] The city, which was defended by a group of Muslim mercenaries, put up a considerable show of resistance but eventually capitulated.[10]

We actually know very little about Peter the Pilgrim and we cannot even be sure that he was a cleric, although given his literacy and education, it is almost certain that he was. He obviously found the siege long and tedious, a feeling shared by many of his fellow soldiers. To while away the time he decided to write a short treatise on what he called 'the indubitable but hidden power of the lodestone concerning which the philosophers have hitherto given us no information.'[11] Lodestone is a natural magnetic iron ore called magnetite. Magnetism was, of course, the archetypical occult property and did not really fit Aristotle's materialist philosophy that forbade action at a distance. It was perfectly obvious that a magnet could affect another object even though the two were not touching. Rather than just worry about the theory, Peter decided to empirically investigate how magnets behave.

Peter was the first to realise that magnets have polarity – north and south. He found that this was always the case even if they are cut into pieces. As for the compass, he correctly deduced that the needle must be attracted towards some giant magnet whose influence could be felt by compass needles throughout the world. He guessed that the heavens themselves were magnetic and that compasses pointed towards the pole star rather than the earthly magnetic north pole. Another of his mistakes had extremely positive repercussions. Because he believed that the whole universe, which was spherical in shape, was a cosmic magnet, he deduced that a sphere would be the best shape for one. This is untrue, as a bar magnet is considerably stronger, but the idea would be picked up in the sixteenth century by a man who believed that occult forces could overthrow the Aristotelian laws of physics. That was William Gilbert (1540–1603), whom we will meet in chapter 18. Peter the Pilgrim's work on magnets had a long and profitable future because it came to the notice of a man who was much more famous than he was. This contemporary praised Peter as 'a master of experiment' and an expert on alchemy, surveying and military tactics.[12] The tribute came from Roger Bacon (1214–92), probably England's most renowned medieval scholar.

The Life of Friar Bacon

Roger Bacon was born into a well-to-do family from the west of England who sent him to Oxford University when he was in his teens. Shortly after 1230, he travelled to Paris where he was lecturing on Aristotle by 1237. Ten years later he resigned his position and devoted himself to private study. He spent his time and plenty of money (some £2,000 by his own reckoning) on alchemical research.[13] He seems to have had little interest in high living, but alchemy was notoriously expensive and soon he had run out of cash.

This was not an unusual situation for a student to be in, and many were forced to leave university because they were unable to pay the fees. Luckily a solution was at hand. The Dominicans and Franciscans were actively recruiting scholars because they needed trained preachers to carry out their evangelising missions. Joining these orders might not sound very attractive to us, as it involved living in complete poverty and in obedience to the commands of superiors. The advantage was that the friars had their own priories at

Oxford that provided food and board. More importantly, the orders paid the university's fees for it to educate its members.[14] We know that Bacon was devoutly religious and in about 1257 he signed up with the Franciscans who ensured that he could continue his academic career.

As a Franciscan, he was moved between the friaries attached to the universities of Paris and Oxford. While he was at Paris, he met Cardinal Guy le Gros de Foulques who later became Pope Clement IV (d. 1268). Bacon made such an impression that after he was elected Pope, Clement asked him to send his prescription for Christian reform. The result of this request was the *Opus Major* (or 'Larger work') which Bacon compiled in a frantic year of activity. He also managed to run off the *Opus Minor* and *Opus Tertium* (the 'Lesser Work' and 'Third Work' respectively) during the same period as summaries of the *Opus Major* that he subsequently expanded with much original material. Unfortunately, Clement was dead before he received the massive tomes he had commissioned. It is difficult to know what he would have made of them if he had ever had a chance to read them.

A recent plausible suggestion for why Clement asked Bacon to write up his ideas was that they shared a concern about the end of the world.[15] In the *Opus Major*, Bacon writes: 'I am writing on account of the perils which happen and will happen to Christians and to the Church of God through unbelievers and most of all through the Antichrist.'[16] Certainly, fears of an imminent apocalypse would explain the urgency that drove him to produce such an enormous body of writing in a single year. The main thrust of Bacon's work is the defence of Christian truth. There is no hint in anything Bacon wrote that he had even slightly sceptical religious views. His concern was that the Church was not using the full armoury of the sciences to further the spread of Christianity. He was particularly concerned that Jews and Muslims should be converted before the doomsday, which he saw rapidly approaching.

In chapter 6, we saw how the university of Paris had initially banned Aristotle's books about nature before finally accepting them as the bedrock of philosophy. Bacon was writing at the time that this dispute was in full swing. Therefore, we should read his three *Opera* as part of the case for using the natural science of Aristotle, not to mention mathematics and linguistics, in Christian education. He expounds in enormous detail exactly how the sciences can be of aid to religion.[17] Doubtless, Bacon was also interested in natural philosophy for its own sake. However, this in itself hardly makes him a subversive critic of religion. He held firmly to the idea, then so common among thinkers, that science was the handmaiden of theology.

Oxford's Franciscans already had a strong academic reputation before Bacon joined them because Robert Grosseteste (c.1170–1253) taught theology at their priory between about 1230 and 1235.[18] Grosseteste had risen from humble beginnings to become head of the university and would later become bishop of the immense diocese of Lincoln. His story illustrates that particularly brilliant men from poor families could use the universities as a path to advancement far beyond their initial standing. Their families could expect to benefit from this, especially if the cleric in question did not take his oaths of celibacy too seriously and had a mistress and children. As far as Grosseteste was concerned, however, the oaths were binding and he became a feared opponent of clerical corruption. Early in his career, he suffered from a near-fatal illness that caused him to give up the trappings of worldly priesthood for the discipline of the Franciscans (although it is unlikely that he ever actually joined the order). We know very little about his early life, but he had probably been awarded his Master of Arts degree at Oxford before 1200 when the writer Gerald of Wales (1147–1223) commended him as skilled in the liberal arts.[19] He was made head of the university in 1214, at least according to a longstanding tradition,[20] and then appointed to the see of Lincoln in 1235. After he became a bishop, Grosseteste's devotion to his ecclesiastical duties led to plenty of political disputes with more worldly clergy.

Although he ended his career as a bishop and theologian, many of Grosseteste's early works are about natural philosophy. He wrote his books on nature in his youth before his brilliant mind turned to what he thought were higher theological matters. Grosseteste did not yet have access to all the newly translated books by Aristotle, but he made considerable progress nonetheless. An important aspect of his thought was his frequent references to *experimentum*. This is the Latin word from which we get our modern term 'experiment', but a more accurate translation would be 'experience'. So, when natural philosophers wrote about how *experimentum* informs us of certain properties of nature, they are referring to what they have observed and not to controlled experiments that they have performed.[21] The distinction is very important and follows from the common belief among ancient and medieval philosophers that natural phenomena could not be expected to perform in a laboratory in the same way as they did in the wild. For Grosseteste, to study nature meant having to observe, passively and unobtrusively, to see how things happened in the real world.

The Legend of Friar Bacon

Roger Bacon was next in the line of natural philosophers, after Grosseteste, who would ensure that Oxford had an illustrious reputation for the subject throughout the late Middle Ages. Notwithstanding the importance of his work, Bacon's reputation today actually hinges on two misconceptions. First, we hear, his writing has a peculiarly modern flavour, with references to experiments and future inventions like cars and planes. Second, there is a persistent myth that because he was ahead of his time, he got into serious trouble with the Church.

We should deal with the second of these allegations first. According to several of the standard biographies, the Franciscan authorities imprisoned Bacon for ten years late in his life. For those looking for evidence of the conflict between science and religion, this was a prime example of clerical intolerance. Some historians had no doubt that the Church incarcerated Bacon for his dangerous scientific opinions. For others, it was his sympathetic view of both astrology and alchemy that doomed him to a dungeon. Today, a fresh look at the surviving sources show that it is difficult to prove Bacon's imprisonment happened at all, let alone that it was caused by his dangerous scientific views.

The origin of the story is the *Chronicle of the 24 Ministers General of the Franciscans* dating from about 1370, a full century after his alleged arrest. This document claims that Bacon was a master of theology and imprisoned for unspecified 'suspect novelties'.[22] As we know that Bacon never qualified as a theological master, it is hard to give this account much credence. Furthermore, the controversy in which the *Chronicle* implies Bacon was involved had nothing to do with science. Rather, a sect of extremely ascetic Franciscans was stirring up trouble. These men were convinced, like Bacon, that the world was about to end and that the Church should, forthwith, divest itself of all property in imitation of the poverty of Christ. The material riches of the medieval church are legendary and bishops certainly had no intention of living as beggars. If Bacon had been a supporter of these spiritual Franciscans – and, given the enormous piety and millennialism evident in his writings, this is plausible – he could have got into a great deal of trouble.[23] However, the allegation that Bacon's science led to his imprisonment finds no support in the historical record.

His reputation as a futurologist and experimenter has a stronger foundation in fact. In chapter 1 we saw how the early medieval period had been an era of rapid technological advance, and this trend continued through the following centuries. Among the most important inventions to reach Europe in the Middle Ages was the Chinese discovery of gunpowder. The earliest reference to this explosive substance in the West comes from a work of Roger Bacon which describes firecrackers, a popular amusement in China. A passage in the *Opus Tertium* reads:

There is a child's toy of sound and fire made in various parts of the world with powder of saltpetre, sulphur and charcoal of hazelwood. This powder is enclosed in a packet of parchment the size of a finger. This can make such a noise that it seriously distresses the ears of men, especially if one is taken unawares, and the terrible flash is also very alarming. If an instrument of large size were used, no one could stand the terror of the noise and flash. If the instrument were made of solid material then the violence of the explosion would be much greater.[24]

It is clear from the last line that Bacon could see how the toy could be adapted to become a weapon. He might have heard about firecrackers from Franciscan missionaries who, like Marco Polo (1254–1324), were taking advantage of the trade routes opened up by the Mongol Empire. The conquests of Genghis Khan (1162–1227) had imposed a bloody peace on central Asia which made travelling from Europe to China practicable for the first time since the fall of the Persian Empire in the seventh century.

Some of Roger Bacon's work reflects the spirit of inventiveness that permeated Europe. In a well-known letter called *On the Marvellous Power of Artifice and Nature*, he speculates on ideas like flying machines and horseless carriages:

> It is possible that a car shall be made that will move with inestimable speed and the motion will be without the help of any living creature … It is possible that a device for flying shall be made such that a man sitting in the middle of it and turning a crank will cause artificial wings to beat the air after the manner of a bird's flight.[25]

Today, we know that Bacon's idea for a flying machine would be completely ineffectual, but Victorian historians proclaimed him as a genius ahead of his time on the strength of this document. A more modest view is that he was equipped with an active imagination. We should remember how much he got wrong as a result of the shared attitudes of his era. Where Bacon was unusual was in his interest in the work of craftsmen. Most medieval scholars had no time for technology and handiwork. This was partly a reflection of the views of Greek philosophers, like Plato and Aristotle, who thought that any kind of trade was beneath the dignity of intellectuals. This is an area where Christianity provided a useful counter to pagan chauvinism. After all, Jesus himself had been a carpenter. But old prejudices were slow to die out and most university-educated men did not involve themselves with trade.

In his *Opus Major*, Roger Bacon tried hard to convince the Pope of the importance of 'experimental science', but this also does not have quite the meaning we might expect. A large element of Bacon's thought was clearly magical and the experimental work he did carry out appears to have been largely devoted to alchemy, the pursuit of which had swallowed up so much cash. Although he probably did carry out a good deal of meddling in magical practices, he was not putting forward the kind of research programme that today we would recognise as scientific.

It is possible that the association of practical skills with alchemy and magic helped frighten off university scholars. Bacon himself ended up with a reputation as a magician. In the sixteenth century, he was thought to be a necromancer of the same kind as Faust. Inevitably, he was said to have acquired Gerbert's brazen head, which we last encountered in the hands of Albert the Great. The magic head supposedly gave its name to Brasenose College in Oxford, and some say that the brass doorknocker preserved in the college dining hall is the same one that once belonged to Bacon.

On a more positive note, Robert Record (1510–58), an important Tudor mathematician and writer, credited Bacon with the invention of a glass 'in which men might see things that were done in other places'.[26] This sounds like magic, but Record went on to explain that Bacon had used his knowledge of optical theory and natural philosophy to build the glass. In fact, Bacon never says he actually produced such a device, but he did suggest that it was possible to do so. 'From an incredible distance we might see the smallest letters ...' he wrote, 'so also might we cause the sun, moon and stars to descend in appearance here below and similarly to appear above the heads of our enemies.'[27] Nevertheless, Record's comment certainly proves that the idea of the telescope had been around long before it was officially invented in 1608.

Divine Light

Roger Bacon's ruminations on the magnifying qualities of lenses did not arise in isolation. By the mid-thirteenth century, Arabic and Greek books on the science of light were available in Latin. Bacon digested this daunting body of work and produced an impressive synthesis of all the advances up until his time.

Many of his fellow Franciscans were also fascinated by light. To Robert Grosseteste, it was divine. When God created the world, his first words were 'Let there be light.' This implied that light was a fundamental property of the universe. Grosseteste imagined it as emanating from God, filling the universe with his glory so that his presence was everywhere. Divine light did not just illuminate the physical world; it also had a metaphysical component that allowed mankind to 'see', as in comprehend, the mysteries of the faith. Thus, Grosseteste believed that understanding light would also tell him something important about God.[28] Unfortunately, he was writing before Europeans had access to the relevant Greek and Arab knowledge. Without a proper appreciation of what had gone before, he could make little progress beyond inspiring the next generation of scholars.

Among undergraduates studying physics, optics is one of the subjects with which they seem to encounter the most difficulty. They might be reassured to learn that the nature of light posed similar problems for even the most brilliant of minds in the past. The central problem was to explain how humans can see. Leaving aside psychological questions about the subjective experience of vision, natural philosophers wanted to understand the way in which the eye organises rays of light coming from every direction into a coherent image. By rights, we should not expect to see anything except a formless blur, because incoming rays from different angles should all interfere with each other and prevent us from forming a picture. Yet somehow our eyes reinterpret this chaos into a representation of the world around us.

Vision certainly left the Greeks stumped. Some of them thought we must see objects as whole entities. Thus, if we are looking at a dog, we receive an invisible signal of 'dogness' into our eye that transmits the whole image to our minds.[29] How each signal could fly though the air without getting mixed up with others was harder to explain.

Other Greek thinkers, especially Euclid and Ptolemy, tackled light from a different perspective. They had figured out that light travels in straight lines and that mirrors reflect light rays according to a fixed law. This meant that they could start using their skills in geometry to draw diagrams showing how rays propagated.[30] But this left wide open the problem of how the eye produces a coherent image from disparate rays. Euclid apparently solved it by suggesting that the eye launched rays outwards which interacted with the environment. This makes no difference to the law of reflection, which works in either direction. He may have imagined that the rays emitted by the eye bounced off objects and returned via the same path. This is quite an insight, because it is very similar to the way in which both sonar and radar detection work. In both cases, pulses are sent out which are deflected when they encounter something large and dense enough to form an obstruction. Some of the deflected pulse returns to the detector so that a picture of the environment can be formed. Bats actually do 'see' by listening for the echo of their high-pitched squeaks, and they can navigate through complete darkness as a result. So as a model for vision, Euclid's emission theory is nothing like as far-fetched as it sounds; it is just not right for humans.

Euclid's theory allowed him to use geometry to model the reflection and refraction of light. But his idea had many problems. The most pressing was that luminous objects, like the sun or a torch, were obviously giving out light. This contradicted the theory that we see by rays emitted from our eyes. This led Bacon, following Aristotle and the work of Muslim scholars, to reject Euclid's theory but without abandoning his geometrical work on reflection.

Unfortunately, he was unable to make progress with the central problem of vision: how the eye sorted light rays, coming from different directions, into a coherent image. The answer to this problem that he adopted, which had been developed by the Muslim scholar

Alhazen (965–c.1039), was to assume that only light rays that penetrated the eye head-on went into making up the image that we see.[31] Any light rays that hit at a glancing angle were so weakened, he said, that they were unable to impose themselves on the retina. Of course, this raised the question of why rays that arrive just a few degrees from perpendicular do not produce a weak image that would give us blurred vision. The theory demanded that any ray that strayed even slightly from straight-on should be completely disregarded. As such, although it was the best idea available at the time, later theorists were never entirely happy with it.

Roger Bacon's work on optics gives the lie to some of his rhetoric about the importance of experimental science. Although he seems to have played around with lenses, he betrays no sign of a systematic investigation of light's properties. His theories look like book-work and an attempt to join together the various authorities into a coherent whole.[32] He was not even prepared to completely abandon the theory that our eyes send out rays; since it was supported by respected Greeks like Euclid, Bacon had to find a place for it.

The subject of optics, known as 'perspective' in the Middle Ages, attracted several other thirteenth-century natural philosophers. John Peckham, the archbishop of Canterbury shortly after Robert Kilwardby had been, was a Franciscan like Bacon and shared his fascination with light. He wrote his own textbook on optical theory, largely taken from Alhazen, which remained on university syllabuses until the sixteenth century.[33] Bacon's own work also informed a massive treatise by the Polish theologian Witelo (fl.1250–75) who was working in Rome. This combined all the Greek and Arab ideas together with the insights of his own era.[34] Witelo's book remained the last word on perspective until the sixteenth century when, as we will see in chapter 18, it formed the foundation of modern optical theory.

Perhaps Bacon would have solved the mystery of vision had he lived a little longer, because just as he reached his last days, a new invention appeared in Italy that provided the missing link – spectacles. The way that the eye can resolve an image is through its lens, which refracts all of the light rays emanating from a particular point and focuses them onto a single spot on the retina. Spectacles, with their own lenses, might have given Bacon the clue he needed.

The earliest mention of spectacles is found in guild regulations from Venice dating from 1300.[35] In 1306, the Dominican Giordano of Pisa noted in a sermon that he was preaching in Santa Maria Novella in Florence:

> It was not twenty years since there was discovered the art of making spectacles that help one see so well; an art which is one of the best and most necessary in the world. And that is such a short time ago that a new art that had never before existed was invented. I myself saw the man who discovered it and practised it, and I talked with him.[36]

So the evidence places the invention of spectacles in the late thirteenth century. Several writers patriotically sought to name the inventor as a denizen of their own cities, but these sources are extremely unreliable.[37]

The first spectacles were reading glasses with convex lenses (those that bulge in the middle and are thinner around the edge). These require much less precise grinding than the concave lenses needed to correct short-sightedness, besides also being easier to manufacture in other respects.[38] Spectacles meant that monks and other scholars could continue working even after their eyesight began to deteriorate with age. And as we shall see in the next chapter, mechanical clocks started to tick in England and Italy at about the same time. These inventions catapulted medieval Europe into first place in the race to become the most technologically advanced civilisation on earth. Although he did not know it, medieval man had already surpassed China, Islam and the ancient world.

CHAPTER 10

～※～

The Clockmaker:
Richard of Wallingford

Among the earliest memories of Richard of Wallingford (1292–1336) would have been heat and flames from his father's forge. As the blacksmith of a small town, Richard's father had been an essential part of the community, producing horseshoes, repairing ploughshares and hammering iron utensils into shape. It was hard physical work, but also highly skilled. The young boy inherited a keen interest in the mechanical arts and an appreciation of craftsmanship.

The family of a smith would be reasonably well-off, but there was very little in the way of insurance if tragedy struck. This meant that on his father's death when Richard was only ten, too young to take over the trade, he could have been left destitute. As often happened, though, it was the church that acted as a medieval welfare state. Richard was adopted by a local Benedictine monk who saw to it that he received a proper education in reading and writing Latin.[1] Richard turned out to be a precocious student far more suited for a clerical career than his father's trade. When he came of age, he was sent to study at the college that the monks maintained at Oxford.

Oxford University and the Foundation of Cambridge
Studying at a university during the Middle Ages was not for the faint-hearted. The congregation of so many young men in a single town

without parental authority was a recipe for trouble. Drunkenness, violence and prostitution were facts of life, with the students acting as both the victims and the instigators. The records show frequent complaints about riotous students from the put-upon townsfolk. In 1269, we hear that

> a frequent and continual complaint has gone the rounds that there are in Paris some students and scholars … who under the pretence of leading the scholarly life, more often perpetrate unlawful and criminal acts relying on their weapons, by day and night, to atrociously wound or kill many persons, rape women, oppress virgins, break into inns, also repeatedly commit robberies and other enormities hateful to God.[2]

At Oxford, the students did not stop at oppressing the local virgins. In 1209, the murder of a young woman by a member of the university sparked a major crisis. The circumstances are obscure and it is not clear whether the victim was her killer's mistress or had denied him her favours. The attack happened on the outskirts of the town and the guilty party quickly made good his escape. A mob of outraged residents converged on his lodging and, finding the culprit gone, contented itself with stringing up his housemates. The university was up in arms at this lynching and demanded that the ringleaders be brought to justice.[3]

The dispute was exacerbated by an ongoing conflict between King John (1167–1216) of England and the Pope, Innocent III, over who should be archbishop of Canterbury. The Pope wanted Stephen Langton (1154–1228), a formidable politician and churchman, to get the job. King John required someone altogether more amenable to his will and refused to allow Stephen into the country. In response, the Pope excommunicated the king and placed England under an interdict. Even if John was not worried about his own excommunication, his subjects were horrified at the effects of interdiction. This was the papacy's ultimate weapon and effectively excommunicated an entire territory. Church services were cancelled, the holy sacraments withdrawn and the festivals that punctuated the calendar abandoned. Only the monks in their monasteries, who claimed they had nothing to do with King John's religious policies, received any sort of concession. They were allowed to perform divine service behind closed doors once a week, as long as they were quiet about it.

The crisis at Oxford became mired in the larger dispute. The interdict meant that most clerics had already left the town by the time of the murder, and the hostility of the locals drove the rest away shortly afterwards. For five years, there was no teaching at the university. A group of the masters and their students migrated to another town to the east called Cambridge. It was not a very prepossessing place, being quite small and located in the middle of a bog. But the scholars were welcomed and set up shop there. The resulting university never matched Oxford's prestige during the Middle Ages, but in the sixteenth century royal patronage enabled Cambridge to achieve parity of esteem with the older foundation. During the Reformation, Cambridge enjoyed such prominence that scholars came from as far away as Hungary to hear lectures there. The universal language of Latin meant that this kind of intellectual cross-pollination was common, but misunderstandings were still possible. Our Hungarian found himself wandering around Canterbury in Kent in search of the lecture hall. The Latin word for Cambridge is *Cantabrigia*, so his confusion is understandable.[4]

In 1212, Innocent III declared King John deposed and invited the king of France to invade England. With his people clamouring for the churches to be reopened, John capitulated to the Pope's demands. He handed over his entire kingdom to the Pope who graciously installed him as its monarch. Stephen Langton finally took up residence in Canterbury and promptly sided with the barons in the movement that culminated in John signing the Magna Carta two years later.

As part of the deal between John and the Pope, the townsfolk of Oxford had to make restitution to the teaching masters and invite them back. This part of the agreement, which Robert Grosseteste probably helped to negotiate, gave the university the privileges and rights that were the foundation of its later success. The local people became second-class citizens in their own town and had to pay an annual fine to the university.[5] Thus, by the time Richard of Wallingford began his studies in the early fourteenth century, the university effectively controlled the town of Oxford.

The Trivial Syllabus

Wallingford is only about fifteen miles to the south of Oxford, but Richard's journey to Gloucester College (nowadays subsumed within Worcester College) on the edge of the town may well have been the furthest he had ever travelled. At sixteen, he was slightly older than many of the first-year students who would have begun their careers at fourteen. All of them were already treated as adults for most purposes and, when they joined the student body, they also become clerics governed by canon and not common law. This is why students were known as 'clerks' – a contraction of 'clerics'. Chaucer put his *Clerk's Tale* into the mouth of a virtuous Oxford student who diligently studied the books of Aristotle.[6]

The first step of Richard's academic career was to gain a bachelor of the arts degree, which usually took three or four years. The syllabus he had to follow would have been similar no matter which university he attended. During the course, he would have covered three introductory subjects – grammar, dialectic and rhetoric – together known as the *trivium*. This is the origin of the English word 'trivial'. It is an unfortunate piece of etymology because all three subjects were vitally important.

Grammar involved developing a rigorous style of written Latin. As Richard had done, students were expected to have mastered reading, writing and speaking in Latin before starting their higher education. It was the only language recognised at the university, and the authorities could levy fines for talking in English even in private conversation. Universities tried to limit grammatical tuition on the grounds that new students should already be able to express themselves on parchment, but remedial teaching was often required.

Dialectic means 'logic' in everyday parlance and it is one of the facets of medieval intellectual life that seems most alien to us today. Far from being irrational, thinkers of the Middle Ages were obsessed by logic to an extent that seems completely unreasonable to their modern critics. Students had logical constructions called syllogisms hammered into them until they could repeat them by heart. As a learning aid there were plenty of mnemonics to help students along, but ultimately logic required a great deal of rote learning. The standard textbook was Peter of Spain's *Summulae Logicales*, often called the *Little Logicals*. We first encountered Peter in chapter 6, because as Pope John XXI he had ordered the investigation into Averröism at Paris. During the sixteenth-century backlash against medieval scholarship, Sir Thomas More (1477–1535) could unfairly quip that the *Little Logicals* 'was probably so-called because it contained little logic'.[7] Like so much medieval learning, dialectic derived from the works of Aristotle but it was taken to extremes of which the ancient Philosopher could never have dreamed.

The third element of the *trivium* was rhetoric. This was a much broader subject than learning to speak well in public. As well as oratory, students learned how to construct arguments and the correct form for writing letters.

Once Richard had mastered grammar, dialectic and rhetoric, he was ready for the 'determination' of his bachelor of the arts degree. Determination is simply the term used for the oral examination he had to pass in order to earn the degree. The examination was a highly formalised debate or disputation with the examiner who would then report to the university authorities on his success. A certificate of good character and morals was also required for the bachelorship, something some students must surely have found more difficult to obtain than a pass in the examination.

The Scientific Syllabus of the Middle Ages

When he attained the rank of bachelor, Richard had another three years of study to look forward to before becoming a master of the arts. During this period, he covered the remaining four of the seven liberal arts, called the *quadrivium*, as well as the basics of the three branches of philosophy: ethics, metaphysics and natural philosophy. The seven liberal arts date from the late Roman period when they were regarded as the subjects fit for a free man, rather than a slave, to study. The *quadrivium* was the mathematical component, made up of arithmetic, geometry, music and astronomy. This meant that all students who wanted take their studies beyond a basic level had to study maths.

Basic knowledge of calculations was derived from Arabic sources collectively known as *algorismus*, which is a Latin garbling of the name of the Muslim mathematician al-Khwarizmi (c.780–c.850). The originality of his work is disputed, but it was through Latin translations of it that the principles of algebra reached the Christian West. The word *algebra* itself is a corruption of the title of his best-known book the *Al-jabr*.[8]

Al-Khwarizmi used what we call Arabic numerals (which actually originate from India) and introduced the vital concept of the number zero. Previously, most westerners had used the system of Roman numerals they inherited from the ancient world. This was fine for counting and for dates, but these numerals made it very difficult to do even simple sums. For basic arithmetic, it was far easier to use an abacus but this involved a completely different system of counting. There is some skill involved in efficient calculation with an abacus, as anyone who has been to a market in India will attest. As we have seen, Gerbert had first introduced Arabic numerals to the West in the tenth century to improve this device. However, it was not until the thirteenth century that al-Khwarizmi's *Al-jabr* led to their use becoming widespread.

In recent years there have been persistent claims that the Church resisted the introduction of Arabic numbers and especially of zero. In fact, professional abacus users were the ones who really felt threatened by the new system, as it seemed to make their skills redundant. There was also a problem in agreeing which symbols should correspond to which numbers. As late as 1299, the Bankers' Guild of Florence banned the use of Arabic numerals in official documents because they were causing too much confusion.[9] When a system was eventually settled upon, it was similar to that used in Muslim Spain. This differed markedly from the version used by other Arabs, and those differences remain today. For example, the European zero is the symbol for five in the Middle East. However, with the *algorismus* becoming part of most university syllabuses by 1300, Arabic numbers gradually came to predominate. Most university students were expected to handle addition, subtraction, division and multiplication as well as extract roots and perform basic algebra.

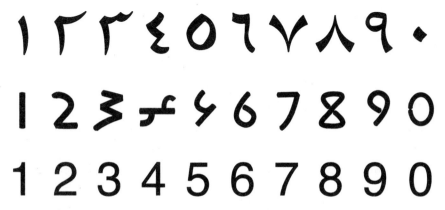

8. Eastern and western Arabic numerals compared to modern western numerals

In addition to learning to do calculations, medieval students mastered theoretical mathematics. This material came from the book *Arithmetic* by Boethius. It featured the properties of prime numbers, perfect numbers and what modern mathematicians call 'number theory'. The idea was to give the student the equipment necessary to understand the mysteriousness of figures and provide him with a vision of logical perfection. Anybody trying to read this book today will find it heavy going. For a start, Boethius's book does not tell you how to do sums and the word 'arithmetic' meant something quite different from what it does now. Rather, this subject dealt with the properties of numbers and ratios in a way that was supposed to train the mind for the higher philosophies. Arithmetic, Boethius said, 'holds the principal place and position of a mother to all the rest' of

Clement of Alexandria said "The druids taught mathematics to Pythagoras"

the arts.[10] This was a Greek tradition that ultimately derived from Pythagoras (c.569–c.475BC) and Plato. To Platonists, a number was the best example of something that transcended the material world and existed on a higher plane. By contemplating numbers, a philosopher could comprehend eternal truths rather than sully his thoughts with the mundane. Closely related to arithmetic was the subject of music. We might think this involved learning to sing or play an instrument. Far from it. Students covered the theory of harmony and an appreciation of the rhythms of the universe, rather than anything as practical as holding a tune.

For geometry, the key text was the *Elements* of Euclid as translated by Adelard of Bath. Most students would probably only have had to make it through the first three books, but even these contain a great deal of elegant and inspirational material. What medieval scholars loved about Euclid was the way that each of his demonstrations derived logically from the previous ones without any possibility of error.

The last of the four subjects of the *quadrivium* was astronomy. For this, students had *The Sphere*, the excellent little handbook by the Englishman John Sacrobosco that Cecco D'Ascoli had used for his lectures. It contained all the essentials they needed in order to understand how the medieval cosmos worked – with the earth at the centre, then the seven planets (including the sun and moon) orbiting around it, thence to the sphere of the fixed stars and finally out to heaven itself. After the seven liberal arts, the prospective Master of Arts would learn something of ethics, metaphysics and natural philosophy. Aristotle was the main authority for all philosophy, although his works were so difficult that most students tackled them through a commentary.

Modern critics of medieval universities have accused them of concentrating too much on useless and obscure logic at the expense of real knowledge. Logic was certainly an important part of the syllabus and it became increasingly complicated through the later Middle Ages. As an intellectual exercise, scholars would invent absurd situations and try to reason their way out of them. Every now and again, the universities would host a special session where students could put their most fiendishly difficult questions to a senior professor. No doubt they went to considerable trouble to come up with the most convoluted riddles they could think of in order to tax the minds of their superiors. The professor gained a chance to show off his mental dexterity by dealing elegantly with whatever his students threw at him. The result was a very rarefied formed of intellectual entertainment. Questions preserved for posterity include 'Should someone born with two heads be baptised as one person or two?' and 'Can a bishop who is raised from the dead return to his office?' Even Thomas Aquinas had had to find an answer to the question 'Is it better for a crusader to die on the way to the Holy Land or on the way back?'[11] The medieval logical conundrum that everybody knows is 'How many angels can dance on the head of a pin?' Sadly, this turns out to be the invention of a seventeenth-century Cambridge academic satirising the admittedly rather abstruse theology of Thomas Aquinas.[12] If a medieval scholar had really asked this, he would have meant it as a joke.

Richard of Wallingford's Career

Richard had entered the university of Oxford in 1308 and by 1314, before he had received his Master of Arts degree, the money ran out. It is not clear whether his studies to date had been financed by Benedictine charity or a legacy from his late father, but neither source was infinite. Without cash, Richard was faced with the same problem as Roger Bacon and he opted for a similar solution. He was intellectually very gifted and the Benedictines would be glad to pay for him to complete his studies if he would commit to joining them. With this aim in mind, the young scholar trekked to St Albans, one of the grandest and richest abbeys in England, and professed as a monk. Three years later, in 1317, he was ordained a priest. Adjudged to have shown sufficient devotion, he returned to Oxford to complete his studies.[13]

Before long, he incepted as a master of the arts. At this point he had to lecture undergraduates for a couple of years before he was allowed to proceed to the theology faculty. Several more years passed before he earned the title of bachelor of theology by composing a commentary on the *Sentences* of Peter Lombard. Although this is a theological textbook, substantial sections in its second book covered the creation of the world and other aspects of physics. This gave students ample opportunity to comment on natural philosophy if this was the subject that most interested them.[14]

We know that this was the case for Richard. While he was ostensibly studying the queen of the sciences, he found time to invent a new astronomical instrument called an Albion. This device is a calculator that allows its user to determine the positions of stars and planets more quickly than using the usual tables. Another advantage was that the Albion did not rely on its user possessing a set of tables that had been accurately copied out. Instead, he recalculated the positions from first principles each time he needed them. This did mean that considerable skill was required to operate the Albion, although mercifully a full understanding of the underlying mathematics was not necessary.[15]

The Albion is made up of brass disks that rotate around a common pivot. Engraved onto the disks are a large number of curved lines from which results can be read off as required. Silk threads hang from the pivot to help the user to take readings as accurately as possible.[16] Richard must have been one of the most accomplished mathematicians of his generation to have had the ability to map out the necessary engravings. He would put these talents to even more spectacular use after he left Oxford in 1326.

Richard returned to the monastery of St Albans to begin his career as a priest and a monk. He had barely settled in when the old abbot died. In the subsequent election Richard, still only in his mid-thirties, was chosen as his successor. It was a huge surprise that, despite his humble origins and relative youth, the returning Oxford scholar had been elected. The abbey's chronicle relates:

The newly elected lord abbot, Richard of Wallingford, having been led into the church with trembling and respect, as is proper, was made to stand before the high altar. The election was announced to the people by ... the archdeacon. It was beyond the expectations of everyone, and especially of the laypeople.[17]

After he had travelled across Europe to have his appointment confirmed by the Pope (with a hefty fee payable in consideration), Richard took up the reins of managing the abbey of St Albans. Although it was ancient and endowed with a vast acreage of lands, profligate abbots had plunged the monastery into debt. The new abbot turned out to be a fearsome administrator. His humble origins did not lead him to deal leniently with the farmers and townsfolk under his rule. Unfortunately for them, he needed their money to fulfil his own ambition of building the most advanced clock the world had ever seen.

The Invention of Time

At roughly the same time as spectacles appeared in Italy, the mechanical clock was invented, probably in England. The earliest mention of such a clock is from 1273 in Norwich.[18] They are recorded in several other documents dating from shortly afterwards, before the invention spread rapidly across Europe. Water clocks and other timekeeping devices had been built by the Romans and Chinese but the mechanical clock was something new. It was made possible by the development of the escapement. The Chinese had produced an escapement of sorts but there is no evidence that this knowledge passed to Europe. Instead it appears to have been independently invented in the West during the late thirteenth century.[19]

An escapement is a mechanism that allows a clock to keep time. Power is provided by a weight hung from a rope wrapped around a horizontal shaft. As gravity pulls on the weight, it tries to rotate the shaft. However, the shaft cannot turn freely because it is attached to a gear wheel whose teeth control its rate of rotation. The escapement itself is a weighted spinning crossbar that allows the gear wheel to turn by only one notch for each one of its rotations. Each time the gear moves on by one notch, there is one tick of the clock and the weight-driven shaft can turn by a very small amount. A series of other gears translate the turning of the gear wheel into the movements of the hands on the clock face.

Initially, mechanical clocks were extremely inaccurate and intended as astronomical models rather than as timekeepers. From these early devices, the idea that the universe itself might be a sort of clock quickly followed. The world as a machine was not a new concept, but because the mechanical clock could power an armillary sphere, which showed the relative positions of the planets, it made the analogy explicit. Before long, the idea that the heavens themselves were like a clock with their own divine inventor took root in the Christian imagination.

The first people to use the new invention to tell the time were probably monks. The divine office of prayers and psalms demanded that monks woke up in the small hours to attend the service of Matins in their church. They were summoned from their beds by the monastery's bells but someone had to be awake to ring them. We still remember this in the French nursery rhyme 'Frère Jacques', a literal translation of which is:

Brother James, Brother James
Are you sleeping?
Sound the bell for Matins.
Ding, dang, dong.[20]

It was easy enough to include a chime on a mechanical clock that could rouse the bellringer but no one else. Before long, clocks spread into secular life and other applications for them soon became clear. Towns installed public clocks, which rang the hours and meant, for the first time, that everyone in the neighbourhood could agree about what time it was. More importantly, an hour became a fixed period of time. Traditionally, there had been twelve hours between sunrise and sunset. However, the duration of daylight varies with the seasons, so the period of one daytime hour was shorter in the winter than in the summer. A mechanical clock chimed the same hours regardless of the season. The modern practice of counting the hours from midday and midnight was introduced by an Italian clockmaker in the 1340s, who included the feature on a famous clock he built for the city of Padua.[21] This is only sensible if all the hours are of an equal length through the day and night.

With the invention of the mechanical clock, time had ceased to be personal and had become a common fact that rules our lives.[22] Now labourers could work to the clock and demand a constant hourly wage, but conversely their employers could insist on their time and not just their product. These developments took centuries to come to fruition but, like so many others, had their roots in the Middle Ages.

Richard of Wallingford built his clock for the abbey church at St Albans. Telling the time was the least of its functions. The face was a giant astrolabe and instead of hour and minute hands, it had an orbiting sun and moon. It also provided the timing of the tides at London Bridge, which was the nearest major port for the monks. Richard carefully wrote down how the clock was put together, and these manuals survive in the Bodleian Library in Oxford where they have been used to create a modern reconstruction. The original suffered neglect over the centuries because after Richard's death, so few people had the ability to maintain it. Nor was its repair the highest priority for the cash-strapped monks. However, it was still in place in the sixteenth century, until it disappeared in the destruction of so much of England's medieval heritage during the Reformation.

Abbot Richard himself did not live to see his clock in action for very long. On his return from visiting the Pope in 1328, he complained of a pain in his eye. Gradually he became more ill and the dreaded disease of leprosy was soon diagnosed.[23] As an important man, Richard could not be exiled to a leper colony and he kept his position as abbot. Leprosy affected all strata of society. It killed King Baldwin IV of Jerusalem in 1185 and perhaps Robert the Bruce of Scotland in 1329. Facing an early death, Richard threw himself into completing his great clock. When he died in 1336, he left a mechanical legacy without equal.[24]

CHAPTER 11

꧁꧂

The Merton Calculators

Richard of Wallingford and Roger Bacon were respected figures in the Middle Ages, but it was not to them that the university of Oxford owed its reputation as a leading philosophical centre. Bacon only achieved the renown he enjoys today when he was rediscovered after the Middle Ages had ended. Instead, a flock of innovative thinkers who were active from the late thirteenth century, many of whom completed their careers in mainland Europe, briefly propelled Oxford to a position of intellectual pre-eminence.

Foremost among them was John Duns Scotus (c.1265–1308), the most important developer and critic of Thomas Aquinas's theology. He acquired the honorific 'the Subtle Doctor' and his work formed the foundation of an entire school of theology. Duns Scotus was educated at Oxford in the aftermath of the condemnations of 1277 and, like Roger Bacon, he joined the Franciscans. Once he qualified as a theologian, his order moved him to Paris and thence to Cologne where he died.[1]

In many ways, Duns Scotus drove forward the train of thought begun by Thomas Aquinas. Like Thomas, he was a great believer in the power of human rationality and insisted that we can prove the existence of God by reason alone. But Duns Scotus also thought that Aquinas had gone too far in subjecting God to the dictates of reason. In natural philosophy, Thomas thought that the workings of the universe had to reflect the character of God. Duns Scotus said that this placed unwarranted restrictions on God's freedom of action. He could make the universe in any way that he pleased and it did not have to reflect anything else at all.[2] Likewise, in ethics, to oversimplify a complicated argument, Thomas thought that God willed what was good and Duns Scotus thought that it was good because God willed it. As Duns Scotus put it, 'the divine will is the cause of good and so by the fact that it wills something, it is good.'[3]

Thus, in both ethics and natural philosophy, Duns Scotus echoed the 1277 condemnations in his emphasis on the freedom of God to do as he pleases. Good is what God says it is and the universe works in the way God says it ought to. For science, this was a positive move. There was now no reason to assume that everything worked in the rational way that Aristotle said it did. Natural philosophers were free to speculate on all sorts of possibilities that they had previously ruled out.

Ockham and his Razor

The trend towards granting God absolute liberty of action and sovereignty continued with the work of another British Franciscan, William of Ockham (or Occam). He was born in Surrey, in south-eastern England, and entered Oxford University around 1300. He probably joined the Franciscans at the same time. In about 1320, he left the university before he had completed the theology course and gained his doctorate.[4] He took up residence in a Franciscan convent for three years, which was when he enjoyed his most productive period as a philosophical author.[5] The reasons behind his departure from the university remain obscure but may revolve around a commentary he had written on the *Sentences* by Peter Lombard. Putting together such a commentary was part of the standard training for theologians. If it was found to contain errors of doctrine, the candidate would have to either correct them or else drop out from the course. Simply making the required amendments would mean that the heretical statements need have no negative effect on the theologian's future career. The procedure was no different from today's students having to give the correct answers in order to pass their exams. However, arguing that it was the examiners who were wrong was riskier. This seems to have been what William of Ockham did, and the matter ended up before Pope John XXII whom we encountered in chapter 8. As well as banning fraudulent alchemy, he was responsible for finally crushing the ascetic wing of the Franciscans with whom Roger Bacon may have been involved.

At the time, the papacy was based in the city of Avignon, in the south of France. Pope Clement V (1260–1314), who was French, had removed himself there in 1309 to avoid the constant conflict and risk of being murdered in Rome. The king of France was delighted to have him and effectively turned the papacy into a branch of the French monarchy.

Pope John XXII, at least, had an independent spirit but he was happy to remain in Avignon rather than return to chance his luck with the unruly Romans. He summoned William of Ockham to Avignon where a committee, headed by John Luttrell (fl.1317–23), the ex-chancellor of Oxford, heard his case. Some modern scholars have suggested that Luttrell lost his job at the university as a direct result of censuring William's work. If so, relations between the two would not have been cordial to start with. Others think it unfeasible that Luttrell would have been involved in a case if he had such a personal interest in it.[6] We will never know for sure. Regardless of the political machinations, the committee declared that 51 propositions in William's commentary on the *Sentences* were heretical. Rather than accept this and amend his work, William fled with a couple of other academics and sought protection from the prospective Holy Roman Emperor. The papacy, allied with France, was continuing its struggle with the Empire for mastery of Europe. Allegedly, William made the Emperor an offer: 'Do you defend me with the sword and I will defend you with my pen.'[7] Thus for the rest of his career, William was part of the imperial court and wrote political propaganda on behalf of his master.

Ockham wrote several important books in the period between leaving Oxford and travelling to Avignon in 1324. The gist of his most controversial work was that human reason is not a sufficiently powerful tool to discover very much about God or the world. We cannot even use reason to prove most statements about cause and effect – experience is the only way to know things.[8] This makes science as well as theology very difficult. As Albert the Great had said, natural philosophy was all about trying to find out causes. William held that we can know many statements about God and the soul only through the light of faith, whereas Thomas Aquinas had thought them provable by reason. Clearly, William's ideas were a serious challenge to the works of rational theology that were championed by Thomas and Duns Scotus. No wonder the old guard were upset.

For a younger generation, of course, William's radical scepticism was exciting and novel. It seemed to honour God by placing him well above the deliberations of human reason. For the rest of the Middle Ages, the theology based on Duns Scotus was called the 'Old Way' while that which followed William of Ockham's ideas became the 'New Way'.[9] Students in Paris were especially excited by the new innovations and their professors responded by trying to ban them in 1339.[10] Theologians argued relentlessly over the relative merits of the Old and New Ways until the Reformation swept away the whole debate in the sixteenth century. At that point, the Catholic Church reverted to the work of Thomas Aquinas while Protestants disowned many of the achievements of medieval theology altogether.

Much of the argument between the supporters of Ockham and Duns Scotus was a proxy for the philosophical dispute between realism and nominalism. In chapter 3, we examined this briefly. To recap, the controversy was over the status of universals. A universal is a term used for a group of things, such as 'dog' for all dogs (including imaginary dogs). Realists believed that universals have a real existence, while nominalists considered universals to be merely names that humans have invented for convenience. William of Ockham, as you might expect, was a forthright supporter of the nominalist position. He insisted that we can only perceive individual things and that any connections we make between them are down to us. There is no need to postulate about the existence of real universals when we can explain the world in terms of the actual individuals it contains. Inventing new concepts, like universals, is unnecessary.

This is an application of the celebrated principle known to posterity as 'Ockham's Razor'. It is deservedly famous and often invoked as a reason for preferring elegant scientific theories over complicated ones. However, the term 'Ockham's Razor' is another nineteenth-century coinage and was never used by William himself. His real point was rather different. He actually said: 'Multiple entities should never be invoked unnecessarily.'[11] What this means is that we should reject physical and philosophical explanations that posit the existence of things, like universals, of which we have no direct experience. From the anachronistic point of view of modern science, this is not a terribly good idea. Species, elements and electrons are all universals that have real and specific properties. The element carbon really does have a unique atomic structure and really does combine with oxygen (another universal) during combustion (yet another). Carbon is not just a collection of black lumps to which we have arbitrarily given a particular label.

On the other hand, medieval nominalists, by rejecting generalisations, tended to be more empirical than their realist rivals were. Because nominalists dealt only in particular real instances, no amount of rationalisation from first principles would convince them that something was so if they could not see it with their own eyes. As we will discover in the next chapter, this attitude meant that they were more than ready to reject Aristotle's conclusions, however reasonable they might be. This rejection was important because very often the Philosopher's conclusions were plain wrong.

The Errors of Aristotle

Does a heavy object naturally fall faster than a light one? Many people believe that it does. They will point out that a feather and a hammer, dropped from the same height, will not land at the same time. The hammer, of course, will hit the ground almost at once while the feather will meander gracefully to the floor. But what would happen if you were to do the same experiment with a pea and a ball bearing? Will the heavy ball bearing fall faster than the pea? If you are not sure of the answer, try releasing two objects of different weights but similar shapes, perhaps a teaspoon and a large serving spoon, from head height. You will find that they both hit the ground at almost precisely the same moment. The reason a hammer falls faster than a feather has nothing to do with their respective weights, but rather their reactions to air resistance. The American astronaut David Scott (b. 1932) carried out a famous demonstration of this on the moon, where there is no atmosphere, during the Apollo 15 mission. He dropped both of these objects and they fell at the same speed, albeit rather more slowly than they would do on earth due to the moon's inferior gravity.

Does nature abhor a vacuum? Can a vacuum suck matter into itself? It is a staple of science fiction that the vacuum of space can suck people out of faulty airlocks. The truth is very different. A vacuum is literally nothing and so cannot do anything. All it does is provide empty space into which matter can move. It cannot suck you out of the window of a spaceship, but all of the air escaping through the hole might blow you out with it.

Aristotle was convinced that a heavier object naturally falls faster than a light one.[12] He also insisted that nature abhors a vacuum. In fact, he did not think a vacuum could exist at all.[13] According to him, it was completely impossible for a space to contain absolutely nothing. Although Aristotle justified his views with careful arguments, he often sided with common sense. For example, he thought the earth was stationary and located at the centre of the universe. This is a sensible position to take when we cannot perceive that the earth is moving and that the stars and planets all appear to move around it. Aristotle said the heavens were incorruptible and unvarying because all the records of his time showed that the movements of the stars and planets never changed. Neither did he believe in atoms. This was partly because a belief in atoms presupposes a belief in a vacuum for them to move around in. There was also no good reason to think that there was a minimum size of object or that matter was not endlessly divisible. Besides, no one had ever seen an atom or any direct evidence of one.

One of his most significant and long-lasting mistakes, baldly stated in *Physics*, was the belief that no object could continue moving without some other object moving it.[14] Often, this is true. If you stop pushing a chair along the floor, it stops moving. But equally, on many occasions things keep going after you have stopped touching them. The best example is throwing a ball. Aristotle was convinced that something must be pushing it after it has left your hand. The only thing he could think of was that the air behind the ball was propelling it forward.[15] This idea is easily refuted. A very powerful blast of wind would be needed to keep a ball moving through the air, and presumably we would notice this gust as we threw an object. Air, we know, actually resists motion, which would be impossible if it was also supposed to be providing the motive force to keep the object moving.

Such was Aristotle's prestige that even his harebrained ideas had to be taken seriously. The trouble was, although critics were unconvinced by the air-pushing concept, they still accepted Aristotle's fundamental law that a moving object must be moved by something else. This made it very difficult to come up with alternative theories. One writer who did take a different tack was John Philoponus (c.AD490–570), a Christian philosopher attached to the famous school in Alexandria. He suggested that the act of throwing a ball impressed a force onto it. This impressed force was then responsible for moving the ball forward but was gradually used up in the process.[16] In this way, Philoponus could maintain the dictum that a moving object (in this case the ball) had to be moved by something (here, the force impressed onto it).

William of Ockham also realised there was a problem with the theory that a moving object had to be moved by something else. Unfortunately, his alternative was hardly more enlightening. William suggested that a thrown ball moves itself, so that it provides its own motive force. He also deconstructs the very concept of motion, claiming that it is simply an object occupying successive places. Movement as a real entity is another idea cut out by William of Ockham's ubiquitous razor. Unfortunately, William's whole discussion of motion is simply an aside while he is talking about a completely different subject.[17] This meant that he never really developed his radical ideas.

Despite their critics, the combination of apparently sound common sense and cogent argument made Aristotle's theories extremely attractive. They also formed a consistent whole that gave a full description of reality. Taken together, Aristotle's philosophy makes for a deeply impressive package. This is the reason, more than any other, why it took such a long time for natural philosophers to realise that he was wrong about so many things.

The trouble is that it is impossible just to tinker with Aristotle's natural philosophy at the edges. It goes much deeper than that. His was a complete theory of reality and rejecting any significant chunk of it would cause the whole edifice to collapse. It is hardly surprising that both ancient and medieval commentators very often gave Aristotle the benefit of the doubt rather than habitually challenging what he said. Reforming natural philosophy had to happen from the ground up. Even if someone wanted to suggest a new theory, the language of philosophy was the language of Aristotle and so that was how he had to express his ideas.

Nonetheless, in the fourteenth century medieval thinkers began to notice that there was something seriously amiss with all aspects of Aristotle's natural philosophy, and not just those parts of it that directly contradicted the Christian faith. The time had come when medieval scholars could begin their own quest to advance knowledge; criticising and correcting their predecessors and striking out in new directions that neither the Greeks nor the Arabs had ever explored. Their first breakthrough was to combine the two subjects of mathematics and physics in a way that had not been done before. The setting for this most essential of steps towards modern science was the quadrangles of Merton College in Oxford.

The Mathematical Archbishop of Canterbury

Merton is among the oldest and grandest of the colleges of Oxford. It was founded in 1264 by Walter de Merton (d. 1277), bishop of Rochester, to provide a home for scholars studying theology for the many years that were required to complete the course and become a doctor of divinity. The college still holds a collection of hundreds of medieval manuscripts housed in a library that has been in continuous use for over six centuries. Also on display in the library is a fourteenth-century astrolabe traditionally known as Chaucer's Astrolabe. As well as *The Canterbury Tales* and his other poems, William Chaucer wrote an instruction manual in English on how to use the astrolabe to tell the time of year and measure the positions of the stars.[18]

During the fourteenth century, Merton College was the scene of some of the most important work on natural philosophy and mathematics in the Middle Ages. A succession of its scholars was famous throughout Europe for pushing back the boundaries of physics. Collectively, these men are known as the Merton Calculators and their influence was still being felt in Italy as late as the sixteenth century. In fact, as we shall see, they almost certainly beat out the path later followed by Galileo and the other founders of modern science.

The earliest of the Merton Calculators, Thomas Bradwardine (c.1290–1349), entered Merton College in 1323 and stayed for about twelve years during which time he became a bachelor of theology. On leaving, he was appointed chaplain to Edward III who was engaged in the Hundred Years War against the French. Bradwardine followed the king on his expedition to France. He may even have been present at the Battle of Crécy in 1346, where English longbowmen slaughtered the heavily armoured French knights. Shortly afterwards, Bradwardine was nominated as archbishop of Canterbury and, like Richard of Wallingford, he had to make the arduous journey across war-ravaged France to the papal curia in Avignon to have the appointment confirmed. He had barely arrived home when he died in 1349.[19] Today his tomb lies in Canterbury Cathedral, just a few feet away from the shrine of the twelfth-century advocate of reason, Saint Anselm. The esteem in which his contemporaries held Bradwardine is illustrated by Chaucer mentioning him in the same breath as Boethius and St Augustine of Hippo in the *Nun's Priest's Tale*.[20] He could hardly have bestowed any higher praise.

All of Bradwardine's important work on natural philosophy was done while he was at Merton. His lasting achievement was to take a step that in retrospect seems blindingly obvious. But at the time it was a radical departure from the accepted norms of scholarship that had been inherited from the ancient Greeks. We saw in the last chapter how students had to spend a couple of years studying maths. Then they moved on and studied natural philosophy. The two subjects were kept separate. Although it was accepted that the stars moved according to predictable geometrical patterns, the use of formulae to produce physical theorems had been frowned upon by Aristotle. He did not believe that it was possible to make deductions in one subject, say mathematics, and use them to prove something in another subject, say physics.[21] This means that there is remarkably little maths in his books of natural philosophy. His account of motion is explained with a few examples and generalisations, but with no attempt to produce a universally valid formula.

Bradwardine, in common with his colleagues, took the opposite point of view. He said that numbers were a vital ingredient of a successful natural philosophy. Mathematics, he wrote,

> is the revealer of every genuine truth, for it knows every hidden secret and bears the key to every subtlety of letters. Whoever, then, has the effrontery to pursue physics while neglecting mathematics should know from the start that he will never make his entry through the portals of wisdom.[22]

Translating this poetic language into practice, Bradwardine wanted to show that if what Aristotle said about how objects move is true, there must be some way to describe their movement as a mathematical function. Furthermore, he reasoned, for the formula to work it had to be valid for all situations, including for very large and very small numbers. All previous efforts to put Aristotle on a mathematical footing had failed because they could not be made to apply in every case. For Bradwardine, this meant that they could not be right.

He decided to concentrate his efforts on finding the correct formula linking the force exerted on an object to its speed.

As it turned out, Bradwardine did finally come up with a formula that properly described Aristotle's laws of motion in all circumstances. We should be clear, though, that the formula was completely wrong. This wasn't Bradwardine's fault but Aristotle's. The basic laws of motion that Bradwardine was describing with mathematics were badly flawed to start with. He may have accurately modelled how things moved in Aristotle's universe, but this was not how things worked in the real world. That does not stop Bradwardine's work being an important step forward and it was recognised as such at the time.[23] He had shown that it was possible to mathematically describe the laws of motion, but more importantly he had gone a long way towards demonstrating that any physical law worth its salt had to be expressible in numerical terms.

Modern historians studying Bradwardine's work have noticed something strange about his formula of motion. Translated into the notation that we employ today, it uses a special function called a logarithm. According to the official histories of mathematics, logarithms were invented by a Scot, John Napier (1550–1617), who published his work in 1614.[24] That they were being used in a limited sense for 300 years previously came as something of a surprise. In fact, Napier almost certainly made his discovery independently of his medieval antecedents, and the applications that he finds for logarithms never occurred to Bradwardine or his contemporaries.

Another area where Bradwardine made headway was in the question of falling bodies. Aristotle, of course, had thought that a heavy boulder falls faster than a pebble. The earliest record we have of someone categorically rejecting this is from the work of John Philoponus back in the sixth century. He wrote:

> If you let fall from the same height two weights, one of which is many times heavier than the other, you will see that the relative times required for their drop does not depend on their relative weights, but that the difference in the time taken is very small.[25]

In this passage, Philoponus is clearly referring to an experiment that he has tried himself. Seven hundred years later, Bradwardine, who probably never did an experiment in his life, considered the hypothetical situation of objects falling in the absence of air resistance. Since the 1277 condemnations, natural philosophers had been considering how a vacuum might behave if God deigned to create one. Bradwardine fruitfully speculated that, in certain circumstances, a light and a heavy object in a vacuum would drop at the same speed.[26] So, by the fourteenth century, it had been shown that objects of differing weights do fall at the same rate both in practice (by Philoponus) and in an idealised situation (by Bradwardine thinking about a vacuum). It would be a while yet before anyone would put these two results together and come up with a general law.

The Mean Speed Theorem

The most talented of the mathematicians at Merton was Richard Swineshead (fl.1340–55) who was probably still a fresher when Thomas Bradwardine left the college. Swineshead's achievement led to his being granted the honorary title of 'The Calculator' by later authors, but we know almost nothing about him. His *Book of Calculations* took contemporary mathematical ideas as far as was possible and pressed them into service in a wide array of applications.

Swineshead was particularly interested in how he could analyse a situation where a quality like heat or speed was increasing and decreasing at various rates. In one chapter of his book, he uses Bradwardine's formula to analyse what would happen if you drilled a hole through the earth and dropped a weight into it. Swineshead's first attempt at a detailed mathematical account of the weight falling to the centre of the earth gives him the absurd result that it would slow down as it fell, so that it could never quite reach the centre.[27] He rejects this and tries a simpler solution that does allow the weight to end up at rest at the centre of the earth. Actually, this is wrong too. We would expect the weight to fall faster and faster, reaching maximum velocity at the centre. Then it would slow down as it came up the other side, and finally come to rest on the far side of the earth (at least if we assume no air resistance in the hole). As this example shows, Swineshead and his fellow Merton Calculators were quite happy to apply their minds to imaginary situations. They set up a fictitious problem and then tried to work out the mathematical consequences. They were still a long way, however, from applying mathematics to a real-world situation and then trying to verify the result experimentally.

It is possible to produce mathematical equations for all kinds of situations. A general law of motion might be too difficult to solve at the first attempt, but simpler situations can be modelled in the same way. That is exactly what the last of the Merton Calculators, William Heytesbury (c.1313–73), did. Again, we know little about his life beyond his sojourn at Merton College and the fact that he wrote *Rules for Solving Logical Puzzles* which was published in 1335. In this book, he derived the most significant result of fourteenth-century physics. There is some doubt as to whether he actually discovered the formula himself or merely had the good fortune to be the first to write it down. For in contrast to Bradwardine's function, Heytesbury's has stood the test of time. If you have ever studied elementary mechanics, it will be familiar to you.

The problem he posed was: what happens when a moving object accelerates at a constant rate? Like his fellow Mertonians, Heytesbury was not very interested in empirical testing. He simply set up a problem and tried to solve it mathematically. The best way to illustrate this is to work through the question of uniform acceleration using some very simple numbers.

Suppose you drive your car at 50 miles per hour for an hour and a half. It is very easy to calculate how far you will have travelled – 50 times 1½ equals 75 miles. However, to use the jargon of the car trade, suppose you accelerate at a constant rate from nought to 60 miles an hour, in ten seconds. How far will you have travelled? First, we should convert 60 miles per hour into 88 feet per second. William showed that the correct method is to calculate how far you would go at your average speed of 44 feet per second over ten seconds, which is 440 feet. As he put it:

A moving body will travel in an equal period of time, a distance exactly equal to that which it would travel if it were moving continuously as its mean speed.[28]

This result, dubbed the mean speed theorem by historians, is central to physics because it describes the motion of an object, any object, falling under gravity. Note that it makes no mention of how much the object weighs. (Nor does it make allowances for air resistance, and so strictly speaking applies only to motion in a vacuum. That is why the feather and hammer fell at the same speed on the moon.)

Unfortunately, William Heytesbury and his contemporaries had no way of knowing the significance of the mean speed theorem. Nonetheless, they found it sufficiently intriguing to discuss many possible applications. Thanks to the 1277 condemnations, they could even talk about motion in a vacuum. For some reason, though, most of the natural philosophy that comes after Heytesbury took place in Paris and Italy rather than Oxford. The year 1350 marks a watershed after which no further mathematicians or philosophers of note emerged from England for almost 200 years. For this reason, we must now leave for mainland Europe and follow Roger Bacon and John Duns Scotus to the university of Paris.

CHAPTER 12

❦

The Apogee of Medieval Science

By the time of the Merton Calculators, Greek philosophy had been fully assimilated into Christian theology. In the twelfth century, western scholars had complained about 'the poverty of the Latins'.[1] They had lacked the best mathematics and natural philosophy – and they knew it. By 1300, though, riches had replaced dearth. The most advanced ancient thought had been discovered and translated. Arabic writers contributed arithmetic and algebra, adding new branches of mathematics to the geometry and number theory known in the ancient world. Thomas Bradwardine and his colleagues started to use mathematics as a tool to generate discoveries in physics and begin the process of combining numerical analysis and natural philosophy. Although Oxford ceased to produce original thinkers to match this golden generation, their ideas spread to France where they were taken up by the most accomplished natural philosophers of the Middle Ages. It was in Paris that medieval science reached its peak.

The Rector of Paris

John Buridan (c.1300–c.1361) was the most remarkable philosopher of the fourteenth century. He was born in Arras, northern France and, as far as we know, never left his home country. He spent his entire career at the university of Paris where he received his Master of Arts degree in about 1320 and was elected rector twice.[2] Unlike most of the individuals we have encountered so far, he did not train as a theologian. This was unusual at the time because theology was such an important subject. People who had mastered the requisite philosophy wanted to go on and study the science of the divine. Buridan also declined to join either the Dominicans or Franciscans, although he did enter the priesthood. Staying out of the religious orders meant that he was something of a free agent, but it also left him without the support of a powerful organisation. Luckily, he was perfectly capable of making his way in the world without any outside help.

Buridan's life as a scholar was not very eventful and so, as usual, the mythmakers have stepped in to liven up his biography. One story tells of how the king of France ordered him thrown into the River Seine for having an affair with the Queen.[3] Another rumour claims that the university expelled him for his support of William of Ockham. As we saw in the last chapter, there was a half-hearted attempt to ban William's teaching at Paris in 1339, which is probably where this tale derives from. What is clear is the high esteem in which he was held. He was called a 'very distinguished man' and a 'celebrated philosopher' by his contemporaries.[4]

Buridan's first love was logic, where he was firmly in the 'nominalist' camp that followed William of Ockham in rejecting the reality of universals. Buridan's nominalism meant that his natural philosophy had an empirical bent. The principles of science, he wrote, are accepted because we frequently observe them to be true and never come across a counter-example.[5] He believed that the job of physics was to explain things based on how they normally appeared, 'assuming the ordinary course of nature'. It did not matter to him that God, by his absolute power, could bring about miracles. As Buridan explains, 'it is evident to us that every fire is hot and that the heavens are moved, even though the contrary is possible by God's power. And it is evidence of this sort that suffices for the principles and conclusions of natural philosophy.'[6]

This logic led him to reject Aristotelian ideas about violent motion because they did not correspond to what he saw in the real world.

Instead, Buridan formulated an alternative theory around the concept of 'impetus' that had had its genesis in the work of John Philoponus in the sixth century.[7] Buridan was familiar with this (albeit possibly only at second hand) and combined it with the insights of William of Ockham.

Contrary to Aristotle, it was clear to Buridan that nothing, especially not the air, pushes a thrown ball after it leaves the hand that threw it. It was equally clear that the ball moves as a direct result of the movement of the hand. Buridan suggested that the hand gives the ball a quality, which he called impetus. To hurl a heavy rock, you would have to give it more impetus than to lob a pebble the same distance. Likewise, faster-moving objects have more impetus than slow ones. Thus, Buridan decided that impetus must be a quality whose magnitude was proportional to both weight and speed.[8] This makes it very similar to, but not quite the same as, the concept of momentum in modern physics.

According to Buridan, an object falling due to gravity will gain impetus as it speeds up. Conversely, once a thrown ball has left the hand, it expends impetus to overcome air resistance. When it has used up all of this impetus, the ball stops moving forward and falls to the ground.

He realised that this led to a radical implication of his theory: 'Impetus', he said, 'would last forever if it were not diminished and corrupted by an opposing resistance or a tendency to contrary motion.'[9] Therefore, if there is no air resistance, such as in a vacuum, then an object will continue moving forever. Looking to the heavens, Buridan suggested that this might be the case for the planets orbiting the earth. He did not believe that they moved in a vacuum. Rather, he agreed with Aristotle that the heavens were a perfect world of 'quintessence' without decay and without friction. This meant that the planets would not meet any resistance and should, therefore, keep moving forever. Buridan had solved a problem that had exercised the minds of both pagan Greek and Christian thinkers: what keeps the planets moving in their orbits?

Recall that when we met Adelard of Bath in chapter 4, he was discussing with his nephew whether or not the stars needed to eat. He thought that they may be living creatures in some sense because they moved of their own accord. In *Timaeus*, Plato had suggested that the planets might possess some sort of soul, which caused them to move.[10] It was a short step for medieval thinkers to replace these pagan souls with Christian angels.[11] This, together with some fanciful iconography, is the source of an unfair caricature of Christian thinking. Medieval theologians did not have an image of angels, complete with wings and haloes, pushing the planets around the sky. Rather, they thought that angels were immaterial spiritual beings who certainly did not need wings to get around. All the theologians were trying to do was stick to Aristotle's law that everything that moves has to be moved by something else. It was obvious that the planets were in motion, so something had to be pushing them.

Now, with his theory of impetus, Buridan had shown that the planets did not need angels or anything else to get around. Bradwardine had already specifically compared the universe to a clock.[12] Buridan may well have been inspired by this analogy, and imagined that God had wound up the 'world machine' at the beginning of time. As he explained:

> In the celestial motions, there is no opposing resistance. Therefore, when God, at the creation, moved each sphere of the heavens with just the velocity he wished, he then ceased to move them himself and since then those motions have lasted forever due to the impetus impressed on the spheres.[13]

God, he was suggesting, had endowed each celestial sphere, and the planet embedded in its rim, with a certain amount of impetus at the creation. Because there was no resistance in the heavens, this impetus would be sufficient to keep the sphere rotating and the planet moving on its circular course until doomsday. This comes quite close to the modern principle of inertia. However, it is not quite the same. No one is suggesting that Buridan had mechanics done and dusted

300 years before Newton. And while Buridan came close to describing inertia when he realised that the planets would keep moving if there was nothing to stop them, he never formulated it in the right way. Aristotle said that if a moving object is left undisturbed, it will stop. The modern principle of inertia states that a moving object will keep going at the same speed in a straight line until it is subjected to another force. Buridan thought the planets would keep moving in circles but he never extended his principle to movement in a straight line. Nevertheless, he had successfully challenged Aristotle's natural philosophy and laid the foundations for the new science of mechanics.

John Buridan's other great achievement was to ignite discussion of the subject that is emblematic of the beginning of modern science – the motion of the earth. Although he never thought that the earth orbited the sun, he did give serious consideration to the possibility that the earth might be turning.

According to almost all Greek cosmologists, the earth did not rotate each day. The entire heavens turned full circle every 24 hours while the earth remained stationary at the centre. The problem that Buridan had with this was that it seemed rather ugly. The heavens were very large and causing them to turn had to be less efficient than rotating the earth, which was, relatively speaking, minute. Like many medieval Christians, Buridan expected God to have arranged things in an elegant way, always allowing that he could do as he pleased. However, although there was a presumption towards elegance, you still had to check the empirical facts to see if God really had operated in that way. Buridan knew that the night sky appears to turn around, but realised that this could just be an illusion caused by the motion of the earth. To explain why we cannot directly observe the earth turning, he compared the situation to a boat sailing down a river. Imagine, he said, that you are on board the boat. If someone else is watching you from another boat moored to the riverbank, they can see that you are moving relative to them. But without also observing the surrounding landscape, they cannot tell whether they are the ones in motion or standing still. We are all in the same situation with the earth's rotation. Unless we have some exterior viewpoint, it is impossible to tell whether the earth is rotating together with everything on it, or if it is stationary.[14] Buridan needed another way to determine whether the earth itself was turning.

Ptolemy had tried to settle the argument by stating that there would be a great rushing wind as the surface of the earth moved while the atmosphere stayed put.[15] Not so, said Buridan. The atmosphere must be rotating together with the earth and so there is no reason for us to feel the air being left behind as the world turns under it.

At that point, unfortunately, he made a mistake by following Aristotle too closely. Surely, Buridan said, if you fire an arrow straight up into the air, it will eventually fall back on top of you. But if the earth really was rotating, you should have moved with it and thus have avoided the falling arrow. Of course, today we know that the earth does turn and we do not see this happening. As we will see, it was Buridan's most brilliant student, Nicole Oresme (c.1325–82), who first correctly analysed this problem.

Nicole Oresme: The Bishop Philosopher

Unlike his master, Oresme qualified as a doctor of theology and forged an impressive career outside the university. He was tutor to the future Charles V of France (1338–80) for whom he later translated a number of scientific treatises into French. As a reward, the king elevated him to the bishopric of Lisieux in 1377.[16] Despite this busy professional life, he wrote several important treatises on mechanics and mathematics that built on the work of Buridan and the Merton Calculators. Unusually for his time, he was also an implacable opponent of astrology.[17]

Oresme is most celebrated for his contribution to the argument about the rotation of the earth. Taking forward Buridan's analysis, he explained that when we fire an arrow into the air, it already shares the earth's rotational motion. As the atmosphere is also moving with the earth, there is no reason that the arrow should not fly straight up and down relative to the archer. In proposing this, Oresme was actually challenging Aristotle's law that different kinds of motion are contrary.

In his scheme, the arrow must simultaneously move naturally with the rotation of the earth and violently from the bow that fired it. He concluded that we cannot use either reason or observation to conclusively determine if the earth moves. This left Oresme with an unanswered question and he tried to find the solution in the one book he considered completely reliable – the Bible.

Christians already realised that the Bible cannot always be taken literally. They believed that the Holy Spirit had inspired the biblical authors to write in the everyday language of the man in the street, and that it was not a scientific text. In that case, Oresme said, it was hardly surprising that the Bible assumed that the earth is stationary, because that is our everyday perception of the matter. He could treat a passage that says the earth does not move as figurative 'by saying that this passage conforms to the normal use of popular speech just as it does in many other places ... which are not to be taken literally.'[18] Only an extra-terrestrial view of the earth could determine the truth about its motion. God, of course, has such a viewpoint and Oresme examined each of the relevant biblical passages to see if they provided evidence either way. Eventually, he came across Psalm 93:1, which reads: 'The world is established that it cannot be moved.' This, he decided, was good evidence that the earth was not rotating. However, he could easily have employed his reasoning to deal with this verse too if he had had any good reasons for thinking that the earth was in fact moving. We cannot blame him for eventually supporting the common-sense position of Aristotle and all the other ancient and contemporary authorities. What Oresme had done was prepare the groundwork. He refuted most of the objections to a moving earth two centuries before Copernicus had suggested it might actually be in motion.

Oresme's other major achievements were in mathematics. The Merton Calculators had derived the mean speed theorem but Oresme found an elegant way to prove it. One of the best ways to represent motion is on a graph. If you plot a graph with speed on the vertical axis and time on the horizontal, you can see relationships that you previously had to imagine. For example, the slope of a line that plots the change of speed over time represents acceleration. A constant slope means that acceleration is constant. Oresme also realised that the area underneath the line must represent the distance travelled. Simply by cutting and pasting, he could show that the mean speed theorem was true. The area under a horizontal line at the average speed was the same as the area under a line with a constant slope.[19]

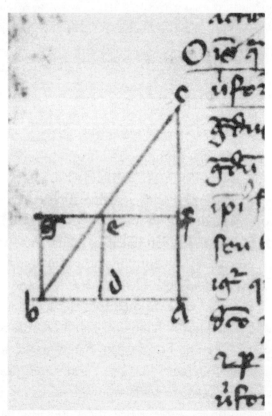

**9. An excerpt from a fifteenth-century manuscript showing
Nicole Oresme's proof of the mean speed theorem**

Using graphs to illustrate motion is an extremely powerful technique. The next step would be to describe the general mathematical relationship between the area under the graph, the line itself and its gradient. Oresme lacked the mathematical tools, now called calculus, to achieve this but his establishment of the meaning of the components of a graph plotting speed against distance was an essential first step.

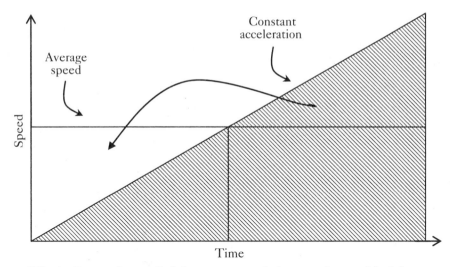

Nicole Oresme's proof of the mean speed theorem in graphical form

The work done by the Merton Calculators and the natural philosophers of Paris became widely diffused across western Europe. The free movement of scholars and common knowledge of Latin meant that a pan-Catholic republic of letters existed in the late Middle Ages. For example, at the same time as Buridan and Oresme were working at Paris, a German by the name of Albert of Saxony (c.1316–90) arrived to teach philosophy. He began his academic career rather late and did not receive his Master of Arts degree until he was in his mid-thirties. He probably studied under Buridan himself and followed him as rector of the university in 1353. Nine years later the Pope called him to Avignon. Albert took advantage of his closeness to the Pope to get permission to found a new university in Vienna before he was appointed a bishop in 1366.[20] Thereafter his career was more political than academic, but he carried the work of his master John Buridan far beyond the confines of Paris.

Albert's own books are less advanced than either Buridan's or Oresme's, which arguably made them more popular. In one small way, though, he did strike a blow against Aristotelian physics. Recall from chapter 9 how Aristotle implied that a cannon ball would drop out of the sky like a cartoon character running off a cliff when it ran out of speed. Albert could see that this was not so. Instead, he drew a diagram showing a cannon ball shooting out of the barrel of the gun in a straight line but then curving gently to earth when it reached the limit of its range.[21] This is not mathematically accurate, but it is the earliest picture we have of a curved trajectory.

Another travelling philosopher, Paul of Venice (1369–1429), carried the new ideas to Italy. He had spent three years of his early academic career at Oxford but, in 1393, returned to the Veneto to take up a post at the university of Padua. Venice ruled the Mediterranean Sea and made its money by trading with the Byzantine Empire and the Turks. The city also controlled a large chunk of northern Italy including, after 1404, the city of Padua, home to the famous university. For the Venetians, having their own university was a matter of considerable pride. They supported it by providing fat salaries for the professors and protection for the students. After the Reformation, Protestants still continued to attend the university of Padua even though Catholic Italy officially considered them heretics.

Paul of Venice brought the mathematical work of the Merton Calculators back with him from Oxford.[22] His colleagues realised that it had a wide range of applications, especially in the field of medicine for which Padua was most famous. They began to consider how they could best calculate the effects of combining drugs with different properties using the techniques developed by Thomas Bradwardine and Richard Swineshead.[23] By 1400, tracts by all the Merton Calculators could be found in Padua's university library.[24] The new science spread to Germany too where, in 1425, a document from Cologne refers to the era as the 'age of Buridan'.[25]

Bread, Wine and Atoms

The Church looked benignly upon all this speculation. Almost all the practitioners were members of the clergy and many, like Nicole Oresme and Albert of Saxony, were in senior positions. Mechanics and mathematics did not cause any concern. Even though Aristotle's natural philosophy was almost the official position of the Church in many areas, thinkers could still challenge and reform it. In one field, though, matters were considerably more delicate. Back in the eleventh century, Lanfranc had used Aristotle's theory of substance and accident to explain transubstantiation. This linked natural philosophy closely to the Eucharistic miracle when bread and wine turned into the body and blood of Christ. In 1215, transubstantiation had become the official dogma of the Catholic Church. From then on, an attack on Aristotle's theory of matter looked like an attack on a key Christian doctrine. Anti-Aristotelian reformers had to tread carefully. One idea that was completely incompatible with the theory of substance and accident was atomism. Medieval philosophers did not know very much about this alternative ancient Greek theory of matter beyond Aristotle's criticisms of it. His predecessor, Democritus (c.460–c.370BC), had suggested that all matter consisted of tiny indivisible particles swirling around in a void. Aristotle had no time for this idea because he rejected the void and believed that everything was infinitely divisible. Most of his medieval readers concurred with his opinion. Thomas Bradwardine even went to the trouble of writing a refutation of atomism called *On the Continuum*.[26]

A few scholars did think that atomism was at least worthy of consideration. Nicholas of Autrecourt (c.1300–69) went further. After gaining his Master of Arts degree, he joined the Paris theology faculty and began to study for his doctorate. Along the way, he produced a commentary on the *Sentences* of Peter Lombard, the standard theological textbook of the time. Again, as was usual, the faculty subjected this to a test of Christian orthodoxy. It failed spectacularly. The case ended up before the Pope and did not reach a final settlement until 1347. The Pope found Nicholas guilty of holding heretical opinions and ordered him to recant. The offending commentary was consigned to the flames. His academic career was over but he received a lucrative appointment as dean of Metz Cathedral in compensation and lived out his days in peace.[27]

Nicholas was the ultimate anti-Aristotelian. He thought that the world would be a better place if people stopped reading Aristotle and rejected almost all of his philosophy. It is possible that he was espousing atomism simply because Aristotle did not agree with it. The trouble was that, for medieval theologians, atomism left transubstantiation without any philosophical foundation. Lanfranc had explained how the underlying substance of the bread and wine could change while the accidents such as taste and appearance stayed the same. But the concepts of 'substance' and 'accident' came from Aristotle. According to atomism, they were meaningless. Instead, the properties of matter were due to the shapes of the individual atoms. For example, fire atoms were sharp and prickly which is why they burn. Water atoms were round and smooth so they flow across each other easily. With the atomic model, there did not seem to be any way to allow the bread and wine actually to be the body and blood of Christ. Either the host contained 'bread' atoms or it was made of 'flesh' atoms. There was no room in atomic thinking for the underlying substance to be different from the outward appearance.

Modern readers might find it hard to understand what all the fuss was about. After all, transubstantiation is accounted a miracle and trying to explain it scientifically is a fool's errand. This misses the point of just how rationalistic medieval theology was. One consequence of this was the near-universal belief among theologians that logic limited God's absolute power. He could not bring about a logical contradiction. The answer to the old canard 'Can God make a weight so heavy that he cannot lift it?' was a straightforward 'no'. More interesting was the idea that God cannot make a human being sin. Theologians thought that this was a logical contradiction because doing God's will can never be a sin. So if God makes you sin, you are not actually sinning.[28] Likewise, not even God could make the host into bread and flesh at the same time. Aristotle's substance and accident concepts had allowed Lanfranc to perform some mental gymnastics to separate the idea of flesh from the idea of bread. By assigning these two ideas to different concepts, he gave the impression that no contradiction existed. Atomism did not provide any such way out, so it seemed to make transubstantiation logically impossible. If that was the case, not even God could perform it. Thus Nicholas was required to specifically abjure his ideas about the Eucharist.[29]

Admittedly, the philosophical arguments for atoms were not very strong. There was no evidence that they existed and they explained nothing that Aristotle could not handle just as well. In fact, atoms were exactly the kind of superfluous concept that Ockham's razor was supposed to dispose of. With hindsight, we now know that atomism turned out to be an extremely fruitful theory of matter. Perhaps it would have been better if the Church had not stamped so hard on the idea. Certainly, this was a clear-cut case of theological orthodoxy curtailing philosophical enquiry. But this happened so rarely that we cannot maintain that the Church held back science in general. In Nicholas of Autrecourt's case, he specifically linked his ideas about matter to the all-important doctrine of the Eucharist and so the authorities felt compelled to act. Supporting atomism without pursuing the theological implications would not have provoked such a hostile reaction.

The popular image of the medieval Church as a monolithic institution opposing any sort of scientific speculation is clearly inaccurate. Natural philosophy had proven itself useful and worth supporting. It is hard to imagine how any philosophy at all would have taken place if the Church-sponsored universities had not provided a home for it. But the price of having a rich sponsor is having to bend to their interests and avoid subjects that they find controversial. Modern scientific researchers competing for funding from big companies have exactly the same problem. The Church allowed natural philosophers a much wider dispensation than many corporate interests allow their researchers today. They were free to speculate as much as they pleased as long as they avoided religious controversy. Even atomism would make a triumphant comeback in the seventeenth century.

Nicholas of Autrecourt disagreed with John Buridan as well. Their dispute was over the empirical status of natural science. Buridan, as we have seen, thought that if nature always behaves in one way and never in another, that is quite enough evidence for the truth of a natural law. Nicholas disagreed. He demanded incontrovertible proof before he would believe any scientific proposition. This placed the burden of proof so high that science could never have advanced based on his method.[30]

Buridan took a more cautious line with the theologians of Paris. In one treatise, while he was discussing whether a vacuum can exist, he suddenly states that his colleagues in the theology faculty have accused him of straying into forbidden territory. Buridan responds by asserting that he is allowed to consider the question as long as he settles it in an orthodox fashion.[31] He agreed with Aristotle that a vacuum was naturally impossible, but admitted that God could create one if he wanted to. This was an example of the 1277 condemnations in action. They stated that it was heretical to limit God to what Aristotle said was possible. So, in following the condemnations to the letter, Buridan was not saying anything controversial. Rather, the oath he had made when he received his Master of Arts degree compelled him to settle the dispute on the side of orthodoxy. In chapter 6 we saw how, following the arguments surrounding Siger of Brabant, philosophers at Paris had promised not to encroach on the theologians' territory, or at least to ensure that they did not make any heterodox claims. The deal still held and so immediately God came into the question, Buridan thought it best to make clear he was sticking to the rules.

The Decline of Medieval Science

As it turned out, the ban on natural philosophers covering theological questions was not much of a handicap to science in the Middle Ages. Most of the writers on natural philosophy were also theologians and so did not suffer from any restriction. Thomas Aquinas, Nicole Oresme and Albert of Saxony, among others, could all discuss the interaction between God and nature to their hearts' content. And they did. In fact, there was so much natural philosophy in medieval theological books that some clerics started to complain about it.[32] Questions about the nature of infinity, what lies outside the universe and how God created the world all appear frequently. Theologians used their training in Aristotle's ideas together with their knowledge of Christian doctrine to try to come up with some answers.

Buridan and Oresme represent the high water mark of medieval natural philosophy. Some interesting work was done through the fifteenth century at the university of Padua where Paul of Venice had brought the discoveries of the Merton Calculators. And we will shortly meet two more innovative thinkers who combined fruitful speculation in natural philosophy with their job of cardinals in the Catholic Church. Furthermore, as we will also see in the next chapter, the invention of spectacles, the compass and the mechanical clock were followed by even more impressive developments in the fifteenth century. But given how close Oresme and Buridan had come to some of the key concepts of modern science, it seems a disappointment that their immediate successors did not take their insights forward. What went wrong?

The deadly incursion that stopped all of Europe in its tracks entered Italy through Venice in 1348 and swept through the continent, leaving between a third and a half of the population dead. This first wave of the Black Death took Thomas Bradwardine with it. The university towns, crowded with students from far and wide, were especially badly hit. As the students fled from the plague, they took it with them back to their homes and monasteries. One noted theologian and astrologer, Richard Holcott (d.1349), had used his art to confidently predict a peaceful death for himself. Maybe, lying in his pallet as the Black Death ravaged his body, he wondered where his calculations had gone wrong.[33]

The initial assault of the plague had burnt itself out by 1350 but it returned in 1360 – when John Buridan was probably among its victims – in 1369, in 1375 and intermittently for the following three centuries. Even those who survived one epidemic could never be sure that they would live through the next. The effects of the Black Death went beyond the enormous death toll, which included many of Europe's finest natural philosophers. It mocked the ability of man to control his destiny and made fools of the doctors. Perhaps the scholars of Europe lost their nerve. In finding it again, they would discard almost the entire legacy of medieval philosophy.

CHAPTER 13

⧸∾⧸

New Horizons

By the late fourteenth century, the Catholic Church had serious problems. The popes had taken up residence at Avignon in 1309, but in 1377 Pope Gregory XI (1330–78) moved back to Rome. The French cardinals were very unhappy about this and elected their own pope as his successor to remain at Avignon. Thus began the schism, when there were two and sometimes even three popes all reigning at the same time. Attempts to reunite the papacy started almost at once and eventually the need for a General Council of the Church was agreed. The first attempt, in 1409, was a failure but in 1417 the Council of Constance appointed a single Pope about whom almost everyone could agree. The dispute now transmuted into a debate about papal power. Above all, no one knew what should happen if the Pope himself was corrupt or a heretic. The argument ran out of steam in the 1430s with the Pope having suffered no diminution of his prerogatives. This was extremely unfortunate, because the next century would see a succession of popes who were obsessed by military power or massive ostentation. This meant they were always in need of money. One of their demands on the pockets of the faithful in Germany would lead to the Protestant Reformation.

Despite the vexations of the Church, the fifteenth century produced several philosophers of note. More significantly, European civilisation recovered its poise after the disaster of the Black Death and enjoyed another spurt of technological progress, the consequences of which would involve the entire globe.

Nicholas of Cusa

The most original thinker of the fifteenth century was Nicholas of Cusa (1400–64) from Germany. His interests included mathematics, philosophy and theology. For his education he travelled to the university of Padua, where he received a doctorate in law, before leading an eventful life mixed up in ecclesiastical politics. In 1448 he was made a cardinal.[1] Despite his busy professional career, he still found time to write important books on theology and philosophy.

While returning from Constantinople on church business he conceived *On Learned Ignorance*, the book that made him famous.[2] The title suggests that that book was not meant to be taken seriously, but learned ignorance is actually a method of discovering truths about God based on accepting what we cannot know. Nicholas's views are not always easy to comprehend but to us, they can seem quite inspired. He argued that in order to reflect God's majesty, the universe he created would have to be limitless, if not quite infinite. He continues: 'Therefore, the earth cannot be in the centre … and just as the earth is not at the centre of the universe, so the sphere of the fixed stars is not its outer border.'[3] He continues that the earth must also be moving although, and this comes straight from John Buridan, we do not notice because we are riding along with it. Most radically, he reduces the earth to just another star (albeit the most important one) and suggests that alien life forms could exist elsewhere in the universe.[4] At the time, no one would have objected to this kind of speculation as long as it stayed hypothetical. As Nicholas could not prove anything, he simply postulated ideas to see what they looked like. He probably never dreamed that within two centuries of his death, his speculations would be found to have been uncannily accurate.

A further short work of his is worth noting. In *On Measuring*, he set out a method of doing natural philosophy that would soon come to define modern science. We have seen how empirical observation was becoming increasingly popular and how magic encouraged some dabblers to get their hands dirty. But these proto-experiments tended to be observational and qualitative. Nicholas realised that to effectively marry natural philosophy to mathematics required not just careful observation but exacting standards of measurement. Only by carefully weighing, timing or measuring the material from experiments could sufficient data be generated to build a mathematical description of nature. He also appreciated that this sort of painstaking research was both difficult and time-consuming, but urged his contemporaries to apply themselves to the task.[5] Sadly, his own official duties left him with no spare time for experiments and no one else took much notice of his proposals. Truly experimental science would have to wait but in this area, like so many others, a medieval thinker had seen the way ahead.

The Crisis of Ancient Geography

Nicholas of Cusa was not the only cardinal with a sideline as a natural philosopher. A central figure at the Council of Constance had been Pierre D'Ailly (1350–1420), a graduate of the university of Paris who was made a cardinal in 1411.[6] He was unusual for a senior churchman in that many of his opinions were on the outer rim of orthodoxy. Unlike Nicole Oresme, Pierre was an enthusiast for astrology. Building on the speculations of Roger Bacon, he even managed to predict the French Revolution or, at any rate, a serious upheaval in 1789.[7] His other act of prophecy was even more prescient. In a book called *Picture of the World*, Pierre explained why it would be feasible to sail west in order to reach the East Indies. He made this suggestion based on a very fruitful disagreement between ancient Greek geographers about how large the earth really is.

The best-regarded figure at the time for the circumference of the globe was 252,000 *stades*, calculated by Eratosthenes (who died around 200BC), a librarian of the famous library in Alexandria. He noted that when the sun was directly overhead in the town of Syrene, it was at 7.2° from the vertical at Alexandria. 7.2° is one fiftieth of the full circle of 360°. Correspondingly, this meant that the distance from Alexandria to Syrene was one fiftieth of the entire circumference of the earth. All he had to do then was measure that distance. He did this by pacing it out, quite a feat for a span of over 500 miles.

Alexandria to Syrene turned out to be 5,000 *stades*, so the earth's circumference is 250,000 *stades*.

Unfortunately, no one knows exactly how long Eratosthenes took a *stade* to be, because the unit comes from the length of a stadium. Its precise length depended on which stadium was being measured. However, we know that the figure only varied between about 170 and 205 yards, which means that his figure for the earth's circumference was between 24,000 and 30,000 miles. The most likely figure is 202 yards (the length of an Athenian *stade*) which gives a circumference of 29,000 miles.[8] The true figure is 24,900 miles, so Eratosthenes's calculation gave him a figure that was slightly too large. Nevertheless, the Roman encyclopaedia of Pliny the Elder (AD23–79) used it and thus transmitted it to the Middle Ages. During the thirteenth century, John Sacrobosco reports the same number in his *Treatise on the Sphere* and it became widely accepted over the next few centuries.[9]

There the matter rested until the rediscovery and translation of Ptolemy's *Geography*, another ancient Greek scientific masterpiece, in 1406.[10] Western scholars had long known about Ptolemy's astronomical compendium, the *Almagest*, but they previously had had no access to his work on geography.

Like Ptolemy's other works, the *Geography* is a technical treatise. It deals with how to make a map of the world using a system of latitude and longitude to plot the locations of cities, islands and other features. The most difficult aspect of the job is how to represent the curved surface of the earth on a flat piece of paper. The bulk of the *Geography* is taken up by tables of cities and other locations together with their grid references. Taken together, they allow the reader to produce a series of maps, including one of the whole world.

Ptolemy's map of Europe looks very familiar, except that Scotland's coast takes a sharp turn eastwards and juts out 300 miles into the North Sea. His map of the world is not entirely alien either, but does have a few features that raise modern eyebrows. India appears to be an island and things get a bit vague around the edges. Naturally, there is no trace of America or Australia. However, the most striking

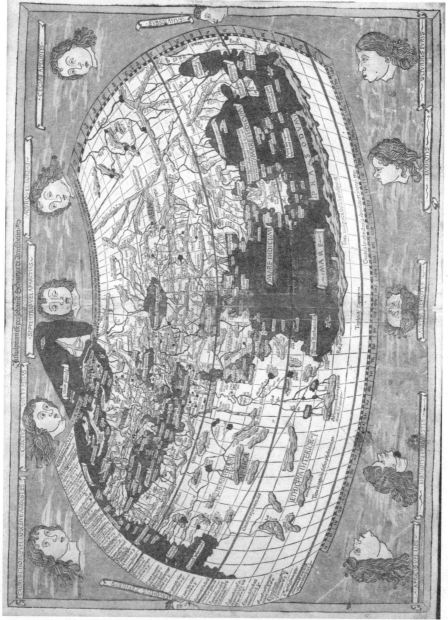

10. A map of the world according to Ptolemy from an atlas printed in 1482

feature is Ptolemy's belief that the southern end of Africa loops eastwards and joins on to the far end of Asia. This has the result that land completely encloses the Indian Ocean and means that it would be impossible to sail from Europe to India around the Cape of Good Hope.

The *Geography* also contained a very significant mistake. Ptolemy estimated that the earth's circumference was only 180,000 stades, 70,000 stades less than Eratosthenes's figure.[11] This potentially meant that the eastern tip of Asia was not all that far from western Europe. Pierre D'Ailly, taking his inspiration from Ptolemy, suggested that sailing west might be practicable. At about the same time, a popular book of travellers' tales under the name of Sir John Mandeville included a story about a Danish sailor who had circumnavigated the earth.[12] Whoever wrote Mandeville's book had probably never travelled further than his local library, but the book was fantastically popular and much more influential than Marco Polo's more truthful account.

Ptolemy's work on geography decisively demonstrated that the ancient authors completely disagreed on basic questions. There was just no way to reconcile the various opinions that they expressed. For instance, Aristotle divided the earth into five different climatic zones. The northern and southern polar regions are too cold and icy for anyone to live there. The tropics are also uninhabitable because they are too hot.[13] Pliny described the equator itself as a ring of fire that completely cut off the northern hemisphere from the southern. Between the poles and the tropics were two temperate regions. According to him we lived in the northern temperate region and there was another in the south, although the ring of fire meant that we could never travel there.[14]

This raised some tricky theological problems, and we saw in chapter 2 how Virgil of Salzburg may have encountered trouble over the issue in the ninth century. Christians wanted to know if anyone lived in the southern temperate region and if so, how they had got there. After all, every animal on the planet should be a descendant of one of the breeding pairs which Noah loaded onto his Ark during the Great Flood described in the book of Genesis. When the flood waters receded, the Ark came to rest on Mount Ararat in Turkey from where all the animals went their way. However, if there really was a ring of fire at the equator, then there was no way for the animals from the Ark to reach the southern hemisphere. Worse still, if humans were living there, how were Christians supposed to go and save their souls if the equator formed an impenetrable barrier?

Ptolemy presented a different view. He made no mention of the tropics being uninhabitable and clearly believed that reaching the southern hemisphere was simply a matter of sailing far enough south. Then, in 1473, the argument was settled decisively in Ptolemy's favour when a Portuguese ship, exploring the western coast of Africa, crossed the equator. It was a point against Aristotle and raised the question of where else he might be mistaken. The Portuguese soon proved that Ptolemy was fallible too. Bartolomeu Dias (d. 1500) rounded the Cape of Good Hope in 1488 and showed that it was possible to reach the Indian Ocean by sea. A route to the Spice Islands of the East Indies lay open.[15]

The spice trade was a great deal more diverse than we might imagine today. Although the export of nutmeg, cloves and pepper was an extremely lucrative venture, the word 'spice' included any imperishable high value, low volume cargo. Ivory, resin, amber, glue, perfume, wax, dye and sugar all came under the category of spices.[16] Along the old Silk Road, threading its way across central Asia, tollbooths took a cut at each national border. That is, if the route was open at all. Alternatively, Arab traders could ship the spices to the Red Sea and carry them by camel train the short distance to the Mediterranean. Once the goods had reached the east coast of the Mediterranean Sea, Venetian ships transported them to western Europe. For Portugal and Spain, the only way to get a share of the trade was to find another route to the Far East. The Portuguese controlled the route around Africa and so the Spanish would have to consider another direction.

The Discovery of the New World

Christopher Columbus (1451–1506) was born in the mercantile republic of Genoa and decided to seek his fortune on the sea rather than enter the family wool business. He was self-taught and must have been a man of exceptional talent to persuade monarchs to let him sail off into the unknown at their expense. Columbus had a different idea about the best sea route to the Spice Islands of the East Indies. He decided that the quickest and safest direction would be to sail west.

Columbus's conjectures received support from Pierre D'Ailly's *Picture of the World*. His own copy still survives and contains many marginal notes in the navigator's handwriting. D'Ailly, writing in 1410, still believed that the tropics were too hot to live in, but Columbus noted in the margin: 'The torrid zone is not uninhabitable because today the Portuguese navigate through it. Indeed, it is very populated.'[17] D'Ailly also gathers some evidence from various ancient authors, emphasised by Columbus in his annotations, that only 'a small sea' lies between Spain and the Far East which should be 'navigable in a few days.'[18] Finally, D'Ailly uses Ptolemy's figure for the circumference of the earth – which is only about 21,000 miles.[19]

Based on his research, Columbus was completely convinced that the far east of Asia was only 2,500 miles from the west coast of Spain. He must have reached this figure by taking the largest possible figure for the length of Asia and combining it with the smallest for the earth's circumference. The true distance between Europe and Japan is closer to 12,500 miles and so Columbus was very, very wrong.[20] He was also very lucky. Imagine his fate if the Americas had not existed. Columbus and his little fleet would have sailed on in the endless expanse of ocean with no chance of ever being able to reach the East Indies before succumbing to thirst, starvation or disease. His sailors were well aware of this and it was for this reason, rather than any concerns about the edge of the world, that they were unwilling to sail west. In Columbus's defence, he may have been aware of travellers' tales from northern Europe about Viking discoveries to the west of Greenland. He claimed to have visited Iceland, or Thule as it was then known, in 1477 and to have sailed 100 leagues beyond it.[21] As it was, Columbus made landfall in the Caribbean exactly where he expected to find the East Indies. It is hardly surprising that he went to his grave convinced that his mission had been a success.

By 1550, the new geography was settled and the map of the earth no longer looked anything like the ancient Greeks said it should. The cornerstones of ancient geography had been the division of the earth into five climatic zones and the existence of only one great land mass. Both were wrong. Theological arguments about the inhabitants of the antipodes were also rendered moot. As a contemporary writer noted, 'not only has this navigation confounded many ancient writers about earthly things, but it has also given some anxiety to the interpreters of the Holy Scriptures.'[22] The continent of America threw up exactly the same theological problems that had previously been raised by the antipodes. People realised this but could hardly argue with the brute facts. As Robert Record put it in his book *Castle of Knowledge* (1556), the existence of new lands was as obvious as the existence of those testified by the Bible and, whatever the theological concerns might be, there was no getting away from it.[23] The remarkable thing is how easily the new geography was accepted and how little effect it had on natural philosophy. Initial hopes that the Americas might be small islands or contain sea channels through to the Pacific Ocean quickly evaporated.

The great navigations of the fifteenth century represent Europe expanding its boundaries well beyond those it had inherited from the ancient world. At the same time, the last vestige of the Roman Empire itself was finally conquered in an event that seemed to be nothing less than a cataclysm: the extinction of Byzantium.

The Fall of Constantinople

By 1453, the territory of the Byzantine Empire had been reduced to the area enclosed by the walls of the city of Constantinople, and now the new Ottoman Sultan Mehmet II (1432–81) was determined to snuff it out. His army numbered 100,000[24] and his fleet controlled the harbour of the Golden Horn. The only thing standing between

Mehmet and the city's motley group of a few thousand defenders under the leadership of Emperor Constantine XI Paleologus (1404–53) were the mighty land walls. These walls had been built in the fifth century and had never been breached.

The land walls still stand today. To reach the centre of Istanbul from the airport, you have to drive through them. The ruined sections are imposing enough, but in the sectors where they have been restored they present a formidable obstacle, far higher and broader than any other castle or city walls from the Middle Ages.

To penetrate this barrier, Mehmet needed the latest technology from western Europe. It arrived in the guise of a Hungarian engineer called Urban who promised Mehmet that he could build a cannon large enough to blow down the walls. To be fair, he had already offered his services to the Byzantines but had been rebuffed over his fee. The bombard he constructed for the Turks was said to be 27 feet long and able to fire a 1,340lb ball a mile.[25]

These figures are doubtless exaggerations, but the weaponry available to the Sultan was surely formidable.[26] A siege gun from precisely this era survives in Edinburgh Castle and allows us to deduce some rather more sensible figures for the size and performance of Mehmet's weapon. The Edinburgh gun, known as Mons Meg after the Belgian city of Mons where it was first tested, was made in about 1449 and given to the king of Scotland by the Duke of Burgundy to stir up trouble with the English. It weighs over six and a half tons and takes a 400-pound ball. Its range from the top of Castle Mount was about two miles.

In 1680, Mons Meg's barrel exploded and it was never fired again. The burst gun allows us to clearly see how it was constructed. Thirty-seven iron hoops were heated and slotted onto a thin-walled inner tube made of 25 iron staves. The tube formed the inside of the barrel while the thick iron rings provided a reinforced exterior. As the iron cooled, it shrank and formed a tight fit around the inner tube.[27]

By the day of the final assault, the siege of Constantinople had lasted two months but in the end, the Sultan's cannons did the job. The vast volume of stone that made up the land walls had been slowly chipped away. During the small hours of the morning of Tuesday, 29 May 1453, with the walls breached, the final assault began. The defenders manned the ramparts and resisted long after the cause had become hopeless. The first two waves of attackers were beaten back, but as a last resort Mehmet sent in his elite forces, the Janissaries. Finally, they managed to scale the wall. The Emperor himself was seen where the fighting was thickest. When the battle was over, Mehmet ordered that his body be recovered for an honourable burial, but it was so mutilated that it could not be recognised. A decapitated corpse was eventually identified on the basis of the imperial regalia on its socks.[28]

As was the custom in war, when the city fell, the Turks fell on it in an orgy of pillage and destruction. The exquisite sixth-century cathedral of Hagia Sophia was saved by its immediate rededication as a mosque, but little else has survived of Christian Constantinople. Ironically, the finest remaining collection of Byzantine art is that looted by the fourth crusade in 1204 and still on display at the Cathedral of San Marco in Venice. The Venetians had led the crusaders who sacked Constantinople during an abortive attempt to retake the Holy Land. Even though the Byzantines had managed to retake their capital a few decades later, the crusaders had fatally weakened them and made their eventual conquest by the Turks inevitable. The fall of Constantinople in 1453 to an Islamic army caused consternation in western Europe. But as Catholics bewailed the loss of eastern Christendom, they only had their own crusaders to blame.

By supplying Mehmet with the weaponry and expertise he needed to breach the mighty land walls, the West had also contributed more directly to his victory. Without the cannons of Urban, taking Constantinople would have been much harder, if not impossible. Although the knowledge of gunpowder had been around for centuries, it took a long time to produce reliable weapons that used it. In the 1420s, the technique of granulation was developed to make gunpowder safer and more reliable. Mixing it with brandy or another alcoholic liquid caused the powder to coagulate into course lumps. This made for easier handling and a more even combustion.[29]

Advances in metallurgy, principally the development of the blast furnace, meant that iron became a strong enough material for cannon barrels and provided an alternative to bronze. Blast furnaces also turned out metal with a lower melting point than traditional means of forging iron so it could be cast into cannon balls or other shapes using clay moulds. The Turks had used a huge bombard during the siege of Constantinople, but smaller-calibre cannon turned out to be a more effective battlefield weapon.

At the smaller end of the scale, handguns required a failsafe firing mechanism before they could displace the crossbow. Such a weapon appeared shortly before 1500 in the form of the matchlock musket, which had an enclosed mechanism and a spring-loaded trigger.[30] Guns meant that Europeans could dominate almost every other people that they came up against. The English longbow, so instrumental during the Hundred Years War against France, was probably a superior weapon to the musket. The problem was that a bowman needed exceptional strength in his right arm to pull back the drawstring. Training began at an early age and maintaining a company of archers was expensive. The early gunpowder weapons may have been less effective in combat, but any fool can fire a gun. This made musketeers cheaper and consigned the longbow to obsolescence.

The Invention of the Printed Book

The development of guns made men better at killing other men. Printing is a pacific invention, but probably changed the world even more than the black powder. Like gunpowder, print originated in the Far East where people had used carved woodblocks and metal type for centuries. In Europe, short religious pamphlets, consisting of little more than a devotional picture and a short explanatory text, appeared in the late fourteenth century. These too used woodblocks. Stationers found that they were only economical for large print runs of standard documents, because they had to cut an entire block that they could not reuse for a different job.

The secret of paper had also travelled from China to Europe during the Middle Ages. Europeans had had to use parchment or vellum as their main writing medium since late antiquity. This is made from animal skins and was ruinously expensive. A paper mill is recorded in central Italy by 1276 and in France by 1348.[31] Although it was hardly cheap, paper was already well on the way to replacing parchment by the fifteenth century.

The father of printing was Johann Gutenberg (c.1398–1468), a metallurgist from Mainz in Germany. While the idea of printing was not new and paper already existed as a suitable material, Gutenberg had to make several intuitive leaps that mean we really can celebrate him as the inventor of the modern printed book. His greatest advance was moveable metal type. He cast the type for every letter separately so that the typesetter could rearrange them for each new page. Next, Gutenberg needed an ink that adhered to paper and did not smudge. He experimented with various oil-based inks, before settling on a mixture of soot and turpentine suspended in walnut oil. Finally, he adapted the old-fashioned wine or olive oil press into a device for bringing type and paper together under pressure.[32]

Gutenberg's first printed book was a short Latin grammar, but his great Bible of 1455 irrefutably demonstrated the potential of the new technology.[33] The Latin alphabet gave European printers a significant advantage over many Asian countries. Because European languages employ only about 25 letters each, a printer needs only a small assortment of type to be able to print absolutely anything. Chinese and other oriental languages have a different symbol for each word, necessitating a vast collection of different type. This rendered the capital cost of setting up a printing operation prohibitive and increased the time required to typeset a page.

The advantages of the printed book went well beyond cost. Of course, they were cheaper than manuscripts, but they were also much more legible. Better still, they had the potential to be more accurate. Unfortunately, poor editing blighted many early printed books. 'There is always the carelessness of printers to contend with', one noted writer moaned in his introduction to the *Natural History* of Pliny the Elder in 1525.[34] This meant that errors tended to proliferate through many copies rather than staying confined to a single manuscript. At least this gave publishers a useful marketing tool when no copyright existed. They could always insist that their version of a book was free of the gratuitous errors that allegedly plagued rival editions. 'Newly corrected and amended' was among the most common blurbs on the covers of early printed books.

The earliest of books printed included the works of the scholastic natural philosophers whom we met in the last chapter. Several of the most important books by John Buridan and Nicole Oresme were printed in Paris. Richard Swineshead was published in Venice, the city that dominated the early market in scientific texts. The 1494 Venetian edition of William Heytesbury's *Rules for Solving Logical Puzzles* included not just the mean speed theorem but also Nicole Oresme's diagrammatic proof. Most popular of all was Albert of Saxony's *Book on Proportion*, which went through no fewer than nine editions and included plenty of references to the work of his masters Buridan and Oresme. All this activity means that it would be quite wrong to think that the discoveries of medieval thinkers were 'lost', hidden in impenetrable manuscripts and gathering dust in monastic libraries. The fact that printers were willing to invest the capital outlay on many editions of these works shows that there was demand for them. It is true that some of these early books are a struggle to read, making use of arcane abbreviations and an ugly gothic typeface. But at the time they were a joy compared to the effort of deciphering a manuscript.

The invention of printing meant that the scientific advances of the Middle Ages were preserved even when the intellectual tide turned against them. The books of Albert of Saxony and William Heytesbury were about to become deeply unfashionable, but because they were already in print they remained available to people who knew where to look. In the remainder of this book, we will hear the story of the men who built on this vital but unloved legacy to bring modern science into being. First, though, we must meet a group of scholars and gentlemen who turned the word 'medieval' into an insult and sought to forget everything they owed to their immediate past in favour of glorifying the residues of the ancient world. These men have become known as the humanists.

CHAPTER 14

⋙⋘

Humanism and
the Reformation

By the time Constantinople fell in 1453, the Pope was back in
Rome after his sojourn in Avignon and better still, there was
only one of him. However, the popes of the late fifteenth
century were among the worst in history. They battled for secular
power in Italy while using the wealth of the Church to fund lavish
parties and building projects. The Vatican was a den of vice and corruption. On the plus side, the profligacy of the papacy and the rest of
the Italian nobility provided the funds necessary to produce the artistic jewels of *quattrocento* Italy as each city tried to outdo the others in
flamboyance. The best painters, sculptors and architects were held
in high esteem and the need to produce more and more remarkable
art drove innovation. Filippo Brunelleschi (1377–1446) designed the
spectacular dome that tops Florence's cathedral and was the first to
codify the technique of perspective.[1] Painting was further improved
when vegetable oil became the medium of choice for artists. Oil
paints allowed for a greater depth of colour than the earlier egg-based
pigments. All these innovations resulted in art so dazzling that later
generations dubbed the entire period the Renaissance.

The Rise of Humanism
As we have already seen, historians have a tendency to give value-laden
names to historical periods. These help to fix popular perceptions of

the period in question. And, if the name catches on, other historians co-opt it to their own periods. We have met two renaissances so far, the Carolingian renaissance under the Emperor Charlemagne and the resurgence of scholarship during the twelfth century. The period that everybody means by 'the' Renaissance lasted roughly from the late fourteenth to the early sixteenth century. The French historian Jules Michelet (1798–1874) first coined the term in the 1850s, but it took Jacob Burckhardt (1818–97) from Switzerland to turn it into common currency in his seminal work *The Civilisation of the Renaissance in Italy* (1860).[2] Michelet and Burckhardt both strongly contrasted the rebirth of culture in the fifteenth century with medieval stagnation. As Burckhardt claimed: 'In the Middle Ages, both sides of the human consciousness lay dreaming or half awake beneath a common veil.'[3] He went on to suggest that the Renaissance was the period in which individuality conquered the herd mentality of the previous era. It need hardly be said that much of this analysis was inaccurate, and recent scholarship has largely discredited it. The Renaissance was as much an age of faith as the Middle Ages and, if anything, even more superstitious and violent.

The desire to look back to Greece and Rome was the true mark of the Renaissance, which in many ways was a conservative movement attempting to recapture an imaginary past rather than march forward. It was a time when, in order to be up to date in writing or architecture, artists had to model their work on a prototype that was over 1,000 years old.

In literature, humanism typified this trend. 'Humanism' is yet another word that was invented in the nineteenth century, but it derives from a historical group of people in the Renaissance who studied the humanities.[4] Modern historians have piled all sorts of concepts on top of humanism and almost succeeded in turning a helpful term into a useless abstraction. Even more recently, non-believers have further muddied the waters by hijacking the word 'humanist' to mean a softer version of 'atheist'. A fifteenth- or six-teenth-century humanist was simply someone who was interested in classical Greek and Latin literature. They felt that medieval Latin was an ugly and barbaric tongue best replaced by the pure Latin of the ancients. Their paragon was the Roman orator Cicero (106–43BC) whose language they believed to be the most cultured and stylish.[5] In fact, by insisting on maintaining Latin in its fossilised classical form, the humanists went a long way towards killing it as a living language. Medieval Latin was untidy precisely because it was a spoken language that could be adapted to new situations as they arose. No one had ever spoken formal Latin as Cicero wrote it. In seeking to turn back the clock, humanists thought they were at the cutting edge of innovation, but they were really incorrigible reactionaries.

Some humanists were so convinced that nothing good could have come out of the early Middle Ages that they mistook ninth-century manuscripts in caroline miniscule for genuine classical artefacts.[6] These products of the reign of Charlemagne often represent the earliest extant copies of Latin literature. It never occurred to Renaissance humanists that there might have been any interest in preserving them during the so-called Dark Ages.

The one positive achievement with which the humanists are indelibly associated is the reintroduction of ancient Greek into western Europe. After an absence of 1,000 years, mere translations were no longer enough for westerners. They wanted to drink Greek literature from the original fountain. To do this, they needed teachers. Luckily, the Byzantine Empire's collapse precipitated an influx of Greek speakers in search of work.[7] After a slow start, knowledge of Greek became *de rigueur* for anyone who wanted to appear cultured. The universities started to teach it and printers cut new type so that they could mass-produce the Greek classics with the right letters. A great argument began, still ongoing, about the correct method of pronunciation. For those unable to master the language, almost the entire body of classical Greek literature was rendered into Latin. Although a good deal of Greek philosophy and science had been translated during the twelfth century, much of it remained unknown in the West. The epics of Homer (fl. eighth century BC), the *Iliad* and *Odyssey*, became widely available for the first time. By 1600, nearly the entire corpus of surviving Greek and Latin writing was in print.

The most significant writer to become available to a Latin-reading audience for the first time in the fifteenth century was Plato. His influence on Renaissance thought was profound and long-lasting. Throughout antiquity, it was Plato whom the Greeks regarded as their leading thinker.

In the Middle Ages, both Islamic and Catholic scholars preferred 'the Philosopher' Aristotle. This changed in the fifteenth century when manuscripts of the dialogues of Plato arrived in Italy. His brilliance was immediately obvious to the humanists. Furthermore, his works are polished and complete examples of classical Greek prose. Unlike Aristotle's surviving rough drafts, they are literary as well as philosophical masterpieces. This made them all the more appealing to the humanist scholars, who were interested in style as well as substance.

The greatest of Renaissance Platonists, Marsilio Ficino (1433–99), not only translated the dialogues into Latin, but also developed an entire theological system that tried to demonstrate how Plato was the heir to an ancient tradition of mystical wisdom that anticipated many of the doctrines of Christianity. We saw in chapter 2 how early Christian fathers like Augustine of Hippo had been aficionados of Plato because his philosophy seemed easy to combine with Christianity. Because Ficino's theology owed much to the Platonic philosophy of late antiquity, a craze for so-called neo-Platonism swept Europe.[8]

The subject in which Ficino thought that Plato most clearly surpassed Aristotle was the immortality of the soul.[9] At the start of the twelfth century, Amaury and his followers had denied personal immortality on the strength of Aristotle's book *On the Soul*. Later in the same century the Averroïsts, also basing their ideas on Aristotle, suggested that the immortal element of each person was not unique to them. Instead of having our own souls, all of mankind participated in a hive intellect to which we returned after death. Any concept of posthumous judgement by God evaporated if people had no individuality after death. However, despite their religious beliefs, Christian scholars were unwilling to throw away Aristotle's *On the Soul* because it was an extremely interesting analysis of the psychology of all living things. It is more a book about how the mind works than what the soul is. Once again, Thomas Aquinas had come to the rescue with an effective combination of Christian and Aristotelian views. Aristotle's science of the mind was separated from his dubious theology and took its place as one of the key texts of natural philosophy. This left the immortality of the soul exclusively within the purview of faith.

Ficino thought he could do better than this. He wanted philosophical proof of life after death and thought that, through Plato, he had found it. Unfortunately, Ficino relied on a dubious theory that Plato had inherited an ancient tradition ultimately derived from Moses. It was an easy matter to tear apart Ficino's assertions and show that whatever Plato believed about the soul, he never proved it.

A professor of philosophy at the university of Padua, Pietro Pomponazzi (1462–1525), wrote a refutation of Ficino's ideas that sought to preserve the traditional demarcation between philosophy and religion. He wrote:

> It seems to me that in this matter, keeping the saner view, we must say that the question of the immortality of the soul is a neutral problem, like that of the eternity of the world. For it seems to me that no natural reasons can be brought forth proving that the soul is immortal, and still less proving that the soul is mortal.[10]

He concluded, 'Plato wrote so many and such great things about the immortality of the soul. Yet, I think that he did not possess certainty.'[11]

To some Christians, this was a grave disappointment. They desperately wanted rational proof for religious claims and looked upon Pomponazzi's conclusions as an affront. In an effort to reassure Christian opinion, the Church had already officially declared that the soul was indeed immortal in 1513.[12] There was even an attempt to trump up heresy charges against Pomponazzi. He survived the assault and Platonism never managed to displace Aristotle at the universities.

For the humanists who had lauded Plato this was a setback, but it did not mark the end of their efforts to reform intellectual culture.

The Destruction of Medieval Scholarship

As far as humanists were concerned, the principal rule of philology was always 'the older the better'. Therefore they cast aside medieval corrections and commentaries, along with plenty of medieval mistakes. They castigated Pliny, writing in the first century AD, as late and unreliable compared to Aristotle who wrote 400 years earlier. The best place to look for knowledge, humanists said, was not in the real world but in the most ancient writings. The French essayist, Michael de Montaigne (1533–92), reported meeting a humanist of Pisa who assured him that 'the touchstone and measuring-scale of all sound ideas and each and every truth, lie in their conformity with the teaching of Aristotle, outside of which all is inane and chimerical: Aristotle has seen everything, done everything.'[13]

Renaissance humanists' obsession with the classics led them to search out the lost works of the ancients. They scoured dusty monastic libraries for forgotten books and sent travellers to the remnants of the Byzantine Empire to bring back Greek manuscripts. The trouble was, they also cleared away the vast bulk of medieval commentaries that had expanded on and criticised Aristotle's thought. They did not recognise that medieval writers had made great advances. As far as humanists were concerned, medieval thinkers were far too recent to have produced anything worthwhile. Scholasticism was undeserving of their attention and so they dumped it. The effect was rapid and nearly disastrous for natural philosophy.

Medieval logic or 'dialectic' was the first subject to go. Some humanists were not sure that logic was worth studying at all. Even if it was, they felt that Aristotle must have covered everything anyone needed to know. During the Middle Ages, logicians had in fact taken dialectic well beyond the achievements of the Greeks towards an advanced logic that was not rediscovered until the nineteenth century.[14] This was very hard to master and so deeply unpopular with students. By the mid-sixteenth century, it was off the syllabus and practically forgotten.

Theology survived little better. Protestants, as we will see, wanted nothing to do with medieval Catholic thought. Catholics themselves did not go quite so far, but the Reformation caused them to reconsider a great deal of their doctrine. As a result, they decided to enshrine Thomas Aquinas as the official theologian of the Church. The writings of most of those who had followed him, such as John Duns Scotus and William of Ockham, became marginalised. Few philosophers have seen their reputations plummet as far as Scotus. In the sixteenth century, humanists condemned him as the worst of scholastic obscurants. Thomas Cromwell (c.1485–1540), the adviser to Henry VIII (1491–1547) who was responsible for dissolving England's monasteries, banned Scotus's work from Oxford in 1535. Cromwell's cronies boasted that they had pulled the manuscripts of his work from the libraries and used them as lavatory paper.[15] Perhaps the worst indignity Duns Scotus suffered is that the word 'dunce' is a play on his name. It seems that the only thing he had done to deserve all this was to be too difficult for later readers to understand. The situation in Italy was just as bad. There, Richard Swineshead was dubbed the medieval bogeyman-in-chief. One humanist moaned about 'inane Swinesheadism', while another ranted 'I am filled with horror even at the mention of his name.'[16]

In natural philosophy, the humanists were again convinced that they would find everything worth knowing in the work of Aristotle and other ancient authors. Medieval writers were neglected as the humanist agenda swept all before it. Oxford, proud of the tradition of the Merton Calculators, resisted the rise of humanism longer than many universities. A student reading natural philosophy at Oxford in around 1510 would most likely have used a textbook written by a Parisian theologian called Thomas Bricot (d.1516). He was a nominalist follower of William of Ockham whose book was obviously popular.[17] It was reprinted at least eight times in three different cities. He was familiar with the work of medieval scholars and included cross-references to his famous predecessors at Paris including John

Buridan. Students using Bricot's textbook would have covered impetus theory and learnt about the defects of Aristotle's mechanics.[18]

Then humanism wiped the slate clean of scholastic authors. After
1523, Bricot's book was not reprinted anywhere again. At Oxford,
Cromwell banned him in 1535 at the same time as he outlawed the
work of Duns Scotus. As for Buridan himself, his natural philosophy
suffered a similar fate and was never reprinted after 1518 until modern times.

In comparison, an Oxford student studying natural philosophy in
1560 would probably have used a textbook by Johann Velcurio of
Germany (d.1534). This is admittedly an easier read than Bricot's
effort, at least for those who can handle Latin. However, its success
was more likely due to Velcurio being a partisan of Luther[19] and to the
fact that he completely ignores medieval writers. Instead, he refers
almost exclusively to Aristotle and other classical authors, together
with the Bible. His book was extremely popular, going through
nineteen editions before 1600. Unfortunately, a student who used
it would have been unaware that there had been any advance on
Aristotle's ideas during the Middle Ages.

To add insult to injury, historians have long assumed that the frequent attacks on Aristotelians in the sixteenth and seventeenth century were aimed at the obduracy of medieval natural philosophy. For
instance, Galileo tells the following story in his *Dialogue concerning the
Two Chief World Systems* (1632), an important book that will feature
again in chapter 21. To understand the story, it is necessary to know
that Aristotle believed that the nervous system centred on the heart
and not the brain. 'One day', Galileo tells us, speaking through a
character in the dialogue, 'I was at the home of a very famous doctor
of Venice.' While he was there, the doctor was carrying out an anatomical dissection. Galileo continues:

He happened to be investigating the source and origin of the nerves.
The anatomist showed the great bundle of nerves leaving the brain
and passing through the neck, extending down the spine and branch-
ing out through the whole body. Only a single strand as fine as a

thread arrived at the heart. The anatomist had been exhibiting and demonstrating everything with unusual care because he knew one of those present was a follower of Aristotle. Turning to him, the anatomist asked whether he was finally satisfied that nerves originated in the brain and not the heart. The Aristotelian thought for a while before he said, 'You have shown me this so clearly that I would be forced to admit that you were right, if only Aristotle himself did not contradict you.'[20]

Generations of historians believed that this story and Montaigne's anecdote were aimed at the natural philosophers of the Middle Ages. Only recently has it become apparent that their targets were one-eyed Aristotelian humanists who had completely lost the medieval critical attitude towards the Philosopher.

The Loss of Manuscripts

In combination with printing, humanism had one other damaging effect. As printed books replaced manuscripts, the old tomes became waste paper. Combined with the changes in taste and the lack of interest in medieval writing, this meant that entire libraries could disappear. Sometime between 1535 and 1558, Oxford University contrived to lose every single manuscript in its collection and even sold off the bookcases.[21] Merton College, home of the Calculators, threw out three quarters of its ancient library, as many as 900 manuscripts, in the same period.[22] These were not burned because they were made of valuable vellum. Instead, the college handed them over to bookbinders who cut them up and used them to make covers for newly purchased printed books. Today, it is still commonplace to find the beautiful calligraphy of a medieval manuscript glued into the covers of a sixteenth-century printed book.

In traditional histories, the rise of humanism is usually portrayed as 'a good thing', but the truth is that the humanists almost managed to destroy 300 years of progress in natural philosophy. By discarding the advances made by medieval scholars together with so many of the manuscripts that contained them, they could have set back the advance of science by centuries. Einstein might have had to do the work of Newton. The reason that progress in science was not so held back (although it arguably didn't move forward as quickly as it might have done) was that the invention of printing had guaranteed that, if nothing else, the old books were preserved. Most people forgot about them but a few, like Galileo, used the knowledge found within.

In the religious sphere, humanists' interest in the ancient sources meant that they encouraged reading the scriptures in their original languages of Greek and Hebrew. The Bible that everyone had read during the Middle Ages was the Vulgate of Saint Jerome (c.340–420) that he had completed during the fourth century AD. The name 'Vulgate' comes from the fact that it was written in the 'vulgar' Latin of the common people. Jerome had deliberately done this to make his work as widely accessible as possible. Of course, 1,000 years later, Latin was no longer an everyday language and only educated people could understand the Bible. By then, however, the Catholic Church had canonised the Vulgate, sanctified by age, as the authoritative version of the Bible. It viewed freelance attempts to work from the original languages with some suspicion. Instead, it set up an official project in Spain where a crack team of academics produced the *Complutensian Polygot Bible*, which included Old and New Testaments in Greek, Latin, Hebrew and Aramaic.[23]

This laborious process was too slow for some humanists. When this project was finally completed, it was overshadowed by the publication in 1516 of a Greek New Testament in Paris prepared by the most famous humanist of them all, Desiderius Erasmus (1469–1536).[24] Erasmus was the illegitimate son of a priest and, like Richard of Wallingford, had been adopted by the Church. He entered a monastery at an early age but, unlike Richard, did not find it to his liking. He escaped by getting a job as a bishop's secretary, then studied at the university of Paris for several years. He did not enjoy himself much there either. It did provide him, however, with a solid Latin education that served him well when he became a writer in need of patrons. He travelled Europe, visiting England three times, before his published books brought him both enormous fame and moderate fortune. Today, his most widely read book is *In Praise of Folly* (1511), a biting satire on sixteenth-century religious and secular society. Erasmus, a pacifist and idealist, wanted everyone to live simple Christian lives and sit around discussing the books they had read. In many ways, though, *In Praise of Folly* is also a typical humanist assault on medieval intellectual and religious life. 'Those subtle refinements of subtleties', he wrote of medieval theology,

> are made still more subtle by all the different lines of scholastic argument, so that you'd extricate yourself faster from a labyrinth than from the tortuous obscurities of realists, nominalists, Thomists, Albertists, Ockhamists, Scotists – and I've not mentioned all the sects, only the main ones.[25]

Erasmus's rhetoric was part of a sustained attack on the Catholic Church that soon split Christianity asunder.

The Reformation

The Protestant Reformation was an even more profound revolution in thought than humanism had been. There is no doubt that humanists, like Erasmus, gave the Protestant Reformation some of its initial momentum. Many of them were critical of the Catholic Church but wanted to reform it from the inside. Once Protestants started a formal schism with the papacy, though, that was going too far for some humanists. Erasmus, for example, fell out with Martin Luther (1483–1546) and never abandoned Catholicism.

We date the start of the Reformation to 31 October 1517 when Luther reputedly nailed his famous 95 theses to the door of the Castle Church in Wittenberg. This was not in itself an unacceptable thing to do, even if Luther's contentions were strongly worded, as long as it remained merely an academic exercise conducted in Latin.[26] Luther was a fully qualified theologian from the local university and his theses were simply items for debate. But when the theses were translated into German and printed as pamphlets that spread rapidly around Germany, matters quickly escalated. Before long, Luther found himself with strong popular support for taking on the Pope.

His original complaint had been that the Church was selling the forgiveness of sins in exchange for hard cash. However, his thinking rapidly developed into a whole new theology of salvation that insisted that only faith was necessary to reach heaven and that the Catholic penitence industry was an offence against the divine. The remission of sins, he wrote, is a free gift from God and not something that can be earned by our own good deeds. At the same time, he appealed to German nationalism in three short books published around the end of 1520, *Address to the Christian Nobility of the German Nation*, *The Babylonian Captivity* and *The Freedom of the Christian Man*. He also launched a full-scale attack on rational theology and the domination of Aristotle. Luther preferred his religion to be based purely on faith and the Bible. Before long his efforts at reforming the Church had taken on a life of their own and become a Reformation. Unlike 'Renaissance', 'Reformation' is a word that people used at the time to describe what they were living through. It is also a highly appropriate term to describe a process of religious change where many of the old certainties dissolved, only to require replacement or re-establishment afterwards.

When Martin Luther came to compose his 95 theses, all of what is now Germany, as well as substantial territories outside it such as Austria and Bohemia, were part of the Holy Roman Empire. From 1519 it was ruled by the hyperactive Emperor Charles V (1500–58), who was also king of Spain. He was fiercely loyal to Catholicism if not to individual Popes. The Empire was a confederation of city-states, duchies and other small principalities which enjoyed a good deal of real independence. That meant that Luther's followers, and Luther himself, could be sheltered by individual local leaders who had religious or political reasons for harbouring them. Once it was clear that Charles V was on the side of the old Church, Protestantism became closely associated with opposition to the Emperor's policies. Luther bravely stood up to the Emperor at the Diet of Worms in 1521 when asked to defend his beliefs. He was granted a certificate of safe-conduct that meant he was immune from arrest and was allowed to leave unmolested once he had stated his case. He was said to have concluded his speech with the famous declaration: 'Here I stand, I can do no other.'[27] As a summary of his sentiments this is accurate, even if the exact words are apocryphal.

After facing down the Emperor, Luther thought it prudent to spend some time in hiding. Concealed in Wartburg under the protection of the local ruler, Frederick the Wise (1463–1525), Luther busied himself translating the New Testament into German. Soon afterwards, inflamed by radical rhetoric, the rural peasants of southern Germany launched a huge rebellion, which Charles V put down with horrifying brutality. After some procrastination, Luther fully supported the suppression. Neither he nor any of his aristocratic sponsors were interested in drastic social reform.

Protestantism was fragmented from the start. In Switzerland another reformer, the French exile John Calvin (1509–64), worked on his own version of Protestantism, known as the Reformed rather than Lutheran tradition. Lutherans dominated in much of northern Germany and Scandinavia as well as having outposts in Transylvania and Poland. Calvinist or Reformed Protestants formed the backbone of the Dutch, Scots and, after much redefinition, English national churches.

Existing beside these larger denominations were a myriad of smaller sects, some of whom no one else considered Christian at all. About the only matter that all the groups agreed on was hostility towards the Pope and the Catholic Church. The most despised radical sects were the proto-Unitarians who denied the Trinity. Everyone else agreed that the Unitarians were heretics that had to be stamped out. The hostility towards them is exemplified by the famous case of a Spanish doctor by the name of Michael Servetus (1511–53). In trouble with the Spanish Inquisition over his extreme views, he fled to France and trained in Paris under a pseudonym. After qualifying in medicine and still under an assumed name, he started living in Vienne in south-eastern France, where he corresponded with John Calvin and published his book *The Restoration of Christianity* (1553).

The book infuriated Calvin, to whom Servetus had foolishly sent an advance copy, and he may have leaked the details of Servetus's secret identity to the local inquisitors of Vienne. Servetus escaped but fled through Geneva where Calvin himself lived. There he was apprehended and, at Calvin's insistence, executed as a heretic. Not all Protestants agreed with this vicious action, especially the fact that he was burnt alive rather than beheaded, as Calvin would have preferred.[28] On the other hand, few people had any sympathy for Servetus's theological ideas. In later years Servetus's work on the heart led to him enjoying a reputation as a martyr for science. This is unwarranted because he was executed for purely religious ideas that had nothing to do with anatomy. What the case of Servetus does teach us is that executions for heresy were not a Catholic prerogative. In England, Unitarians continued to be burnt until 1608 and in Scotland a student, Thomas Aikenhead (c.1676–97), was hanged for blasphemy shortly before the end of the seventeenth century.[29]

The progress of the English Reformation would depend on the opinion of the reigning monarch. King Henry VIII (1491–1547) broke with Rome when the Pope refused to allow him to divorce. Religiously, he remained a Catholic and executed both Protestants and papal loyalists. His son, Edward VI (1537–53), or rather the boy-king's council, completed the move to a Protestant state on the Swiss model but full-blooded Catholicism was restored on the ascension of his sister, Mary I (1516–58). She earned her nickname of Bloody Mary by burning nearly 300 Protestants as well as forcing many more to flee into exile. Her reign was short and when Henry VIII's younger daughter, Elizabeth I (1533–1603), came to the throne after Mary's death from stomach cancer, she enforced her own moderate Protestant settlement despite complaints from extremists on both sides. She ended up executing as many Catholics as Mary had burnt Protestants, but characterised loyalty to the Pope as treachery rather than heresy – a secular and not a religious crime.

Thus, it was the monarch who determined which way the religion of the country would go. This trend was repeated throughout Europe and was eventually codified in the Peace of Augsburg in 1555, the treaty that ended the first in a series of wars between Catholics and Protestants. The treaty stipulated that each ruler could decide the religion of his territory but that he would let dissidents move somewhere more congenial if they wished.

After a slow start, Catholicism began to react against the new movements. The Council of Trent (1545–63) began a process of Counter-Reformation by which the Catholic Church stamped out the abuses of absentee or illiterate priests, simony and bribery, concubines and intemperance. Cracking down made the Church into a more authoritarian institution but also strengthened it enough to combat and then turn back the tide of Protestantism. Whereas it had once looked like it would become a minority church, after the Thirty Years War of 1618–48, Catholicism was again on the rise. The only thing that prevented its complete triumph over Protestantism was the inability of France and Spain to get along.

The Theological Foundation of the Reformation

So, what was the Reformation all about? Simplifying enormously, the disagreements between the two Christian denominations were as follows. On the essential matter of how to get to heaven, Catholics believed that it was necessary to both repent of all one's sins and do penance. The penance could be anything from saying a prayer to going on a lengthy pilgrimage. Luther's initial objection, of course, was that the penance could take the form of a cash payment to the Church. The system also allowed people to pay the penance on behalf of others, even after they were dead. Lutherans rejected all this in favour of a simple formula – forgiveness came from faith in Jesus Christ. Calvinists also cast off the penance industry but believed that God himself elected who would go to heaven. Human beings just had to hope that they were one of the chosen ones, although being a Calvinist was a pretty good indication that they were.

The other major area of disagreement between Catholics and Protestants was the issue of authority. Protestants thought that they could find everything that they needed to know about religion in the Bible. They called this *sola scriptura* which means 'scripture alone' in

Latin. Catholics agreed that the Bible was the most important source of doctrine but said it was not enough. For them, the traditions of the Church were also authoritative, especially those that explained how believers should understand the Bible. Luther countered this by claiming that the meaning of the Bible was obvious to true believers who consequently didn't need any fancy exegesis. Unfortunately, his argument was belied by the fact that Protestants could not always agree on its interpretation among themselves.

Protestant emphasis on the plain meaning of the biblical text meant that they tended to interpret their Bibles more literally. Catholic theologians had built up a rich tradition of symbolic meanings, especially for those parts that did not make much sense with a literal interpretation. Whether or not this affected attitudes to science is unclear. On the one hand, Biblical literalism would seem to make contradictions between science and scripture more likely. However, on the other hand it has been suggested that Protestants carried over their rejection of metaphor in the Bible into the natural world as well.[30] The magical worldview, which saw the world as full of symbols and connections, might seem less convincing to those attracted to literal interpretations. Those who rejected metaphor in the Bible were unlikely to go looking for hidden messages in nature either. Despite this, as we will see in the next chapter, magic was one of the primary beneficiaries of humanism and the Renaissance.

The doctrinal authority of the Catholic Church has led some to suppose that Protestants enjoyed greater intellectual freedom.[31] For instance, there was no Protestant Inquisition or list of banned books like the Catholic Index. However, facts on the ground do not substantiate the claim that Protestants were freer or that freedom is a prerequisite for scientific progress. Religious toleration, when it finally came, was a result of exhaustion rather than any enlightened philosophical programme. In the seventeenth century, the devastation caused by the Thirty Years War as armies marched back and forth across Germany, coupled with a series of civil wars in the British Isles, made the acceptance of other religious points of view a pragmatic necessity. At least everyone could console themselves with the belief that their opponents were bound for Hell in the next world even if they could not be eliminated from this one.

Protestantism and Science

Historians have devoted plenty of attention to the question of how the Reformation affected the rise of science. In the nineteenth century, the pioneer sociologist Max Weber (1864–1920) suggested that the spirit of individualism and the work ethic of the Puritans, members of an ascetic branch of Protestantism, was one of the central factors in the rise of capitalism.[32]

In 1938, Robert K. Merton (1910–2003) enhanced the thesis by suggesting that Puritanism was also a crucial cause of the new scientific philosophy in seventeenth-century England.[33] Merton found that some Puritan beliefs promoted manual work. By valuing craftsmanship as much as scholarship, Puritans encouraged natural philosophers to get their hands dirty with experiments. As we have seen, experimentation had almost no part in medieval scientific practice.

Not all historians have been convinced by this. Firstly, Merton's critics felt that he had drawn the definition of Puritanism too widely and so included many people who would not identify themselves as such. Secondly, many of what he took to be specifically Puritan religious reasons for studying God's creation were actually common to all Christians. If you read seventeenth-century Catholic science or indeed anything from the Middle Ages, you will find exactly the same ideas. The desire to worship God by actively studying his creation is a Christian imperative, not just a Puritan one. For example Robert Record, an English Protestant, noted that astronomy 'leads men wonderfully to the knowledge of God and his high mysteries.'[34] At about the same time, Henry Howard (1540–1614), a Catholic aristocrat, could write:

> Natural philosophy drives us violently to know God, for when we see the causes of things are so orderly, depending on one another, we cannot at length but have recourse to that principle and special cause from which all other things, as if from their fountain head, do flow.[35]

One of the reasons for the illusion that Protestants have had a dispro-portionate influence on the rise of science is the bias of traditional English-language histories. They give an impression that after an ini-tial contribution from Galileo, England was the place where almost all important scientific advances occurred. We might assume from this that Protestant tolerance was behind this alleged advantage.

Another big problem for the thesis is the case of France. A history of science with impartial international coverage would recognise the French achievement to be equal to the English. The Reformation in France led first to civil wars, then to religious toleration and then to absolutist Catholic militancy. But as science in France became more highly developed, the country itself became more solidly Catholic.

The undisputed elite of Catholic science were the Jesuits. The Society of Jesus was founded by the Spaniard Ignatius Loyola (1491–1556) and received its official charter from the Pope in 1540.[36] Initially, missionary work and not science was the Society's driving force. Jesuits also involved themselves in education and the estab-lishment of schools. They set up a college in Rome which provided training for the Jesuit priests, and it grew into an important research establishment. As natural philosophers, the Jesuits have often been criticised for their reliance on Aristotle long after everyone else had abandoned him. This is a reasonable complaint, but it did not handi-cap their experimental work which was admired even by their oppo-nents. Between 1600 and the order's suppression in 1773, Jesuits produced 6,000 scientific papers, including 30 per cent of all publica-tions on electricity.[37]

We can see how little effect religious affiliation had on science in the sixteenth century by examining the cases of Oxford and Cambridge universities. Throughout this period, the universities found themselves under close scrutiny from the English govern-ment. Academics were in a position both to challenge the govern-ment's religious policy and provide it with intellectual ballast. This made them dangerous and useful in equal measure, so the authorities were keen to keep the universities loyal. As the country swung back and forth between Catholic and Protestant, professors of the wrong religious persuasion were thrown out of their jobs and new ones appointed. This allows us to see how members of both denominations treated the study of nature and mathematics. At Cambridge, for instance, we know the identities of the official mathematics lecturers from 1500 onwards. These individuals came from every religious tradition. Some were unable to compromise their principles and ended up in prison or exile. Others were happy to serve whoever was in charge at the time. But it is impossible to find any link between their being Catholic or Protestant and their attitude towards astronomy or geometry.[38] This conclusion should not surprise us. The theological arguments between Protestantism and Catholicism had no bearing on mathematics. They could hardly have led to one side or the other being more adept at science.

The debate over the effect of the Reformation on scientific advance continues to smoulder. English-language history has traditionally had a marked anti-Catholic bias, and once this is filtered out, evidence that the Protestants were better at science is scant. A recent survey has shown that early modern scientific pioneers were divided exactly evenly between Catholics and Protestants.[39] The influence of humanism was much more profound and potentially far more damaging. Humanists chased medieval writers out of the universities with a mixture of invective and satire. If it had not been for the prior efforts of printers, a huge body of knowledge could have been lost, or at least become so inaccessible that it ceased to have any relevance. Thanks to printing, the natural philosophy of the Middle Ages remained at hand for the next generation of scientific thinkers. They plundered it at will and did not always feel that they had to acknowledge the source of the treasures they unearthed. In the last chapters of this book, we will see how the discoveries of the Middle Ages came to be incorporated into modern science.

CHAPTER 15

〜〜

The Polymaths of the Sixteenth Century

One thing with which the Middle Ages is indelibly associated is belief in magic. But, in fact, the sixteenth and seventeenth centuries were when magical thinking was at its most intense and widespread. At first this seems strange; at precisely the period when most now believe that modern science was being born, the occult enjoyed a remarkable resurgence. Even Sir Isaac Newton, it turns out, was obsessed by alchemy and at the end of the seventeenth century, astrology was still big business. Of course, it still is today.

Hermetism

To find out why magic was so popular during the early modern period, we have to travel back to Florence in 1463. There, Cosimo de Medici (1389–1464) lay dying. He had been de facto ruler of the city for 30 years, his family having made a successful transition from banking to politics. His long administration had seen Florence become the jewel of Italy, its status celebrated by the completion of Filippo Brunelleschi's magnificent dome on the cathedral in 1461. Cosimo had been a patron of both the arts and literature. He sponsored the humanists who searched out ancient manuscripts and founded the Laurentian Library to house their booty. This library, for which Michelangelo later built the vestibule, already housed 3,000 manuscripts when it opened to the public in 1571, including many of those that first allowed western scholars to enjoy the surviving literature of ancient Greece.[1]

Before he died, Cosimo wanted to read Plato, whose books had been almost entirely unknown during the Middle Ages. He was unable to understand Greek himself, but he could afford to hire the best scholars to translate newly discovered manuscripts into Latin. During the Council of Constance back in the 1430s, Cosimo had first heard about the philosophy of Plato from a neo-pagan magician. The council had been called to reunite the Catholic and Greek Orthodox Churches so that the Pope would launch a crusade to rescue Constantinople from the Turks. The project fell apart but the meeting of the finest minds of eastern and western Christendom had profound consequences. Among the Byzantine intellectuals at the council was one Gemistus Pletho (c.1355–1452). He was Plato's greatest admirer, but also a sun-worshipper.[2] The young Cosimo de Medici heard Pletho extolling the virtues of neo-Platonism but probably also learned about more esoteric writings.[3]

Marsilio Ficino, Cosimo's most gifted translator, had long been gathering together manuscripts of Plato's dialogues. But the dying Cosimo ordered Ficino to lay aside the work on translating Plato. Cosimo's agents had discovered a manuscript languishing in the library of a Macedonian monastery. It purported to have been written by the legendary sage Hermes Trismegistus of whom he had, perhaps, heard through Pletho.[4]

Ficino and Cosimo both believed that Hermes Trismegistus was an Egyptian seer, a contemporary of Moses and the oldest of the prophets of God. In English, Trismegistus translates into something like 'three times blessed'. Today we hear little about Hermes and might wonder what all the excitement was about. However, during the Renaissance, his writings were thought to contain the long-forgotten secrets of a race of magicians. There were good reasons for believing this. Augustine of Hippo, in a passage otherwise hostile to Hermes, had conceded that he was a great sage, while other Church fathers praised his wisdom and authority.[5]

Ficino and his contemporaries dated the works of Hermes to about 1300BC, roughly at the time of Moses's exodus from Egypt as described in the Bible. Ficino took Hermes's works to be among the original sources that Plato had used to formulate his philosophy, and he also thought them to be fully compatible with Christianity. That meant that the Hermetic corpus, as it was known, was the original pristine theology in which the deepest truths of God could be uncovered. Within it, readers could find intimations of the Trinity, the incarnation and a multitude of neo-Platonic ideas such as the 'great chain of being'. Most exciting of all, the Hermetic corpus laid out procedures that allowed a magician to compel angels and other benign spirits to carry out his orders.

Ficino was one of the foremost scholars of his age and one of the few who were able to accurately translate from Greek into Latin. He was also a Catholic priest so his interest in ancient magic might seem rather incongruous. Nonetheless, he was convinced that it was possible to harmonise the work of Hermes with Christian theology because he believed that, although Hermes was a pagan, he had enjoyed some glimpse of the divine mind that was now lost. It took Ficino only about a year to produce his translation and he was able to hand it to an ailing Cosimo de Medici shortly before he died. The resulting book was hugely influential for 150 years after Ficino published it.

In Ficino's translation, this magic attempted to harness the power of the stars and especially the sun. He called the sun 'the lord of the sky, which rules and moderates all truly celestial things',[6] and revived the old notion that the heavens produce a sound, the music of the spheres. By composing his own astral songs, Ficino believed he could control the power of the stars. His rituals drew on a hotchpotch of ancient traditions including stories about Pythagoras and Orpheus.[7] Pythagoras was important because he formed a link in the chain between Hermes Trismegistus and Plato along which the wisdom of Egypt had been transmitted to Athens.

Anyone visiting a library to read an English translation of the Hermetic corpus will probably be disappointed.[8] It is mainly in the form of dialogues between Hermes and his son about mystical matters that do not make much sense to the uninitiated reader. The language is heavily symbolic and subject to a multitude of different interpretations. Once Ficino had finished his translations the more difficult task was to figure out what it all meant. This spawned a voluminous literature of commentary and discussion that only the most dedicated of today's researchers feel inclined to plough through. Then, in 1614, the whole Hermetic corpus was finally revealed to be a forgery. Isaac Casaubon (1559–1614), a French Protestant exiled in England, exploded the myth. In a massive refutation of the errors of Catholicism, he pointed out that Hermes was actually a phoney invented well after the birth of Christ and, furthermore, the apparent anticipations of Platonism and Christianity were simply plagiarism.[9] After word of this got around, the Hermetic corpus became something of an embarrassment and scholars quietly dropped it.

The revival of magic set in train by Ficino was carefully couched in terms that made it acceptable to Christians. As long as the subject was confined to discussion among educated priests and bored aristocrats it caused little stir in the outside world. The most visible result may have been the proliferation of magical sigils and symbolism in art such as in Botticelli's paintings the *Mystical Nativity* and *Primavera*. Pope Alexander VI (1431–1503) had his private apartments in the Vatican decorated with frescos of astrological figures. These included a picture of Hermes Trismegistus himself, in conversation with Moses and the Egyptian goddess Isis.[10] However, the influence of Hermes went well beyond the safe bounds of Italian high society. With the backing of such esteemed scholars as Ficino, magic now went mainstream. It was no longer the preserve of strange old women and renegade priests. Couched in the elegant Latin of the humanists, the Hermetic corpus was a licence to reinvent astrology, alchemy and other forms of occult knowledge. Soon it seemed that almost every European intellectual was infected with the magical bug, labouring under the impression that the Hermetic corpus opened up exciting vistas of forgotten wisdom.

John Dee Reforms Astrology

Among the biggest beneficiaries of the revival of the occult was astrology. During the Renaissance, astrology's problem was not official censure but the inconvenient fact that it did not work. Almost everyone agreed, whether they supported astrology or not, that the success rate of even the best practitioners left a great deal to be desired. Among those earnestly seeking a solution to this problem was the English magus John Dee (1527–1609).

Dee was a product of Cambridge University. Humanists – led by Erasmus himself, who had briefly taught there – saw off medieval philosophy and were propagating the new learning. Greek, Hebrew and mathematics were the subjects of the day. Dee himself taught Greek at Trinity College, founded in 1546, and benefited from the new surge in maths teaching brought on by the reform of the university syllabus in 1549. Even so, as he arrogantly explained in a short autobiography he wrote late in life, Dee quickly outgrew the ability of anyone at Cambridge to teach him and studied on his own. He travelled throughout Europe, delivered lectures at the university of Paris on geometry and hobnobbed with the most admired natural philosophers of the day.[11] He returned to England determined to launch a thoroughgoing reform of astrology and place it on a firm scientific foundation. It never seems to have occurred to him that astrology might not work at all, but that was not his fault. The art of prediction was completely consistent with Dee's worldview and offered such great benefits that it was quite rational for him to devote himself to perfecting it. Most attacks on astrology from this time are religious rather than scientific. Countering the naysayers, some religious thinkers, like Luther's right-hand man Philipp Melanchthon (1497–1560), were extremely enthusiastic about astrology. He spent many years lecturing on it at Wittenberg University, where Luther was based.[12]

Dee's astrological agenda harnessed mathematics to astronomy so as to provide a scientific explanation of how the stars and planets influenced events on earth. He began by postulating that each planet gave off rays that imparted certain properties to whatever they struck. As the planets moved around the sky, the relative intensities of these rays on earth changed.[13] This explained the variety of astrological effects and why where you were born was as important as when. Thus, if the rays from Jupiter and Mars were especially strong at the moment of your birth, then you would obtain jovial and martial characteristics – that is, you would be jolly but prone to violence. Dee wanted to mathematically calculate the force of the rays from each of the individual planets at a particular place and time to provide an exact horoscope. However, he faced some formidable problems. He needed to know the relative distances to the planets more accurately than was possible at the time. He also could not assume that the rays emanating from each planet were of equal strength. This meant he was unable to calculate their intensity once they reached earth. Finally, he realised that the angle at which the rays struck the earth had a profound effect on how powerful they were. This is most obviously the case with the heat from the sun, which warms the equator (where it is directly overhead) far more than the Arctic region (where it never rises very high in the sky). The necessary calculations were beyond Dee, however, and he lacked the patience to devote enough time to such a complicated problem.

Dee's attempt to reduce astrology to mathematics showed that he was thinking along the right lines. But as a scientific venture, it was always doomed to failure because his underlying theory was wrong and he had no reliable way to experimentally verify his hypothesis. Not many astrologers shared Dee's passion for a reformed astrology, but they all realised that it was essential to use the very best methods available to plot the movements of the planets. This demand for accuracy provided a major market for astronomers' tables and helped drive their standards up. In turn, having the ability to master the complicated geometry necessary to predict the position of the planets improved the professional status of the astrologers themselves. They were no longer treated with the suspicion that had clung to them in the Middle Ages. In the sixteenth century, astrology enjoyed a boom and its practitioners became celebrities. Girolamo Cardano, although virtually forgotten today, was the most famous astrologer and physician of his time. His reputation spread as far as England and even over the border to Scotland. In fact his name was anglicised to Jerome Cardan, which is remarkable in itself. It means that he was as famous as his contemporaries Michelangelo and Raphael, both of whom were also re-christened with English names.

The Life of Jerome Cardan

Cardan was born in Pavia in 1501 amid the turmoil of early-sixteenth-century Italy.[14] He spent much of his childhood in Milan living under French occupation after Louis XII (1462–1515) had invaded in 1499. The city was briefly liberated in 1512 and plague finally drove the French from the city in 1522. Jerome's family was genteel but not wealthy. The shortage of money when he was a child instilled in him determination to get rich and famous as quickly as he could. His horoscope gave him some measure of encouragement but also a dreadful warning. It predicted that he would become extremely well known but would die before he reached old age. 'There were stars which threatened my death from every aspect', he wrote, 'which all declared would be in my 45th year.'[15] From his earliest youth, Cardan believed he was a man pursued by the fates to succeed, burn brightly and quickly fade.

His father, Favio, was a successful lawyer who wanted his son to follow in his footsteps. But Jerome, fascinated by philosophy and mathematics, decided that he would become a doctor. Favio himself had dabbled in these subjects when he prepared the first printed edition of the textbook on optics by John Peckham, the thirteenth-century archbishop of Canterbury.[16] Perhaps that is how Jerome knew there was no money in natural philosophy. There were a series of blazing rows over the matter before Favio finally relented and allowed his son to study medicine at the university of Pavia.[17] Although no money was available to pay for the fees or Jerome's living costs, he was already an accomplished enough mathematician to earn his keep by hiring himself out as a tutor. Jerome had another skill that helped him to make ends meet – he was a gifted gambler.

Despite the disapproval of the church, Italy was awash with gambling dens where men wagered whatever they could afford. Jerome Cardan had an enormous advantage over almost all of his fellow gamblers – he understood the concepts of probability and odds. While playing with dice is a matter of luck, those who know what the odds are can be careful only to wager substantial amounts when the odds are in their own favour. There was no 'banker' with a built-in advantage, so if the skilled gambler's opponents had a less sure grasp of the odds than he did, then he could fleece them. The science of probability did not even exist at the time – Cardan invented it. He was the first to decree that all probabilities can be shown as fractions between zero (for impossible) and one (for a certainty). Thus the probability of a coin landing heads-side-up is ½ or one in two. It tells us something about western civilisation that this most vital field of mathematics was first developed by a student trying to raise enough money for his bar tab. Even so, Cardan's book on the subject was not published in his lifetime and by the time it was, others had established priority.[18]

In 1526, Jerome Cardan qualified as a physician and rode back to Milan to start what he hoped would be a medical career. Unhappily, the Milanese College of Physicians decided to reject his application for membership. Without the College's approval, Cardan could not work in Milan and so had to build up his reputation as a healer in the provinces. He was bitterly disappointed by the snub and spent many years attempting to gain admission to the College until, in 1539, he was such a celebrated doctor that they had to let him in.[19]

Cardan's success as a physician was almost entirely down to his philosophy of non-intervention. His peers tortured their patients with a gruesome regime of bleedings, purgatives and potions guaranteed to weaken the constitution of even the hardiest soul. As we have already seen, doctors of the time based their theories on a mixture of Galen and the Arabic medical writers, with a dose of natural magic and plenty of guesswork. Merely by sporting a professional qualification, doctors could demand hefty fees for accelerating the demise of their clients. This is surely the reason that so many people put their trust in the village wise women and the intercession of the saints. Both were much less likely to kill them before they had a chance to get better on their own.

Whether by intuition or genius, Cardan gave his patients a chance to heal themselves. Rather than the constant round of prescriptions and bloodletting, he suggested a healthy diet and plenty of rest. In this way, he gained a reputation far beyond the borders of Italy as a doctor without parallel.

In 1551, John Hamilton (1511–71), the archbishop of St Andrews in Scotland, was suffering from failing health. Hamilton was not only the leader of the Church in Scotland. He also wielded considerable political power during the minority of Mary Queen of Scots (1542–87), who was just nine years old at this time and safely exiled in France. The archbishop suffered from a lung condition, very possibly asthma, which had defeated the efforts of all the physicians in Scotland. In desperation, he called for Cardan and offered to meet him in Paris where Hamilton expected to travel on diplomatic business. With his affairs in order, the Italian very much liked the idea of an all-expenses-paid expedition to France and set off as soon as he could. He had got as far as Lyon when he learned that Hamilton had not, in fact, left Scotland and was begging him to travel all the way to Edinburgh. A substantial purse of gold convinced Cardan that he could risk the journey and he reached Hamilton's palace in the summer of 1552.[20] Dismissing the assembly of physicians arguing over Hamilton's treatment, Cardan launched the archbishop into a detailed regimen of milk, rich food and relaxation.[21] The new approach worked and the patient began to improve. He lived for another twenty years (before being executed by Scotland's new Protestant rulers) and always attributed his longevity to Cardan.

The case sealed Cardan's status as Europe's foremost doctor. Even before he left Scotland, a messenger arrived from London asking him to come and treat the sickly king of England, Edward VI. Obeying the summons, Cardan found that it was his skill as an astrologer as much as his medical talents that led the English court to consult him. He met John Dee and discussed the properties of a magical gem, as well as producing a chart showing that the king's stars would grant him marriage and a long life.[22] Cardano later claimed that his actual horoscope for Edward VI had shown that the king had just months to live (which in fact he did), but that he lied about this in order to avoid provoking an affair of state. Not everybody was convinced by this, especially as he had already published the king's false horoscope. In an unfortunate coincidence of timing, it appeared shortly after the young man had died.[23]

Astrology was Cardan's lifelong passion. He called it 'the most lofty of the branches of knowledge because it deals with celestial things and with the future, awareness of which is not only divine but also very useful.'[24] Medical practice merely provided an income and a secure reputation that allowed him to indulge his real interest.

Alert readers will have already noted that Cardan did not die at the age of 45 as his horoscope had demanded. Clearly, there was a problem with the prognostication, albeit an extremely welcome one. Like John Dee in England, Cardan launched a project for the reform of astrology.[25] However, his plan was completely different from Dee's attempt to provide a mathematical basis for the art. Despite being the better mathematician, Cardan adopted an empirical approach. Perhaps he was well aware of the shortcomings of Dee's idea and realised that the calculations required to plot the influences of all the planets would just be too difficult. Instead, he gathered together the horoscopes of as many people as he could find and compared the predictions to how their lives actually turned out.[26] This was a difficult business in itself, because he needed accurate information on when and where all his subjects had been born. He hoped to be able to collate the horoscopes so that he could formulate some rules about the ways in which the influences of the planets interacted. His collection of horoscopes went through several editions, some printed by the publisher of Copernicus's *Revolutions of the Heavenly Spheres*, and was a commercial success.[27] But, needless to say, Cardan's great venture failed scientifically and as astrology fell from favour among scholars, it has tended to be written out of history. What is most interesting about the efforts of Dee and Cardan to restructure astrology is how closely they both conformed to the methodology of science as we would recognise it today. The twin tasks of gathering data and generating mathematical theories are both essential parts of almost all modern science. It is ironic that the first concerted attempts to do this were performed in a subject that is no longer considered scientific at all.

Among Cardan's collection of horoscopes was one he prepared for Jesus Christ. He must have known how much trouble this had got Cecco D'Ascoli into and it was rumoured (falsely) that the medieval magus Guido Bonatti had been burnt at the stake for the same reason. Nonetheless, Cardan made no effort to conceal what he had done, and even included a lengthy discussion of Jesus's horoscope in his edition of Ptolemy's book on astrology.[28] This bravado would later bring great misfortune upon him.

Cardan and Tartaglia

In the 1540s and 1550s, Cardan was at the height of his fame and influence. He published on any subject that interested him and wrote a considerable number of bestsellers. His most successful works were two volumes of what we might call 'popular science' called *On Subtlety* (1550) and *On Variety* (1557). They contained a smorgasbord of interesting factoids as well as numerous inventions cooked up by Cardan's resourceful mind. A few of these are still in use today. For example, buried somewhere in the engine of modern cars is a device called a 'cardan joint' that allows a spinning driveshaft to bend. Cardan was also familiar with impetus theory,[29] which he used in many applications, and he accepted the possibility of a vacuum.[30] *On Subtlety* is one of the earliest books in which we find Archimedes (287–212BC) raised to the highest status among the ancient sages. Cardan's discussion of his writing probably did much to bring it to a wider audience.[31]

Archimedes' work was more advanced than that of any other Greek mathematician. Unfortunately, this meant that it flew over the heads of many of his contemporaries and even some of those who came long after. We tend to remember Archimedes for crying 'eureka!' (Greek for 'I've found it!') and leaping out of his bath. The story is probably apocryphal and refers to his discovery of the law of buoyancy. The law states that any object dropped into water will displace an amount of liquid equal to its own weight. If the object has a lower density than water, then it will not displace enough water to cover itself completely and so will float. At the end of his life, Archimedes designed some impressive military equipment to defend the city of Syracuse from the Romans during the Second Punic War in 212BC. Despite the reported use of burning glasses and a giant hook to sink the Roman galleys, the city eventually fell to them and Archimedes was one of the victims of the slaughter that followed.[32] Many of his mathematical works had been translated into Latin during the Middle Ages, but it was not until the sixteenth century that they were intensively studied.[33]

Cardan's most significant book was of a more academic hue – a guide to advanced mathematics, which he called *The Great Art* (1545). This contained the first published solutions to cubic and quartic equations, the former of which occasioned an almighty row with a fellow mathematician, Niccolò Tartaglia (c.1499–1557) who claimed that he had found the answer first.

Tartaglia was born in Brescia and was brought up by his mother after losing his father at an early age. He barely made it into his teens himself. In 1512, the French stormed the city. Many civilians, including Tartaglia and his mother, sought sanctuary in the cathedral but the French were no respecters of the holy building. They mutilated the boy and left him for dead. One blow split his jaw but by some chance he survived. His mother nursed him back to health but his injuries left him with a permanent stutter. Rather than be ashamed of this, he named himself 'the stutterer', which in Italian is Tartaglia.[34] We do not even know what his real family name was.

Although he was not ashamed of his speech impediment, Tartaglia always had a chip on his shoulder over his lack of formal education. He was self-taught and never mastered the elegant Latin that his peers thought was essential to intellectual discourse. Nevertheless, he was a mathematician of genius and was soon earning his living teaching others the subject. In 1537, he moved to Venice and published his first book called *New Science*. This dealt primarily with the mathematics of cannon balls. At the time, Venice was preparing to do battle with the Turks who were following up their conquest of Constantinople by trying to achieve mastery of the rest of the Mediterranean. *New Science*, dedicated to the Venetians' general, was Tartaglia's contribution to the struggle. In the book he showed that a cannon ball should move with a curved trajectory, at least for part of its flight. Later on, in his *Questions on Diverse Discoveries* (1546), he demonstrated that the path of a projectile should be a curve throughout.[35]

In 1539, Cardan invited Tartaglia to visit him in Milan. During his stay, Tartaglia imparted the secret of how to solve cubic equations to Cardan on condition that he kept the method to himself until it was published. Understandably, Tartaglia wanted to reveal the discovery in one of his own books. In the meantime, he had other projects on the go, including his 1543 edition of the medieval Latin translation of the works of Archimedes. This book finally brought the achievements of the great Greek mathematician to a wide audience and is probably how Cardan himself came to hold him in such high esteem. Tartaglia strongly implied that he had translated Archimedes from the Greek himself. He wanted to appear as well-educated as the cultured humanists but the plan backfired after he was found out.[36] And when he had not published his solution of cubic equations after six years of procrastination, Cardan lost patience. He included cubic equations in *The Great Art*, much to his now-rival's fury.[37] Tartaglia died twelve years later in Venice, still convinced he never received the respect that was his due.

Cardan's Family Problems

The last fifteen years of Cardan's life were filled with one terrible misfortune after another. It was his sons who were to bring down disaster upon him. His elder son was called Giovanni and the younger Aldo. He also had a daughter Chiara who, at least, caused him no trouble beyond the cost of her dowry. This expense, which Cardan was still talking about decades later, earned him a valuable confidant in his son-in-law.[38] Sadly, to her father's lasting regret, Chiara had no children.

Aldo appears to have been something of a psychopath. He was a thief and a violent ruffian, continually being arrested for his crimes. Cardan was now rich and spent a fortune paying fines and bail on behalf of his dissolute son. Eventually his behaviour became too much even for this most indulgent of fathers. Aldo was expelled from the family house and then exiled from the city itself.[39]

On top of these family worries, Cardan had all the usual problems of fame. When he became a lecturer at the university of Bologna, his colleagues were even more envious of his extramural success than academics of today would be. They kept up a constant campaign of sniping, tale-telling and poaching of his students.[40]

His books did not please everyone either. *On Subtlety* provoked what one modern scholar has called 'the most vitriolic book review in the annals of literature'.[41] The humanist Julius Caesar Scaliger (1484–1558) wrote a 900-page tirade, twice the length of *On Subtlety* itself, berating Cardan for every aspect of the book. Scaliger was especially outraged about the elevation of Archimedes above Euclid and Aristotle. 'You have put a builder [Archimedes] before Aristotle, who was no less knowledgeable in these arts', he fumed, continuing: 'After Archimedes, you have put Euclid as if the light after the lantern.'[42] The rebuttal proved almost as popular as its target with the book-buying public who then, as now, liked nothing more than a good literary spat.

Cardan decided not to rise to the bait and Scaliger was disappointed not to receive a reply in kind. The silence of his target began to unnerve him and he allowed himself to be convinced that Cardan had read his book and died of shame. Overcome with remorse at causing the death of his adversary, Scaliger composed a gushing obituary. 'Cardan', he wrote,

> in addition to his profound knowledge of God and nature was a consummate master of the humane letter ... A great man indeed! I affirm

that the man who would venture to compare himself to Cardan may well be regarded as one who is lacking in all modesty.[43]

Luckily for his own reputation, Scaliger heard that Cardan was very much alive before his panegyric was published.

Cardan had to endure a far more painful experience than wounds to his pride. His son Giovanni had unwisely married a shrewish woman whose family constantly tried to extort money from his wealthy father. So Cardan must have been shocked but hardly surprised when word was sent to him that his daughter-in-law lay dead from poison and his son was under arrest as the prime suspect. Here was a crime that it was beyond the means even of Cardan to make good with a hefty bribe. The case came to trial in the Palazzo Marino, now Milan's city hall, in February 1560. Giovanni claimed that he had used his medical expertise to manufacture a toxin with which he had laced a cake he baked for his wife. He also admitted that his incompetence was such that he had failed twice in the scheme before finally managing to do away with her. After the confession, the trial judges had only to decide on the sentence. That would have been a formality but for Cardan's celebrity and civic importance. He launched a heartfelt and eloquent appeal for his son's sentence to be commuted to exile or life imprisonment:

> Although I brought forward many arguments and their like, it availed naught except in so far as it was decreed by the court that if I should be able to come to terms with those who had brought the charge against him, his life would be spared. But the very indiscretion of my son forbade this. For he had boasted of riches which I did not possess and the accusers tried to exact what did not exist.[44]

No agreement could be reached and Giovanni was executed in his prison cell. Cardan never recovered.

Even then, his suffering was not over. In 1570, he was arrested by the Inquisition in Bologna. The details of the charges are unknown but probably resulted from the jealousy of his colleagues at the university. The Inquisition, like many other official organisations of oppression, relied very heavily on denunciations from informers to do its job. This provided an easy method of revenge for anyone who wished to take advantage of it. In Cardan's case the sheer volume of his writing, his relatively unguarded pen and, of course, that horoscope of Christ provided plenty of material for the tribunal to wade through. It seems, though, that they could not be bothered to do so. After a few weeks the prisoner was released on bail and eventually he must have been acquitted, for we hear nothing of any trial, let alone a conviction.[45]

At least this marked the end of his troubles. Cardan travelled to Rome and petitioned the Pope over his treatment by the Inquisition. The outcome was a papal pension and a respectable retirement until his death in 1575, 30 years later than his horoscope had predicted. He spent his last four years writing his autobiography, *The Book of My Life*, through which we now know so much about him.

The astounding breadth of Cardan's interests makes him a true renaissance man. His *Great Work* on algebra taught Europe all the new advances in the subject which Italian mathematicians had been working on. His medical practice was both humane and successful. He tinkered and invented constantly and, unlike some authors, he made no effort to keep his good ideas to himself. His obsession with astrology was based on a desire to place this useful art on a firmer footing than hitherto. Perhaps his careful collation of horoscopes had the unintended but beneficial consequence of demonstrating that the stars have no effect on our lives after all. Above all, Cardan and Dee show us that the boundary between the occult and the scientific in the sixteenth century was both porous and indistinct.

CHAPTER 16

≈≈

The Workings of Man:
Medicine and Anatomy

As a doctor, Jerome Cardan had proved more successful than the practitioners of orthodox Galenic medicine by discarding bleeding and purgatives. But he never rejected medical orthodoxy outright. This meant that he could only provide palliative care to help the body heal itself. Any direct treatments he prescribed would have been useless at best. In contrast to Cardan, other dissenting doctors were not afraid to attack conventional medicine vigorously and directly. Foremost among them was Theophrastus Bombastus von Hohenheim (1493–1541) who, despite a chaotic lifestyle and idiosyncratic ideology, managed to found his own school of medicine.

An Unorthodox Physician

Theophrastus von Hohenheim has always been better known by his Latin name, Paracelsus. His family was German, but he came from Switzerland. He led the life of a wanderer, forced to keep moving as his loudly voiced opinions made him unpopular in one place and then another. However, Paracelsus was no quack. He claimed to be a fully qualified physician[1] and, despite the lack of documentary evidence, there is little reason to doubt him. Rejecting Galenism, he used his skills as an alchemist in his medicine. And his work went much further than that. He invented an entire system of thought that blended ascetic Christianity with mysticism and alchemy to produce what came to be called the chemical philosophy.

To Paracelsus, all materials comprised the three elements of salt, sulphur and mercury. He had added salt to the traditional alchemical pairing of sulphur and mercury to act as the third part of the neo-Platonic trinity of body, soul and spirit. As he put it in a typically opaque way:

> All metals can be produced from the three types of matter – mercury, sulphur and salt … Mercury is the spirit, sulphur the soul and salt the body. But a metal is the soul in the middle, between the soul and body, as Hermes Trismegistus says. The soul in sulphur unites the two opposites, body and spirit, and changes them into a single essence.[2]

The Hermetic wisdom translated by Ficino was just one of many sources for Paracelsus's thought. The result is a hotchpotch of magic, religion, natural philosophy and medicine. A constant theme running through it all is hostility towards the Galenic academic medicine of the day. Paracelsus urged the use of the drugs he had devised rather than traditional treatments such as bleeding. He believed that illness was caused by individual diseases rather than an imbalance of the humours. This led him to look for specific drugs to treat particular conditions, but unfortunately he had no idea what might work. Paracelsians, as his followers became known, were especially keen on prescribing antimony and mercury, both deadly poisons. Nevertheless, he scored some notable successes as a healer, in all likelihood because doing nothing would always be more effective than the customary methods.

Paracelsus's reputation peaked when the town council of Basle appointed him to the prestigious post of professor of medicine at their university in 1527. He responded with some scandalously original lectures and an outright rejection of the traditional authorities. On 23 June he burnt a medical textbook – allegedly a copy of Avicenna's *Canon of Medicine*, the enormous tome that formed the backbone of academic study – in the market square.[3] Needless to say, his colleagues ensured that he was quickly expelled from the town.

A companion during those years later painted an extremely unflattering portrait of the man.

> As to Paracelsus, while he was living I knew him so well that I should not desire to live again with such a man. Apart from his miraculous and fortunate cures in all kinds of sickness, I have noticed in him neither scholarship nor piety of any kind ... The two years I passed in his company he spent in drinking and gluttony, day and night. He could not be found sober an hour or two together, in particular after his departure from Basle.[4]

When he died in 1541, Paracelsus was a famous but not influential figure. Few of his works had even been published. Gradually, however, books by him appeared and stories about him multiplied. Soon he had become the founder of a new school of medicine. His followers fleshed out his philosophy and developed treatments on its basis. Some highly placed physicians championed his thought in the universities and it seemed for a while that the chemical philosophy might be able to overthrow Galenism. The Galenists fought back, and a pamphlet war developed that led to a ban on Paracelsian medicines at the university of Paris in 1615.[5]

It would be a mistake to see Paracelsus as a precursor of modern medicine. His thought was suffused with theological and mystical speculation. His main objection to Galen was that he was a pagan. Christianity, Paracelsus said, should have its own medicine that showed a proper understanding of God's work in the world. He reinterpreted the book of Genesis according to his own insights and also subscribed to the doctrine of signatures. He believed that God had shown which herbs and chemicals could be used as cures by giving them a recognisable signature that pointed towards the nature of the ailment they treated. Careful observation of nature was an important aspect of his thought, but the doctrine of signatures meant he often ended up looking for the wrong things. This demonstrates that almost no amount of observation will lead to significant headway in science unless it is buttressed by some sort of valid theory to which the observations can be applied. Meticulous empirical examination was standard practice in Greek and medieval medicine. It didn't stop doctors killing people.

Going Under the Knife

During the Middle Ages, it was the surgeons, whom you might expect to have been the real butchers, who achieved some of the most notable successes in saving lives. Professional physicians looked down on surgeons as manual workers because they used their hands. Even the frequent bleedings demanded by Galenic medicine were usually carried out by a barber-surgeon and not the physician himself.[6] Perhaps this lack of official respect gave the surgeons freedom to innovate. When it came to dressing wounds, they had sole charge of the process and by the thirteenth century were challenging the wisdom of the Greeks over the healing process. The Greeks had believed that inducing putrefaction was a necessary prerequisite for a wound to heal, rather than a way to encourage gangrene or at least leave a nasty scar. Instead, radical surgeons advocated cleaning wounds with wine, drying them and binding them shut immediately.[7] Another good and easily available antiseptic was urine. One doctor recalled how he saw a man's nose cut off in a duel. Picking up the severed organ, the doctor explained that he urinated on it and then successfully bandaged it back onto the victim's face.[8]

In fifteenth-century Italy, surgeons offered skin grafts as a way of healing burns and other serious wounds. Strictly speaking, this technique was a rediscovery because it had been used in ancient Rome. One particularly impressive application was rhinoplasty, perfected by Gaspare Tagliacozzi (1545–99) in the late sixteenth century. The procedure involved the complete reconstruction of the client's nose. There was more demand for this than we might think. At the time, syphilis was endemic and one of its more gruesome effects was to cause the sufferer's nose to rot away.[9] The medical textbooks preferred to avoid any references to the pox and instead noted that rhinoplasty was ideal for when the lost nose had been eaten by dogs.[10]

Tagliacozzi set about repairing a lost nose by cutting a flap of skin from his patient's arm and sewing it over the nasal cavity. The arm was strapped in place until the flap had grown onto the face, whereupon it was cut from the arm and remoulded into a new nose. The whole process could take months.

Not all of Tagliacozzi's surgical interventions were a success. A story was told soon after he died of a case where his patient was a nobleman from Brussels who wanted a new nose to replace the one he had lost in a duel. However, he was unhappy about having the necessary incision made in one of his arms. He suggested, therefore, that the strip of skin be taken from the arm of a servant rather than his own. The treatment progressed as usual except that it was the servant's arm that the patient had tied to his face. When the servant was released from this rather intimate bond to his master, the rest of the operation to reshape the replacement nose continued as usual. Everything seemed to be going well but then, after a year, the slave died from unrelated causes. At the same time the new nose, fashioned from the living flesh of the servant, began to putrefy and before long it dropped off.[11]

Such stories led Tagliacozzi's contemporaries to draw completely the wrong conclusion. Paracelsus's belief in magical threads uniting the world was hardly unusual in the sixteenth century. Like many apparently superstitious beliefs, this was based on impeccable if flawed logic and buttressed by apparent successes. In the case of Tagliacozzi's failed rhinoplasty, the obvious explanation was not that donated organs will tend to be rejected by their hosts, but rather that the new nose remained somehow attached to the servant after it had been detached from his arm. Thus, when the slave died, the nose died too. You could hardly ask for better proof of action at a distance and the attendant doctrine of sympathy. Again, this illustrates just how important it is that empirical results are linked to a reliable theory before they can be extrapolated into a general law.

The Beginnings of Human Dissection

Neither Galenic nor Paracelsian medicine could mount any serious attack on disease as long as doctors had no real idea what caused it. The theories behind both schools of medicine were so erroneous that it is an understatement to call them merely wrong. Among other things, the acquisition of true knowledge about disease required an accurate understanding of how the human body worked. Here, at least, the Middle Ages did provide the setting for a massive step forward, beginning in Italy with the introduction of the practice of human dissection. This would eventually destroy Galen's reputation by showing that even the basic building blocks of his theories were mistaken.

The introduction of human dissection in western Europe is one of the most surprising events in the history of natural science. It was practically unheard of in any other culture due to strong taboos against cutting up bodies and not giving them the proper respect demanded by tradition. In the ancient world, it was briefly allowed in Alexandria in the third century BC, perhaps because embalming bodies for mummification already involved the removal of most of the soft organs. Hence, cutting open dead bodies was not as shocking in Egypt as it might have been elsewhere. If we are to believe one ancient writer on medicine, the practice in Alexandria even extended to the vivisection of live prisoners.[12] However, any sort of dissection of humans was forbidden by the pagan Roman authorities. In the second century AD, Galen had to do all his work on animals, especially pigs and Barbary apes. This caused errors in his work on anatomy because he assumed that the physiology of animals would be mirrored in humans.[13]

When Galen's works were transmitted to the Arab world after 800, Muslims were not about to start dissecting humans. Although there was no direct prohibition against such activity in Islam that we know of, this is probably because it never occurred to anyone that it might be permitted. We do know of a Christian who lived under the Caliphate and dissected monkeys, but there is no hint that he ever got to try his hand on humans.[14] Unable to check the facts for

themselves, the Arabs assumed that Galen was right and did not realise that he too had never got further than using primates.

Once Galen became well-known in the Catholic West, it was
natural for everyone to continue assuming that he was correct in
almost every detail. Then, in northern Italy towards the end of the
thirteenth century, something revolutionary happened – human dissections restarted. Many modern scholars think this had something
to do with Bologna's pre-eminent position as the place to study law.
The first dissections may actually have been post-mortem examina-
tions intended to ascertain cause of death for legal purposes. Even

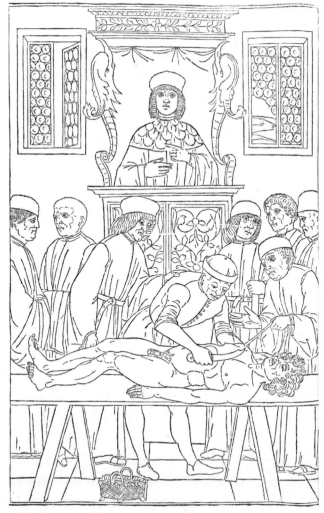

11. Woodcut
showing a dissection
from a fifteenth-
century medical
textbook

Pope Innocent III is on record ordering the forensic examination of a murder victim.[15] These autopsies broke the taboo against cutting up human bodies and, shortly afterwards, the medical faculty in Bologna was carrying out dissections as part of doctors' training.

The anatomy theatre that now stands in Bologna dates from the seventeenth century and incurred serious damage during the Second World War. However, its design is similar to that of the earlier theatres that developed in the fourteenth century for educational dissection. In the centre of the room is a marble slab upon which the cadaver was laid. Surrounding it is an ascending array of benches so that everyone was able to get a good view. Set over the benches is what can only be described as a throne, complete with a carved wooden canopy supported by two statues of flayed men. The professor sat on this seat when he was delivering his lectures and did not actually do the dissecting himself. Instead, his assistants carried this out while he stayed well away from the messy business of actually chopping up the body. As he read out sections of his textbook, the assistants would point out each organ. Alternatively, the professor might be armed with a long stick with which to point out the features of the cadaver's interior as his assistant revealed them.

In 1316 a Bolognese physician, Mondino dei Luzzi (who died around 1326), wrote a manual on how to proceed with human dissections which became a standard work through the rest of the Middle Ages. Mondino had gained his expertise from carrying out post mortems, most interestingly on two pregnant women.[16] He was not universally popular, however. His slight deviations from Galen were too much for at least one German university, which would not allow his textbook to be used.[17]

One of the more prevalent myths about the medieval Church is that it opposed human dissection. As we have seen, most societies and cultures did have a strong aversion to this activity, so it is surprising that the Church allowed dissections to go ahead with barely a whimper of opposition. A papal bull of 1300 entitled *De Sepulturis* is often cited as evidence for a Church prohibition as it forbids the boiling of bodies. This practice had become common during the crusades when those who died on campaign wanted to be interred in the family tomb back home. Boiling separated flesh from the bones so that they could be easily repatriated for burial. The papal bull did have the unintentional effect of preventing anatomists from boiling heads to reveal the structure of some bones under the skull. Mondino admits in his manual:

> The bones which are below the basilar bone cannot be well seen unless they are removed and boiled, but owing to the sin involved in this, I am accustomed to pass them by.[18]

If the Catholic Church had really objected strongly to human dissections, they would not have rapidly become part of the syllabus in every major European medical school.

Human dissection in the late Middle Ages was intended to be pedagogical rather than for the purposes of research. This meant that if any discrepancies with Galen's account of anatomy were noticed, they were not followed up. Instead, students tended to interpret what they saw in terms of what was in their books. Senior professors may have had the experience and prestige to question Galenic orthodoxy but they did not actually get their hands dirty with the business of cutting up bodies. They were paid good fees to teach and to treat live patients so they had no reason to deal with dead people. Then, in the first half of the sixteenth century, several anatomists began to question what they really saw inside our bodies.

Andreas Vesalius

Andreas Vesalius (1514–64) is the most celebrated of all anatomists. He came from the Spanish Netherlands and studied at the new university in Louvain before moving on to Paris to study as a physician. He completed his studies in Italy, then the country in which the premier medical schools were found. Once qualified, he continued to work as a lecturer in surgery. His method of teaching was rather unconventional because rather than seating himself in the professor's throne while flunkies dissected the cadaver, Vesalius got down in the pit and did it himself. Because of his showmanship and enthusiasm his lectures were extremely popular, and they also made him realise that Galen's books were not the last word on human anatomy. He suggested that the key to understanding Galen's omissions and mistakes was that the ancient master had only dissected animals and not human beings.[19] Thus, Vesalius himself could complete Galen's work by carrying through his project to its logical conclusion – investigating man himself, the summit of God's creation. Vesalius did not wish to refute Galen's work but to perfect it.

Vesalius had all the necessary skills to carry out his aims: he had a good eye, he was a fine draughtsman and also an astute businessman. His masterwork was *On the Fabric of the Human Body*, first published in 1543. Vesalius provided the text and the illustrations were based on his own drawings. The actual woodblock artwork was supplied by the workshop of the great Venetian artist Titian (d.1576).[20] As well as their anatomical detail, the pictures showed great flair in arranging the flayed and opened bodies in classical poses as if they were still living. The book's fame and success was as much down to the fantastic artwork as its indubitable scientific value. The entire work was a paean to the magnificent handiwork of the Creator as uncovered by his servant Vesalius.

Of course, *On the Fabric of the Human Body* is not perfect. For example, its treatment of the female genitalia leaves a lot to be desired, being based on observations made with a sixteenth-century eye. Even today, we speak of men producing semen (from the Latin for seed) and women being fertile. This language derives from the ancient view that all a woman did was to provide a place where the man could plant his seed until it grew into a baby. The reproductive organ of a woman was thought of as passively accepting what the man thrust into it. *Vagina* is Latin for the sheath of a sword. Consistent with this view, Vesalius's illustration of the vagina is just an inverted penis designed to furnish a comfortable berth for the man's organ. It was not until the work of his student Gabriele Fallopio (1523–62) that the eponymous tubes were discovered connecting the uterus to the ovaries.[21] A human egg was not observed until the nineteenth century,

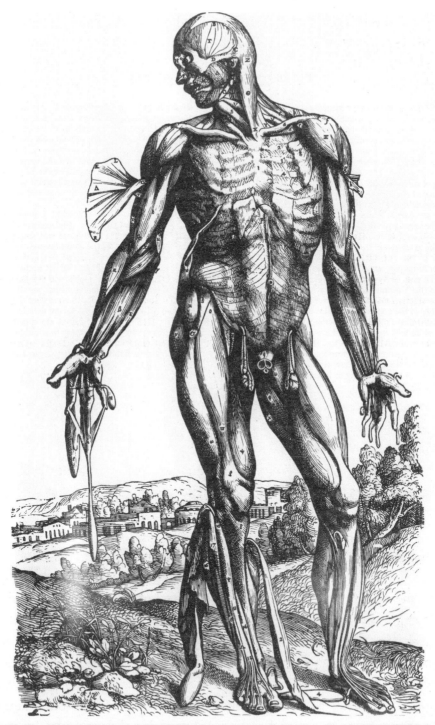

12. A woodcut of a standing flayed figure produced by the studio of Titian for Vesalius's *On the Fabric of the Human Body* (1543)

when it was finally realised that the seed came from the woman and fertilising it was the function of the man's sperm.

13. A woodcut of a uterus from Vesalius's
On the Fabric of the Human Body (1543)

For anatomists, the shortage of bodies was a major problem. Vesalius openly admitted that he resorted to grave-robbing. In one case, he flayed the skin from the body of a dead woman so that her relatives, who had gone to court to retrieve it, would not recognise her. And, notwithstanding the ancient papal prohibition, he had no compunction boiling up bodies to remove the flesh from the bones.[22]

The sixteenth century being what it was, anatomists could count on the occasional free cadaver. Vesalius wrote about how he had discovered the corpse of a criminal who had been burnt at the stake:

> We went to the place where, to the great advantage of students, all those who have suffered the death penalty are displayed on the public highway for the benefit of the rustics … I climbed the stake and pulled away a femur from the hip bone. And when I pulled at the upper limbs, the arms and hands came away. I took the legs and arms home in several secret journeys leaving the trunk and the head. The thorax was tied with a chain high up and in order to take it, I allowed myself to be shut outside the city at nightfall.[23]

Vesalius went on to relate how he climbed up the stake in the darkness to acquire the prizes. He hid the bones nearby until he had an opportunity to smuggle them into the city. Fallopio did not have to sink to such subterfuge. A local ruler gave him a live prisoner to experiment on, although mercifully the anatomist euthanised the subject with opiates before he got to work.[24] It was probably a kinder end than the victim would have expected.

Vesalius's success and popularity with students did not go down well with his fellow faculty members. Nor were some of them pleased to be told that when it came to anatomy, they were not just wrong but had been too lazy to check the facts for themselves. Jacobus Sylvius (1478–1555), a physician from Paris, had a novel reason for rejecting Vesalius's discoveries. He claimed that we should believe the works of Galen were true even when they conflicted with the evidence of our own eyes.[25] Sylvius's extreme humanism had led him astray. Like many humanists, he was a bookworm and an admirer of the classical world who could not accept that Europe had long ago overtaken ancient civilisation.

Eventually, Vesalius left Padua and was employed by Philip II (1527–98) of Spain as his personal physician. This lifestyle did not please the anatomist and he had no opportunity to continue his research. As simply resigning from a royal appointment was out of the question, he contrived to escape by obtaining permission to go on a pilgrimage to Jerusalem.[26] Tragically, his ship was wrecked on the return voyage and Vesalius drowned.

Persecuted Anatomists?

Shortly after Vesalius's death, rumours were circulating that he had been sent on his pilgrimage after a run-in with the Spanish Inquisition and this was later used as further evidence to paint the Catholic Church as an opponent of dissection. In fact, there is no evidence at all that the Inquisition was involved in Vesalius's wish to leave Spain.[27] He was, after all, a man of great piety who saw his life's work as a way of glorifying God. Vesalius suffered from no religious persecution. We have already seen that one of his fellow anatomists, Michael Servetus, was not so lucky, even though his dreadful fate had nothing whatsoever to do with natural science.

Besides producing two highly regarded editions of the ancient *Geography* by Ptolemy, Servetus's scientific fame rests on his study of the heart and lungs. Galen's picture of the system by which blood is transported around our bodies suffered from several major flaws. For instance, he mistakenly believed that veins and arteries were not connected at any point except through the heart. This was hardly his fault. He lacked any sort of magnifying instruments with which to observe some of the tiny capillaries in the muscles and lungs that link the veins to the arteries. Instead, he postulated the existence of pores in the thick interior wall of the heart that separates the left and right ventricles. Because the veins and arteries were not connected in Galen's schema, there was no complete circulation of blood. He believed that blood was being continually created and destroyed by the body. It was produced by the liver, he claimed, and carried around the body where it was used up as fuel.

Historians have long been puzzled about why Galen, with all his experience of the anatomy of the higher mammals, got the function of the heart and circulatory system so utterly wrong.[28] It is a difficult question to answer but serves as a warning that to look at something completely objectively is almost impossible. Galen had his own axes to grind with the philosophical schools of his time and this made him no less vulnerable to misconstruing evidence than Christians might be due to their preconceptions.

Michael Servetus was troubled by the inconsistencies in Galen's system. He noted that Galen's pores in the heart's interior wall must be very small and that the pulmonary artery between the heart and lungs was very large. He correctly deduced that this artery does not just carry the blood that the lungs need to function as Galen thought, but that all the blood in the body passes through it. Air, he thought, might be absorbed into the blood in the lungs before being transported around the body. Unfortunately, he made this observation as an aside to his theological book *The Restoration of Christianity* in order to demonstrate his views about the spiritual qualities of the air. While the theory may be garbled, Servetus is certainly correct in his observation. 'By a very ingenious arrangement', he wrote,

> the subtle blood is urged forward by a long course through the lungs … then in the pulmonary vein it is mixed with inspired air and through expiration it is cleansed of its sooty vapours … The notable size of the pulmonary artery confirms this.[29]

His insight, hidden within what was considered a notorious work of heresy, did not gain much exposure and was later independently reached by the Italian Realdo Columbo (c.1515–59).[30] Even Servetus was only repeating work done by a Muslim physician called Ibn al-Nafis (1213–88) in the thirteenth century.[31] In the end, it was through Columbo's work that the idea became well known among European anatomists.

William Harvey and the Heart

With their exploration of the human body, it was anatomists rather than doctors who chipped away at the Galenic edifice, even if many of them never really meant to damage it. But at the start of the seventeenth century, there was a breakthrough that dealt a crippling blow to Galen's reputation as a reliable source on the workings of the human body – the discovery of the circulation of the blood by William

Harvey (1578–1657). He showed that Galen was not just wrong in the details. His entire system explaining the basic mechanics of life was flawed. And, as Harvey's experiments on the heart were done on animals, Galen lost the defence that he had no access to human cadavers.

William Harvey began his education at Cambridge University where he attended Gonville and Caius College in the 1590s. John Caius (1510–73), the London physician and translator for whom the college was named, had refounded it a generation earlier. Caius was an old school Galenist who devoted most of his scholarly energy to preparing new critical editions of the ancient master's works. As medical instruction at Cambridge was rather thin on the ground, Caius had also endowed a scholarship that allowed a promising student to go to Padua to finish his course. Although Padua was in the territory of devoutly Catholic Venice, the Venetians did not let differences of religion deter fee-paying students, and in any case Harvey had Catholic sympathies. After qualifying as a medical doctor in 1602, he returned to England and used his prestigious degree to start a respected medical practice. He eventually came to count King James I (1566–1625) among his clients.[32]

Harvey confined his research on the heart to his spare time. His wife must have been an extremely forbearing woman who tolerated her husband's habit of vivisecting dogs.[33] His key insight was that both the heart and veins contained valves, which prevented blood from moving in the wrong direction. This might be where Galen was lacking in background knowledge because, while valves were well understood by the seventeenth century, they do not appear to have been common in Roman technology. Harvey also described the heart as operating like a machine. He compared its workings to rapidly turning gears or a flintlock musket.[34] Later in his life, he witnessed a fire engine spraying water through a hose. The water was ejected by a pump powered by the firemen and Harvey noted the analogy with the heart pumping blood around the body.[35]

Despite these modern touches, he couched his analysis within the Aristotelian tradition.[36] He assumed that the heart had been designed by God, and set out to discover what it had been created to do. Echoing Hermetic thinkers he declared that the heart 'deserves to be styled the starting point of life and the sun of our microcosm, just as the sun deserves to be styled the heart of the world.'[37]

Today, our understanding of the heart and circulation is still based on Harvey's work. When blood leaves the heart through a pipe called the aorta, it is full of oxygen which makes it bright red. The oxygenated blood is carried all the way around the body by the arteries, which in turn branch into ever-smaller vessels until they become too tiny to see with the naked eye. These microscopic capillaries are so narrow that the blood cells can only get through them in single file. The capillaries pass the blood through the tissues of the body where the oxygen is unloaded. They then carry the deoxygenated blood, now a purple-blue colour, into broader veins. These connect together like tributaries of a great river, which eventually flows back into the heart.

The heart itself is divided into two halves, separated by a thick impermeable muscular wall. In each half there are two interconnected chambers, an atrium and a ventricle. The atriums provide a reservoir for blood that is waiting to be pumped by the respective ventricles. The blood that has circulated around the body flows into the right-hand atrium and is then pumped by the right ventricle to the lungs through the pulmonary artery. There it is re-oxygenated and returns to the left atrium to be pumped back through the aorta by the left ventricle.

For the reader who can cope with the gory descriptions of vivisections, the logic of Harvey's *On the Motion of the Heart* (1628) remains impressive. Calculating how much blood the heart was moving around the body, he found that it was an amount that far exceeded the quantity of blood that could realistically be produced by the liver. It was clear that the body reused blood and, as the valve meant it could only move in one direction, it must be circulating. Granted, it was still impossible to trace the full circuit as capillaries remained invisible without the aid of a microscope, but Harvey's treatise was convincing enough to persuade many. Not everyone was willing to be swayed though, and he complained of losing patients after the publication of his book because they were not willing to trust their health to a man with such novel ideas.[38]

Sadly, although the authority of Galen had been seriously weakened by the likes of Paracelsus and Harvey, there was as yet little effective improvement in clinical practices. Undermining the theoretical basis of medicine was no good if there were no new theories to replace them which could lead to new treatments. The four humours remained the basis of diagnosis because there was no alternative. Attempts to use Harvey's insights to carry out blood transfusions fell foul of what we now know are differing blood groups. If a patient is transfused with the wrong group, he or she will die. As doctors had no conception of what a blood group was, let alone a way to identify them, the treatment was often fatal for no apparent reason.[39]

Given the demise of the magical worldview in polite society, sympathy and other alternatives also fell from favour. Protestants even tended to reject religious miracles. Somehow, scientific medicine had seen off its main rivals while actually being less effective than either of them. Thus, the history of medicine until the mid-nineteenth century, with the significant exception of smallpox vaccination, is a history of failure. Most doctors were comfortably oblivious to the damage they were doing by bleeding their patients, but there were a few early critics of the practice. For instance, in 1806 Thomas Jefferson (1743–1809), third president of the United States and a notable polymath, wrote morbidly:

Harvey's discovery of the circulation of the blood was a beautiful addition to our knowledge of the animal economy, but on a review of the practice of medicine before and since that epoch, I do not see any great amelioration which has been derived from that discovery.[40]

Thankfully, in the field of physics and astronomy, the natural philosophers of the sixteenth century were able to build much more profitably on the foundations laid in the Middle Ages.

CHAPTER 17

༺∽∽༻

Humanist Astronomy and
Nicolaus Copernicus

There was nothing that humanists enjoyed more than a good squabble. They accused each other not of heresy but of scholarly incompetence, which was much worse. Valuable patronage was at stake and if a humanist could not gain a position at court, he could at least besmirch the reputations of those who had.

One of the most fearsome debates took place between George of Trebizond (1395–1486) and John Bessarion (1403–72) over the question of whether Plato or Aristotle represented the peak of Greek philosophy. Both men were native Greeks who made their living in Italy in the mid-fifteenth century. George of Trebizond, often known by his Latin name of Trapezuntius, actually came from Crete[1] but Bessarion really was from Trebizond, a Greek enclave on the south-eastern shore of the Black Sea.[2] Today, he is best known as a collector of manuscripts. The San Marco Library in Venice holds over 1,000 books which he donated to the city on his death.[3]

Italians were clamouring for translations from ancient Greek and Trapezuntius supplied a steady stream of them. He worked quickly but carelessly, confident that few would have the necessary language skills to catch him out. Bessarion was one man who did. Although the root of their disagreement was over philosophy, Bessarion took great pleasure in compiling long lists of the errors made by his rival. Trapezuntius's shame was compounded by his inability to be diplomatic or admit he had made mistakes. While Bessarion's charm and easy manner saw him raised to the rank of cardinal, Trapezuntius had to flee to Naples to escape the scandal.

The New Astronomy of George Peurbach

When Constantinople fell to the Turks in 1453, Bessarion, who regarded the city as his home, devoted himself to trying to begin a crusade to liberate it. To do this, he needed the support of the Holy Roman Emperor and so travelled to Vienna in 1460 to meet him. To his disappointment, the mission was not particularly successful and no crusade resulted.

However, while he was in Vienna, Bessarion introduced himself to George Peurbach (1423–61), a friend of Nicholas of Cusa and the Emperor's court astrologer. Peurbach was a rare example of a humanist who was actively interested in mathematics.[4] The cardinal's mind was still on the battle with Trapezuntius and he thought that Peurbach's astronomical and literary expertise would allow him to open a second flank in his struggle. In 1450, Trapezuntius had published a new translation of Ptolemy's *Almagest* directly from the original Greek into Latin. It was supposed to supersede the medieval version that had been rendered from Arabic back in the twelfth century. As usual, Trapezuntius's slapdash approach let him down. The *Almagest* is a fearsomely complicated treatise that requires an accurate translation to be of any use at all. Demonstrating neither precision nor mathematical skill, the new version was derided a failure.[5] Bessarion decided to compound Trapezuntius's embarrassment by asking Peurbach, who possessed all the necessary skills in abundance, to produce his own summary of the *Almagest*.[6] Peurbach, who had already written an updated version of the medieval *Planetary Theories*, set to work at once.

He proceeded quickly and had completed six of the thirteen books by the time he died the following year. On his deathbed, he handed the project over to his student and collaborator Johann Müller, better known as Regiomontanus (1437–76). Both master and pupil were Germans who had travelled to Vienna in search of an education and patronage. They shared a desire to make astronomical predictions as accurate as possible and had no trouble in reconciling that aim with being an official astrologer. After all, without accurate tables, astrology was impossible. Peurbach had already found that one lunar eclipse had occurred eight minutes earlier than the standard tables of the time predicted.[7] Given that the tables in question had been prepared in 1252, their accuracy in predicting events two centuries later seems remarkable to us, but for Peurbach and Regiomontanus it was not good enough.

Regiomontanus completed the summary of the *Almagest* in 1463 and, with his master dead, joined the household of Cardinal Bessarion in Rome. He spent the rest of his life trying to correct Greek and Arab astronomy by making new observations and bringing to bear the most advanced mathematical techniques. There were rumours on his death that he had been poisoned by agents of Trapezuntius for his part in Bessarion's schemes.[8] If this is true, it would not be untypical of the politics of Renaissance Italy.

On Triangles, a guide to trigonometry that Regiomontanus wrote in 1464, is often held up as the origin of that discipline. In fact, Richard of Wallingford had composed a similar treatise back in the early fourteenth century drawing on Arab and Greek archetypes.[9] Regiomontanus may not have used Richard of Wallingford's work directly but he certainly owned a rough copy of the Englishman's treatise on how to make his innovative astronomical instrument the Albion, which contains plenty of material on trigonometry.[10]

Looking at the Sky

Ptolemy, in agreement with Aristotle and almost everyone else, thought that the earth was stationary at the centre of the universe and that the planets, including the moon and sun, orbited around the earth. No one in medieval Europe disagreed with this, Nicole Oresme's suggestions about the rotation of the earth notwithstanding. On the other hand, the matter of exactly how the planets were arranged and in what order was open to question.

Today, many people would be hard pressed to identify a planet in the night sky. When it is close to the earth, the easiest to spot is Mars because it is bright and clearly red in colour. Venus is even brighter but keeps lower in the sky and so is often more difficult to find. It appears just after sunset or just before sunrise, giving it the title of the morning or evening star. Jupiter and Saturn are both often visible, although the latter is hard to differentiate from a normal star unless you know what you are looking for. Mercury is quite dim and stays even closer to the horizon than Venus. Until the eighteenth century, these five were the only known planets.

We now know that the reason why Mercury and Venus are only visible around sunrise or sunset is that both planets orbit closer to the sun than the earth does. This means we only see them when we are looking in the direction of the sun, but obviously they are invisible when the sun is actually in the sky. In the ancient and medieval worlds, the way that these two planets were tied to the sun was a mystery. A few people suggested that they did orbit the sun rather than the earth, although this was not a popular view. The perturbations of the outer planets – Mars, Jupiter and Saturn – also share the same duration as the sun's orbit and there seemed to be no reason why this should be so.

Since Pythagoras, the ancient Greeks had believed that the planets moved with an unchanging and uniform circular motion.[11] According to their worldview the heavens were perfect and so the motion of the planets had to reflect this. The planets themselves, oblivious to this theory, did not behave themselves. While they supposedly orbited the earth in circles, they did not travel across the sky with a uniform speed. Worse still, they could clearly be observed to move backwards from time to time. Finally, the brightness of the planets (and the moon's size) varied over the course of months or years. This should not happen if they stayed the same distance away from the earth.

Ptolemy had presented answers to these problems but his solutions were very convoluted. The idea of the planets sailing serenely through the heavens was lost in a fog of fiendish geometry. In essence, his method was to assign to each planet several different uniform circular motions that, when added together, gave a very close approximation to the observed movements. His two principal mechanisms were eccentrics and epicycles. An eccentric orbit was one that did not centre on the earth, meaning that the planet was not always the same distance away from it. An epicycle was a smaller circle about which the planet orbited that was, in turn, carried around the larger eccentric orbit. By manipulating the speeds at which the various circles rotated, Ptolemy was able to model the movements that he observed in the sky, including the backwards motion and changes of speed. He could also explain the differing brightness of the planets and size of the moon by the fact that the eccentrics and epicycles carried the planets closer and further away from the earth. To get a really exact fit between these models and his observations he also had to use a device called an equant which is simply too complex to describe and explain here.[12]

Many natural philosophers hated Ptolemy's system because it made the heavens such a muddle. The original principle of uniform motion in circles disappeared in a blizzard of geometrical constructions. This was a particular embarrassment for Jews, Christians and Muslims because they believed that God had created the world as perfect. How could he have made the heavens when they were in such a jumble?

In Muslim Egypt the Jewish philosopher Moses Maimonides, who would later inspire Thomas Aquinas, wrestled with this matter in his influential *Guide for the Perplexed*. He eventually conceded that astronomers could do no more than find a hypothesis that fits the observed motions of heavenly bodies. He hoped that someone would come after him who could show how the planets actually moved.[13] John Buridan rejected epicycles as physically ridiculous but found eccentric orbits acceptable.[14] Still, this left him unable to reconcile observation with theory. Sadly, no one else could come up with an alternative that did away with Ptolemy's constructions and still reproduced the observed path of the planets in the sky.

The physical construction of the heavens was also a source of debate. The ancient Greeks envisaged a series of shells into each of which a planet was embedded. Peurbach combined this proposal with Ptolemy's epicycles to postulate a universe of solid crystalline spheres. Each sphere had to be thick enough to accommodate its planet at both its minimum and maximum distance from the earth. Assuming no space in between the spheres, Arab astronomers had calculated the total radius of the universe from the centre of the earth to the fixed stars to be 90 million miles (roughly the distance we measure between the sun and the earth today).[15] Figures of around this order of magnitude had been determined in antiquity and were well known in the Middle Ages.[16] This was merely the minimum size of the universe assuming that there were no gaps between the planetary spheres. No one can call the medieval universe small, even if our own is vastly larger.

Peurbach and Regiomontanus revealed the problems with Ptolemy's system but did not suggest an alternative. Peurbach also realised that the differences in the apparent size of the moon were not as great as Ptolemy's model predicted. As thousands of their printed books were sold to students at universities throughout Europe, the insights of Peurbach and Regiomontanus spread. Their summary of the *Almagest* might have had its genesis in a literary dispute and astrologers' need for accurate planetary tables, but that did not stop it being the cutting edge of astronomical theory. The shortcomings of Ptolemy, long recognised, were now clear for all to see and several astronomers began work on alternatives.

The Life of Copernicus

In 1543 Nicolaus Copernicus, a Polish clergyman nearing the end of his life, finally allowed his contribution to the astronomical debate to be published. It was a book called *Revolutions of the Heavenly Spheres*. In it, Copernicus claimed to have demonstrated that the sun was the centre of the universe and the earth orbited around it along with the other planets. The impact of this radical idea was softened slightly by the fact of his friend, Andreas Osiander (1498–1552), adding a fore-word that explained the theory was only meant to be a hypothesis and was not presented as a fact.[17] What Copernicus thought about this late addition is not recorded, since he died as his book came off the press. Osiander had penned his preface because he found the idea of the earth rushing though space at high speed while simultaneously spinning on its axis ridiculous and he knew Europe's intellectual elite would agree. There was no question of ecclesiastical pressure being brought to bear and no chance that the church would seek to suppress the book. After all, it was dedicated to Pope Paul III (1468–1549) himself. This was the done thing at the time when all scholars needed patronage, and a great number of books were presented to princes and kings complete with gushing pronouncements on the royal virtues. Paul III had been the dedicatee of another book presenting a reformed model of astronomy five years before Copernicus's.[18] All the signs are that the Pope appreciated the flattery and read neither of them.

The real problem with stating that the earth is moving was that almost all of the available evidence and all expert opinion was against it. We saw in chapter 12 how John Buridan and Nicole Oresme had suggested a rotating earth in the fourteenth century. Oresme had concluded that there was no physical reason to reject the hypothesis, but no positive confirmation for it either. On the other hand, physical evidence against the earth orbiting the sun did exist. When we look up at stars each night, we see that they are fixed in the same patterns, called constellations. If the earth were in motion, we should expect the stars would change their relative positions as the earth followed its orbit. For example, a star that was directly overhead at midwinter should be off to one side at midsummer.

To grasp this phenomenon, called stellar parallax, a bit better, remember how many children are convinced that the moon is following them when they are travelling by car. This is because, however far the car moves, the direction and size of the moon do not change at all. In comparison, all nearby everyday objects move across our field of vision as we approach and then pass them. The only exception to this rule is another car that is following our own. So it seems to a child that the moon is tracking his or her own motion. Of course, the real reason the moon's size and direction appear fixed is that it is an immense distance away compared to the everyday scale of road journeys.

Substituting the moving earth for a moving car, we note that even as we travel over the entire distance of the earth's orbit, the direction and brightness of the stars remain the same. This must mean that either they are preposterously far away, or else, as everyone in the Middle Ages thought, the earth is not moving after all.[19]

People often imagine Copernicus as a lone revolutionary genius working at the fringes of Europe. Poland, however, was not a backwater at the time but part of a vast commonwealth whose king, Sigismund I (1467–1548), was one of the most powerful monarchs of Europe, a correspondent of Erasmus and patron of the arts. Born Nikolaj Kopernik, Copernicus adopted the common habit of taking a Latin name to emphasise his links with the cosmopolitan and international elite. He was brought up in the house of his uncle, the bishop of Ermeland, before studying at the university of Cracow from 1491. Then, five years later, he travelled to Italy to continue his education. He spent time at the ancient universities of Bologna and Padua before receiving a degree in canon law at Ferrara.[20] In all, he spent ten years in Italy when he would have been exposed to the fashionable Platonic works of Marsilio Ficino. At the same time, the books of the medieval natural philosophers were pouring off the presses in Venice for use at the nearby universities. Copernicus would also have been taught astronomy from Peurbach's textbooks, as well as geometry from Euclid's *Elements*.

On his return to Poland, Copernicus embarked on the comfortable and leisurely life marked out for a well-educated man of his class. His uncle had presented him with a canonry at Frombork Cathedral for which he was paid a substantial stipend and not required to do very much work. These canonries were extremely popular with scholars because they provided a steady income without onerous duties attached. You may recall that Peter Abelard had had one back in the twelfth century. So did Nicole Oresme and Nicholas of Autrecourt 200 years later. Copernicus had the necessary family connections to expect a post of this kind, and it meant he had plenty of time to indulge in his full-time hobby of astronomy.

Copernicus first circulated some ideas about a heliocentric universe in manuscript form some time after 1507.[21] His contemporaries looked upon his proposals as interesting but certainly wrong. After this failure, he went back to the drawing board and tried to produce a viable cosmology that would withstand the scrutiny of his peers. He read a copy of Regiomontanus's treatise on trigonometry and found it contained the mathematical techniques that he needed to perfect his work.[22] The result was *Revolutions of the Heavenly Spheres* which remains one of the great triumphs of human genius.

Unlike some of the famous books in the history of science, this is not one that most people can just pick up and read. The bulk of the book contains a complete reworking of Ptolemy's system using the cutting-edge geometry of its day. Even though geometry was part of the basic university education given to all candidates for Master of Arts degrees, Copernicus's book was still very advanced.

The heliocentric system proposed by Copernicus, despite all his years of effort, was not very much simpler than Ptolemy's had been. Copernicus managed to cut down the number of epicycles, and the heliocentric system also explained why Mercury and Venus always stay close to the sun. As we have seen, they must appear to do so because they are orbiting the sun more closely than we are.

Unfortunately, Copernicus could not provide any direct demonstration that the earth orbited the sun and not vice versa. Conversely, the evidence against the earth moving still seemed strong. The stars remained obstinately stationary. This lack of stellar parallax meant that the universe was either much, much larger than anyone had previously thought (and they already thought it was extremely big), or Copernicus was wrong. He resolved the problem with some intellectual sleight of hand. It was recognised that because the universe was so large, it appeared the same from wherever on earth you looked at it. Copernicus simply said that it was actually so indescribably huge that it appeared the same from wherever in the earth's orbit you looked at it.[23] To do this he had to increase the size of the universe by about a factor of a billion. This explanation offended the principle of parsimony – the idea that nature does nothing unnecessarily. Copernicus was making the universe far bigger than it needed to be just so that it would fit with his theory. We know today that he was right, but his reasons were not convincing.

The Sources for *Revolutions of the Heavenly Spheres*

Where did Copernicus find the idea that the earth orbits the sun? And why was he willing to entertain such an absurd suggestion in the first place? He explained in the preface to his book that he was as dissatisfied with the then-current models of the universe as many of his contemporaries. He wanted a model of the world machine worthy of its Creator whom he called 'the best and most orderly workman of all.'[24] As far as Copernicus was concerned, Ptolemy's system was too messy to have been designed by God. So, he claimed, he read all the books of philosophy that he could lay his hands on in search of an alternative.

One obscure Greek astronomer, Aristarchus of Samos (c.310–230BC), had believed that the earth orbits the sun. It would seem obvious that Aristarchus represented the best chance to impart some ancient legitimacy to Copernicus's ideas. But instead he excised all mention of him from the final version of *Revolutions of the Heavenly Spheres*.[25] Rather, he quoted from various followers of Pythagoras, even though they did not actually support his theory. This was because it was the neo-Platonism of Marsilio Ficino that most strongly influenced him. Platonists looked back to Pythagoras as the font of wisdom, and so Copernicus quoted from his followers rather than from Aristarchus of Samos. Ficino was aware that the sun was a great deal larger than the earth or any of the other planets, 160 times larger by his reckoning. 'All celestial things appear by divine law to lead back to the one Sun, the Lord and regulator of the heavens', he wrote.[26] Besides Ficino, Italy was awash with occult theories about the sun, placing it figuratively, if not literally, at the centre of the universe. One neo-Platonic sage, Francisco Giorgi (1466–1540), referred to the sun as the 'heart of the heavens' in a book published in 1525.[27] This is too late to have stimulated Copernicus directly, but it is indicative of the zeitgeist in which he was educated. We even find the Pole citing Hermes Trismegistus, who wrote that the sun was a visible god.[28]

As for his technical arguments for the rotation of the earth, Copernicus appears to have lifted them straight out of the work of John Buridan. They both suggested that the rotation of the earth is more parsimonious than the rotation of the entire universe. And compare these two passages – here is Buridan writing in about 1350:

> If anyone is in a moving ship and imagines that he is at rest, then should he see another ship, which is truly at rest, it will appear to him that the other ship is moved ... And so, we also posit that the sphere of the sun is everywhere at rest and the earth in carrying us would be rotated. Since, however, we imagine we are at rest ... the sun would appear to us to rise and then to set, just as it does when it is moved and we are at rest.[29]

Copernicus wrote 200 years later:

> When a ship sails on a tranquil sea, all the things outside seem to the voyagers to be moving in a pattern that is an image of their own. They think, on the contrary, that they are themselves and all the things with them are at rest. So, it can easily happen in the case of the earth that the whole universe should be believed to be moving in a circle [while the earth is at rest].[30]

As late as 1516, Buridan 'still ruled the subject of physics at the university of Paris.'[31] And commentaries on his work were produced by several of the masters at the university of Cracow at the time that Copernicus was studying there.[32] True, Buridan and Oresme had only discussed the rotation of the earth, but their reasoning stood just as well to justify the earth orbiting the sun. Even if Copernicus did not have direct access to Buridan's work, exactly the same argument appears in Nicholas of Cusa's *On Learned Ignorance*.[33] Nicholas of Cusa even studied at Padua in the century before Copernicus arrived there.

Theorems developed by earlier Muslim astronomers are also included in *Revolutions*. For example, a geometrical construction of a Persian astronomer, Nasir al-Din al-Tusi (1201–74), which has been dubbed the Tusi couple by historians, was used to generate a linear motion from two circles. Copernicus's diagram of the couple bears a remarkable resemblance to Arabic manuscripts, quite apart from his use of the same theorem.[34] Furthermore, his model for the moon is exactly the same as that developed by the Syrian Ibn al-Shatir (d.1375).[35]

Copernicus may have learnt about these ideas while he was travelling in Italy. Unfortunately, historians have not been able to determine exactly where he came across them. It is unlikely that he actually read the Arabic treatises containing the theorems, because he could not understand the language and there is no record of them being translated. So the route by which Muslim astronomy found its way into this seminal work of western science remains a mystery.[36]

Thus, Copernicus was not a lone genius who rediscovered ancient wisdom. He was part of the long-running European school of natural philosophy that went back to William of Conches and Adelard of Bath, cross-fertilised by the parallel occult and Arabic traditions. That is not to say that heliocentricism was not radical and new, but *Revolutions of the Heavenly Spheres* is written in the language of medieval thinkers and uses their arguments. If John Buridan had picked it up, it would have made perfect sense to him, far more than it does to us today, even though he probably would have respectfully disagreed with its thesis.

The Impact of Copernicus

The reaction to Copernicus's work was initially muted. Its difficult mathematical content meant that not many people read it, and his ideas appeared at first glance to be absurd. Martin Luther, over dinner one day, said he thought Copernicus's idea was just a newfangled theory designed to attract attention.[37] Astronomers were more impressed. Although they rejected the movement of the earth, they realised that Copernicus's system was easier to apply than the existing alternatives. A new set of astronomical tables was produced in 1551, calculated using the methods in *Revolutions of the Heavenly Spheres*, which rivalled the popular medieval version. Even the Catholic Church found that Copernicus had his uses in the area of calendar reform.

Senior churchmen had been worried about the calendar for centuries. Throughout the Middle Ages, Europe had used the 365-day year, with a leap year once every four years, as instituted by Julius Caesar (100–44BC). Unfortunately, this is not precisely correct because it makes the year, on average, about eleven minutes too long. That might not sound like much, but by 1500 the error had accumulated to almost ten days. The immediate concern of the Church was that Christian festivals were being celebrated at the wrong time, but there were other, more practical disadvantages as well. In the fifteenth century, the cardinals Pierre D'Ailly and Nicholas of Cusa had supported the call for improvement. But nothing was done until the aftermath of the Protestant Reformation, when the calendar became one of many things that were changed as the Catholic Church put its house in order.

In the 1570s, Pope Gregory XIII set up a commission to decide on the method and implementation of reform. Because his system made calculations easier, the commission used Copernicus rather than Ptolemy to determine the length of the year. After nearly ten years of consideration, a report was presented to the Pope. The month of October 1582 lost ten days and the calendar was realigned with the solar year. To prevent them drifting apart again, the new calendar omitted a leap year every three centuries out of four.[38] Unfortunately, by the time the Catholic Church had sorted out its calendar, Protestant countries no longer accepted papal dictate. Consequently, they only gradually came to adopt the Gregorian reckoning. Today this causes historians no end of trouble because, for several centuries, different countries used different calendars. A letter sent from Protestant England to Catholic France could arrive before the date on which it had been posted. England did not finally adopt the reformed calendar until 1752, when the month of September lost no less than twelve days. Stories that the move prompted rioters to demand the return of their lost days are, unfortunately, mythical but the change is the reason why the English tax year ends on 5 April rather than the traditional quarter day of 25 March.[39]

Little was heard of the central idea behind Copernicus's book, that the earth is orbiting the sun, for the next 50 years. Astronomers were aware of it and discussed it from time to time, but very few people believed it.[40] It was only when new evidence about the constitution of the heavens came to light that everyone started to talk about Copernicus.

CHAPTER 18

≈≈

Reforming the Heavens

The sun (or earth, depending on your point of view) around whom Europe's astronomers turned was the Jesuit professor Christopher Clavius (1538–1612). He had been a young technical adviser to the calendar reform commission but, by the time the change actually took place, he was considerable less junior and was put in charge of explaining the new system to the Catholic world. He also composed a textbook on astronomy, organised as a commentary on Sacrobosco's *The Sphere*, that he kept up to date throughout his long career.[1]

Astronomy at the time Clavius was active, the late sixteenth century, was in a state of turmoil. Copernicus's radical proposal for a moving earth was initially not the main issue that was preoccupying astronomers. Their problem was rather a slew of celestial prodigies that either refused to obey the theories of Aristotle and Ptolemy or else just should not have been there at all. Clavius's textbook makes an excellent guide to the ferment that engulfed astronomers during his lifetime.

Comets and Supernovas

Two heavenly marvels arose in the 1570s that challenged existing ideas. The first was a nova, or new star, in the constellation of Cassiopeia and the second a comet. Until these events, everyone agreed that the heavens were unchanging and the planets had always followed the same paths since the beginning of time. It was impossible for a new star to appear and as for comets, Aristotle had said that they were confined to the atmosphere.[2] Medieval people found no contradiction in believing that comets and eclipses were completely natural phenomena and also signs from God.[3] When Halley's Comet appeared in 1456, anxiety about the Turkish conquest of Constantinople was still high and the Turks themselves were encamped outside Belgrade. The omen in the sky only heightened fears that catastrophe was imminent. Nonetheless, the story that Pope Callistus III (1378–1458) took the desperate measure of excommunicating the comet is just a legend.

Jerome Cardan was one of the few authorities who disagreed with Aristotle and suggested that comets might originate among the stars rather than in the atmosphere.[4] But his opinions carried little weight with astronomers. Few Europeans had noticed the 'guest stars' in AD185, 1006 and 1054 that were carefully recorded by contemporary Chinese and Islamic astronomers. This was probably because the western belief that the heavens are fixed was completely incompatible with the appearance of new stars. On the other hand, the nova of 1572 (nova, of course, is just Latin for 'new') was impossible to ignore. For the most dedicated skywatchers it presented a remarkable opportunity to test the theory that the heavens never changed.

The 1572 event was what we now call a supernova – a massive star exploding with a brightness millions of times greater than our sun. Even though it occurred 7,500 light years away, the 1572 supernova could be seen in daylight and was brighter than Venus. In all, it was visible for about fifteen months although it faded rapidly after the initial explosion. Astronomers all over Europe carried out observations on it. At the time, the big question that everyone wanted answered was: where was the nova located? According to the traditional view, it should have been an atmospheric aberration, but careful observations proved that this was impossible. Over the fifteen months for which the nova was visible, astronomers made extremely careful measurements of its position relative to its neighbouring stars. Clavius collated all this information and found that the position of the nova was exactly the same for all observers. That meant that the nova had to be beyond the moon and the doctrine that the heavens could not change was proven false. Clavius himself cautiously acknowledged this in the 1585 edition of his textbook.

> If it is true [that the nova is a new star] then Aristotle's followers ought to consider how they can defend his opinion about the matter in the heavens. For perhaps it should be said that the heavens are not made of a fifth element but rather changeable bodies – albeit less corruptible than the matter here on earth ... Whatever it finally turns out to be (and I do not insert my opinion into such matters) it is enough for me at present that the star we are talking about is located in the sphere of the fixed stars.[5]

The result was confirmed by a further nova that appeared in 1604.

Then, in 1577, a huge comet appeared in the sky. Surprisingly, Clavius hardly mentioned it in his textbook although many other astronomers scrutinised its movements. The most famous of these men was an irascible Dane called Tycho Brahe (1546–1601). Tycho was born in southern Sweden, then ruled by Denmark, to a noble family with close ties to the Danish throne. He was brought up by his uncle and sent to the university of Copenhagen at the age of thirteen. There he saw a partial eclipse of the sun which fascinated him – he was impressed not so much by the event itself as by the fact that it had been long predicted. He bought some textbooks on astronomy, including the *Almagest*, and set out to discover as much about the subject as possible. His education continued in Leipzig and he also took the opportunity to travel widely in Germany and Switzerland. By 1572, his father and the uncle who had brought him up were both dead.[6] Tycho returned to Denmark and found that he had inherited money and so had leisure to further his research interests. When the nova exploded in early November of that year he was ready.

Like everyone else, Tycho wanted to measure the parallax of the nova. This was an attempt to calculate the distance to the nova by carefully observing how it moved against the background of the fixed stars. Unlike others, however, he used a potentially more accurate technique. The further apart the observations are made, the better the resulting parallax measurement. Clavius had correspondents throughout Europe so his observations were separated by a couple of thousand miles – from Sicily to Germany. Tycho made his observations 10,000 miles from each other without ever moving from Denmark himself. As the earth is rotating (Tycho thought that the whole sky rotated and the earth remained stationary, but the principle is the same), it followed that taking his observations twelve hours apart meant that he obtained the maximum distance between them. This would not have been an easy task at all, but Tycho was an extremely patient and skilled observer with the money to have his own custom-made instruments built for him. But even with this technique, Tycho could see no parallax of the nova at all. It was simply too far away.[7]

Tycho distributed copies of his work among the intelligentsia of Denmark, although it did not reach Rome where Clavius was working. This advertisement of his genius had the desired effect. The king was keen to keep this adornment to his majesty in Denmark. Tycho wanted a dedicated observatory where he could get on with his work and in 1576 the king presented him with his own island and enough money to build an entire research centre there. Over the next twenty years, the 2,000-acre island of Hven, sandwiched between Denmark and Sweden, was the scene of his life's work.[8]

Tycho had already realised that the medieval astronomical tables were not accurate enough. The date given for one conjunction of Jupiter and Saturn in 1563 was out by a month.[9] With mistakes this large, it was no wonder that astrologers were unable to do their job properly. He set out to complete a set of new observations of planetary movements from his base on Hven. At the same time, he could keep an eye out for any other astronomical novelties as they appeared. He did not have to wait long for the first one – the comet of 1577. Tycho again carried out his observations of the comet's parallax and proved that it must be a heavenly phenomenon as well. He placed it beyond Venus. But unlike the nova, comets are certainly moving and the path of this one took it through the orbits of the other planets. If the planets really were carried on the surface of giant solid spheres, the comet must be passing straight through them. Peurbach's idea of impenetrable shells was by no means universally accepted by astronomers, but Tycho's observations of this and other comets seriously undermined the theory for those who believed it. Tycho himself took a long time to accept this implication and did not even publish his work until 1588.[10]

After twenty years of having free rein over the management of his research centre on Hven, Tycho fell out with the new king of Denmark. He was now so famous that he had to put up with a constant stream of visitors to his island. He had also become an expensive indulgence for the king who cut back his allowance and insisted on economies. Tycho was not happy and neither was he the sort of man to take a perceived insult lightly. In his youth, he had lost part of his nose during a duel over a trivial matter that history does not even record. Annoyed with the king, he departed in search of a new benefactor. Tycho's reputation was such that he was quickly adopted by the richest patron of all, the Holy Roman Emperor Rudolf II (1552–1612). Together with his instruments, students and records, Tycho moved to Prague in 1599.[11] He had hardly settled in when he died without ever publishing his two decades' worth of observations. He had, however, brought out a volume of his preliminary findings in which he suggested a radical remodelling of the solar system.

Recall that Ptolemy and his followers, such as Clavius, believed that all the planets, as well as the sun and moon, orbited the earth. Tycho realised that his observations made this impossible. He continued to place the earth stationary at the centre of the universe with the sun and moon going around it. But the other five planets – Mercury, Venus, Mars, Jupiter and Saturn – he had orbiting the sun instead. This arrangement had one major drawback. In order to get the model to match his observations, Tycho had to accept that the orbit of the sun intersected that of Mars. This was impossible if the heavens consisted of a set of solid spheres as Peurbach believed. Tycho initially despaired, but then he realised that he already had the data to reject solid spheres from his work on the comet of 1577.[12]

Clavius, however, stuck firmly to the solid spheres of Ptolemy. He rejected Tycho's model although he did praise him as 'a distinguished astronomer'.[13] His reaction to Copernicus, whom he called 'that eminent restorer of astronomy in our era', was more complicated.[14] Clavius rejected the idea that the earth moved, but he was happy to use Copernicus's mathematical ideas, suitably adapted, to amend his Ptolemaic picture of the universe. Clavius's textbook goes to the trouble to refute Copernicus at considerable length but does not bother even to mention Tycho's alternative.[15]

The Magnetic Earth

Although Clavius's discussion of Copernicus's theory was unusually full, he was not the only astronomer to refer to it. In England, whether or not the earth moved was a matter debated at Oxford University by the 1570s and the question is also discussed briefly in English-language astronomy textbooks from this period. None of this meant that there were many people who agreed with Copernicus. Before 1600, hardly anyone did. But one of those few was an English doctor called William Gilbert (1544–1603).

Gilbert had entered St John's College, Cambridge in 1558. At this time, St John's was the most mathematical of Oxbridge colleges and even had a paid examiner in the subject. We know that Gilbert himself held this position in 1565 because he had to sign his name in the college's register. He does not appear to have rated mathematics as a useful adjunct to science and went on to earn a medical degree in 1569.[16] In the 1570s he built up his practice and entered the London College of Physicians by 1581. This meant that he was one of the top doctors in the city – wealthy, well connected and with leisure to carry out experiments.

His intellectual development is difficult to pin down because we do not have many documents from his early years. We do know that he rejected the materialist philosophy of Aristotle in favour of something much more like the magical worldview that we discussed in earlier chapters. Gilbert started off with some notes attacking Aristotle's views on the weather, but soon the occult property of magnetism drew his interest. He set out to investigate it as closely as he could.

We have already noted how the compass had appeared in Europe during the twelfth century and was an indispensable aid to navigation. That it worked by magnetism was not doubted. This had led some scholars to suggest that it pointed north because there was a giant magnetic mountain in the Arctic region that attracted compass needles. Others, like Peter the Pilgrim, had thought that compasses pointed towards the celestial pole and that they aligned themselves with a heavenly magnet. In the sixteenth century, these questions began to have an urgent practical significance, because as European explorers moved further away from familiar waters they found that their compasses started doing strange things. Magnetic north is inclined at about 11° to true north. Thus, the deviation between true north and the direction in which a compass points is not the same from place to place, a phenomenon known as 'variation'. Worse still, variation was not regular because, as we now know, the earth's magnetic field is not regular either.[17]

We have also seen how medieval thinking about magnetism was limited by the fact that it did not cohere with Aristotle's materialistic philosophy. No one could deny that magnetism existed, but explanations for occult properties still tended to be marginalised as magical. That is why natural philosophers failed to take forward the work of Peter the Pilgrim. His letter on magnetism was printed in 1558 in an edition by Jean Taisnier (1508–62). Taisnier, although a priest and noted scholar in his own right, passed off Peter's work as his own in one of the most notorious cases of plagiarism of the sixteenth century.[18] The remarkable thing is that he thought that the knowledge of medieval scholarship was now so slight that he would be able to get away with such a blatant act of intellectual larceny.

Gilbert knew Peter's treatise well and based some of his own experiments with a spherical magnet on its suggestions. Even so, he dismissed the earlier writer as 'fairly learned for his time' and did not acknowledge the source from which he had borrowed his ideas.[19] That said, Gilbert took matters much further than Peter and coupled his experiments with some intriguing theoretical suggestions. He showed that he could model compass variation with his spherical magnet. This meant that the earth was possibly a giant magnet too, suffused with the occult property of magnetism.

Gilbert's painstaking work has become a paragon of experimental science. It is no surprise that his experiments were geared towards the investigation of an occult property ignored by Aristotle. The laboratory tradition that inspired Gilbert belonged to alchemists and magicians, not bookish natural philosophers. The ancient Greek prejudice against men who used their hands to make a living still caused university men to sneer at artisans, just as doctors looked down on surgeons. Gilbert took the laborious experimental techniques handed down by generations of alchemists, stripped out their bogus metaphysical speculations and emphasised their virtues of patience, manual skill and meticulous record-keeping. He was well aware that experiments could be extremely hard work. He repeated the same procedure over and over again, explored every possible modification and warned his readers not to despair if they could not repeat his results on their first attempt. Finally, he carefully distinguished magnetism from static electricity and kept his empirical data segregated from theory. This last point is important because Gilbert's theoretical ideas are considerably less impressive than his experiments. The earth, he believed, was not only a giant magnet, it was alive. Magnetism was the soul of the earth, its 'astral magnetic mind'.[20] Unlike Aristotle, for whom the earth was dead, inert and still, Gilbert saw it as moving, active and aware. He thought that the earth turned itself so that all of its surface would be warmed by the sun's rays. The whole solar system was an arrangement of magnetic beings that were propelled through space by the force of magnetism.

Published in 1600, Gilbert's book *On the Magnet* covered his investigations and discoveries in exhaustive detail. It was the first work of natural philosophy by an Englishman to achieve European renown since those of the Merton Calculators. Gilbert was also one of the few convinced Copernicans of the sixteenth century. By ignoring his debt to the medieval Peter the Pilgrim while emphasising his experiments and Copernicanism, historians have portrayed Gilbert as a thoroughly modern man of science. However, we can only maintain this picture if we ignore his place in the tradition of occult properties and his own overarching philosophical position. To be fair, he only wrote explicitly about all of this in his later treatise *The New Philosophy*, which never enjoyed anything like the popularity of *On the Magnet*. However, the clues are there, even in the earlier book.

Gilbert's theories especially appealed to another inheritor of the European medieval tradition who used them extensively in his own cosmological speculations. That man was an assistant of Tycho Brahe called Johann Kepler.

Kepler's Early Career

Born in 1571 near Stuttgart in Germany, Johann Kepler was initially raised by his grandfather while his father, a soldier, was away on campaign. The comings and goings of his parents meant that he had a disjointed youth with little money available. However, he was a bright child and managed to win a scholarship to a secondary school and thence to the university of Tübingen. His eventual aim was to become a priest.[21] Religion was the central theme of Kepler's life – both his own beliefs and the social climate in which he lived. He was an unusually devout man at a time when strong religious views were commonplace.

In the aftermath of the Protestant Reformation, the religious geography of the Holy Roman Empire was a mess. Germany, which made up the bulk of the Empire, was a patchwork of statelets each independently ruled but owing nominal allegiance to the Holy Roman Emperor. In 1600, this was Rudolf II whose seat was in Prague. It was Rudolf who had offered Tycho Brahe a new job when the cantankerous astronomer left Hven.

Each ruler determined the religion of his state, as well as how rigorously he would enforce the local orthodoxy. Although basically a follower of Luther, Kepler did not subscribe to all the official doctrines of the Lutheran church. This made him potentially unwelcome in both Catholic and Lutheran jurisdictions, although in fact he spent most of his life employed by Catholic rulers who found him too useful to persecute.

At Tübingen, Kepler received the standard mathematical education that Lutheran universities offered. The astronomy professor who taught Kepler was called Michael Maestlin (1550–1631), who was part of the international correspondence network that included Tycho and Clavius. As well as the official classes, Maestlin kept his best students, including Kepler, informed about the latest happenings in astronomy. From his professor, he learnt about Copernicus's hypothesis of a moving earth as well as the comet and supernova that had challenged received wisdom.[22] Kepler was so obviously gifted as a mathematician that the university offered him a job teaching the subject at one of its satellite seminaries in Austria for a couple of years before he started his theological training. As it turned out, he never did return to theology because of his disagreements over Lutheran dogma.

Kepler was a theorist and not an observer. His eyesight had been damaged by a childhood illness, which made him completely unsuited to skywatching. As a mathematician, however, he had very few equals and could successfully manipulate vast amounts of data. In 1596 he published his first book, *The Mystery of the Universe*. For Kepler, the most important fact about the world was that God had created it. Like Copernicus, he was convinced that the structure of the heavens had to reflect the perfection of its creator. This perfection, he thought, would reveal itself best through the precision of geometry.

Also like Copernicus, Kepler put the sun at the centre of the universe. He then arranged the planets around the sun, but with the size of their orbits determined by the five basic solids – the tetrahedron (four triangular sides), cube, octahedron (eight triangular sides), dodecahedron (twelve pentagons) and icosahedron (twenty triangular sides).[23] In his model, Kepler imagined that the orbit of Mercury would just fit within an octahedron and that the orbit of Venus, the next planet out, would just fit outside it. The orbit of Venus, in turn, was encased by an icosahedron outside of which orbited the earth.

Next he placed a dodecahedron, then a triangle and finally a cube, with a planet between each one. Saturn, the most distant planet, he believed orbited outside the cube. The neatness of this arrangement seemed to provide additional evidence of the providence of God but was not precisely correct. For Kepler, even a small error rendered the model not quite good enough. No one else was convinced by his idea of using the five regular solids either, although the book did receive some good reviews.

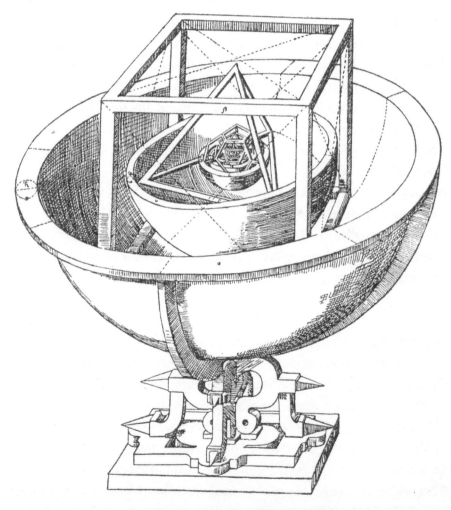

14. Kepler's first model of the universe based on the five basic solids from *The Mystery of the Universe* (1596)

To complete his work and find the perfect fit for the planets' paths, Kepler needed the astronomical data that Tycho Brahe had devoted a lifetime to collecting. Tycho himself was notorious for guarding his figures but after some troublesome negotiations, Rudolf II accepted Kepler as Tycho's official assistant. They were to collate all the data and prepare it for publication as the *Rudolphine Planetary Tables*.[24]

The working relationship between Kepler and Tycho was not good but within six months, the older man was on his deathbed. During a lucid moment before a fever carried him away, Tycho handed over the project of publishing the tables to his assistant. Shortly afterwards, the Emperor appointed Kepler as his new imperial mathematician.[25] The real treasure, Tycho's twenty years of observations, remained the property of the Brahe family, who caused endless trouble for Kepler with demands for money and a say in how the tables were published. After 25 years, the conditions attached to the use of the observations could still not be agreed and so Kepler went ahead and published them anyway.[26]

Kepler's Model of the Solar System

In the next few years, Kepler finally cracked the problem of how the planets move, although no one took much notice. Now that he had access to Tycho's observations, Kepler could test his ideas with data of unprecedented accuracy. Concentrating on the orbit of Mars, Kepler found that his best model, based on Copernicus and including eccentrics and epicycles, was out by eight minutes of arc, which is about two fifteenths of a degree.[27] Kepler was nothing if not pedantic and an eight-minute error would not do. So important was this small difference to Kepler's success that he later called it a 'good deed of God's'.[28]

We can summarise the system of astronomy that had been inherited from the ancient Greeks with three axioms: the immutability of the heavens, circular orbits and uniform motion. Thanks to the supernova of 1572, immutability had already been dismissed by Clavius and Tycho. Many sixteenth-century natural philosophers also thought that the other two axioms were either untrue or irrelevant. A rising tide of scepticism made out that it was impossible to accurately map the planets and certainly to know how they really moved. Astronomy was just a matter of trying to construct the best mathematical model of the planets' observed paths across the sky.

Kepler rejected the defeatism of the sceptics although his reasons were religious rather than scientific. As far as he was concerned, the heavens must reflect their maker. 'For a long time, I wanted to be a theologian', he wrote, 'now however, behold how through my effort God is being celebrated through astronomy.'[29] As the Bible itself states: 'The heavens declare the glory of God; and the firmament shows his handiwork.'[30] There was no imprecision about God and he did not make eight-minute mistakes. Nor was he the capricious sort who would make the heavens into an unsolvable puzzle. If the paths of the planets were ordained by God, then they must be simple and elegant. In keeping with his faith, Kepler was absolutely unwilling to abandon the axiom of uniform motion. But the need for the planets to move in circles was a Greek addition to that basic principle and he could drop it without compromising his belief in God's fidelity.

Copernicus had also derived his astronomical ideas from his theology. However, for all his calculations, he had failed to achieve total empirical accuracy because he was committed to movement in circles. Now Kepler completed the chain between religion and science. His ideas about God provided his hypothesis, he had the mathematical ability to turn his ideas into a system and, at last, Tycho's data meant he could check to see if his system was actually true.

And so, after a great deal of work, Kepler found exactly how the planets move. Like Copernicus, he realised that a planet's orbit could not centre on the sun and that its uniform motion was not its absolute speed. In fact, planets move faster when they are close to the sun than when they are further way. Kepler found that the axis of a planet's orbit swept through an equal area in any given time. This was the uniformity that he was looking for. His greatest insight was that orbits are not circles, or even based on circles, but ellipses. An ellipse is a kind of oval that has two focuses. The sun is always found at one of them. Kepler later found a mathematical relationship between the length of time it takes planets to orbit the sun and their distance from it. These three rules – that orbits are ellipses, that the axis of an orbit sweeps through a uniform area and that the period of time an orbit takes is related to its size – are now called Kepler's Laws.[31] With them, astronomers could work out anything they wanted to about the movements of the planets.

Publicising his ideas was not Kepler's strongest point. He wrote voluminous tomes but they are practically unreadable. His laws are not hard to explain but you would never guess this from his convoluted explanations. Part of the trouble was with Kepler's religion. He saw his science as a religious duty and wrote as if it was a complicated piece of theology. His notebooks are even worse. Sheet after sheet of calculations are punctuated with mystical speculation and prayers. Nevertheless, it remains true that Kepler cracked the mystery of the planets' movements because of his faith in God's creative power.

The difficulty of Kepler's work meant that it took a long time to penetrate Europe's consciousness. His theories only became accepted later in the seventeenth century, not so much because of his books but because of what he did with Tycho's data. In 1627, he finally published the new astronomical tables of planetary positions that he had promised Tycho he would complete. Their accuracy was way beyond anything previously seen and they quickly became the standard for astrologers all over Europe (astrology, of course, being the main market). It was this success that sold Kepler's laws to the public. His tables, based on his theories, were found to predict the future positions of the planets so precisely that it seemed inconceivable that they did not reflect reality.[32] Gradually, it became accepted that the planets, including the earth, did indeed orbit the sun in elliptical orbits.

Of course, Kepler was fallible. He stuck with his hypothesis about the regular polygons even though they did not fit the data. And he adopted Gilbert's theory that magnetism provided the motive power of the solar system.[33] We now know that it is the force of gravity and not magnetism that keeps the planets in their tracks, but at least Kepler realised that there had to be something holding it all together.

Explaining Vision

Kepler's achievements went beyond the field of astronomy. He also solved one of the central problems of optics. We saw in chapter 9 how Roger Bacon had combined Greek and Arabic optical ideas in the thirteenth century. Although he had not really advanced beyond the theories of the Islamic scholar Alhazen, his work had been incorporated into Witelo's *Ten Books on Perspective* which was the main university textbook by Kepler's time. It set out the flawed medieval theory that only non-refracted rays hitting the eye head-on are impressed on the retina and contribute to what we see. In 1604, Kepler published his *Supplement to Witelo* that amended the theory to what we believe today. Rather than non-refracted rays being ignored, Kepler said that the lens in the eye bent all the rays from a particular point so that they all ended up at the same spot on the retina. Perpendicular rays were not bent at all, but the lens refracted all the others exactly as much as necessary in order to produce a focused image.[34]

Spectacles had now been around for three centuries and the way they worked was well understood. What is special about the lens in our eyes is that it grows thicker and thinner to focus on objects at varying distances. Kepler also realised, as a corollary of his theory of vision, that we actually see everything upside down. Luckily, our brains rectify the image that appears on our retinas. An experiment quickly confirmed this result using a bull's eye lens in a *camera obscura*. This too produced an inverted image.

Kepler's interest in light had much to do with its central position in astronomy. Tycho had pioneered optical techniques in his observations and Kepler could also draw upon the medieval mathematical tradition. His insights later went towards an improved design of the telescope which, like the eye, produced an inverted image. However, his fascination with light was also related to the role it played in religious speculation. Roger Bacon had been convinced that light was the emanation of God and Kepler agreed. It was, after all, the first of God's creations. 'Let there be light', he had said. And there was light.

Following these scientific triumphs, Kepler's life fell apart. In 1611, all three of his children caught smallpox and his six-year-old son died. He was followed to the grave by his grieving mother, Kepler's wife, within months.[35] At the same time, he lost his comfortable situation as imperial astronomer when Rudolf was deposed as ruler of Prague. The Emperor died the following year.

Although Rudolf's successor as Emperor was a zealous Catholic, the Protestant Kepler initially kept his job as imperial mathematician as a result of his fame and skill as an astrologer. Kepler's precise views on astrology are opaque but he certainly found the vulgar practice of casting horoscopes for every occasion unpleasant work.[36] However, astrology was where the money was and he went through the motions so that he could support himself and his family. Politics and religion continued to buffet him throughout his life and he was never able to settle down in one place for long.

In 1615, another tragedy blighted his family when his mother was accused of witchcraft. Kepler travelled to her home town of Leonberg to defend her, but the case dragged on and on as the accusers vainly tried to find enough evidence to make the charges stick. She could not be tortured unless strong evidence was produced against her and eventually in 1620 the case was ruled inconclusive. Even then, she was kept imprisoned for another year and died shortly after her release in October 1621.[37]

Kepler and the Occult

There seems little doubt that Kepler's religious beliefs supplied him with scientific ideas as well simply acting as an inspiration. We have already seen how the central concept of placing the sun at the centre of the universe came from occult sources such as Marsilio Ficino and the Hermetic corpus. Magicians like John Dee thought that mathematics offered a key to understanding the natural world, while Nicholas of Cusa had made exactly the same observation in his theological work *On Learned Ignorance*. All these various influences, swirling around in the sixteenth century, make picking out the story of modern science very difficult. Polymaths were able to keep a foot in more than one camp. Ficino was both priest and magician, Cardano a physician, mathematician and astrologer, Gilbert both natural philosopher and occult theorist.

However, this kind of dual facility was becoming increasingly difficult to maintain as the seventeenth century dawned. As an intellectual pursuit, magic became marginalised. Part of the reason that it was less acceptable was that it had become less plausible. Protestants took a much more sceptical attitude towards religious miracles than medieval Catholics had, and this spilt over into how they thought about the occult.[38] Furthermore, as the Counter-Reformation progressed, the Inquisition became increasingly worried about the heretical ideas that the pagan Hermetic corpus had spawned. Ficino's heirs, as we will see in the next chapter, had to watch their step or incur the wrath of the Church. As for the witch trials themselves, they too reflected a decline in the belief in magic, at least among the upper classes. If people had stopped believing in magic, no one yet doubted the existence of the devil. It was not credible that witches could have gained occult powers by themselves, and so they must instead have acquired them from a diabolic source. Witches went from being old wise women to instruments of Satan. The resulting panic swept up Kepler's mother and cost the lives of as many as 60,000 people over the two centuries up to 1700.[39]

Despite the cross-pollination that had occurred between them, magic and science also suffered from an estrangement. The Hermetic corpus was exposed as a fake and magicians lost the cachet that they had enjoyed when they seemed to be peddling genuine ancient wisdom. The work of Gilbert had shown how an occult property like magnetism could yield to scientific analysis, but the magical worldview itself could provide no testable hypotheses. Finally, although men like Dee and Ficino had been fascinated by numbers, the separation between their numerology and practical mathematics became ever clearer. A remarkable document of the divorce between the fantastic and the mundane is the pamphlet war between Kepler and an English alchemist called Robert Fludd (1574–1637).

With his flair for self-publicity, Robert Fludd was the pre-eminent London astrologer of the seventeenth century. Born near the

Domesday village of Otham in Kent where his family had lived for generations, he received his degree from Oxford in about 1598 and travelled Europe training to be a doctor. On his return to England, he started to practise medicine and joined the influential College of Physicians in London on his second attempt.[40] However, during his time abroad, Fludd had imbibed the teaching of Paracelsus, which he mixed with mysticism and the quest for ancient knowledge. In 1616, a series of pamphlets called the *Rosicrucian Manifestoes* appeared in Paris. These were a forgery, perpetrated by a German Lutheran.[41] The documents allegedly revealed the existence of a secret society called the Brotherhood of the Rose Cross that had guarded wisdom through the centuries and would soon make itself known. The brotherhood, which claimed to include many of the most learned and influential men of the time, would bring about a reform of society and usher in a new era of peace. This caused a good deal of excitement, especially in Germany, and provided a shot in the arm for the whole mystical movement. Inevitably, for someone like Fludd, who believed implicitly in forgotten secrets, the Rosicrucian deception was completely convincing. He fell headlong for the hoax and wrote his own pamphlets defending the brotherhood.[42] He even hinted that he was a member of the imaginary secret society himself.

In his pamphlets, Fludd insisted that his mystical speculations were both true science and true religion. It is true that he praised the use of mathematics to uncover the secrets of nature. But there is nothing scientific about these ramblings. At base, they are the purest gobbledegook. This contrasts with Kepler's work, which read like a stream of consciousness but contained nuggets of epoch-making science. That Kepler knew what he was doing and Fludd did not have a clue comes through very clearly in the tracts that they wrote against each other. Kepler explained that when he used mathematics it was to describe how the universe really was in ways that could be empirically tested. Fludd, he claimed, had no solid grounds for his own mystical associations between numbers and physical things, and neither could he provide demonstrations of his ideas in action. According to

Kepler, he dealt only in 'enigmas and hermetism'. Needless to say Fludd vehemently disagreed, calling Kepler 'vulgar'.[43]

Ultimately, history has vindicated Kepler. The magical systems of the Renaissance were so flexible and complicated that they could be twisted to fit anything. Thus, they explained nothing. Kepler also struck a serious blow against the holistic picture of the world favoured by magicians. His planetary model was intended to refute the idea that the universe was somehow alive. 'My goal is to show that the heavenly machine is not a kind of divine living being', he explained, 'but similar to clockwork.'[44]

Despite his failure to convince as a natural philosopher, Fludd was a very successful physician and his magical works continued to be read for decades after his death. His family set up an impressive memorial to him in a church near Otham. Fludd still stares out from the wall of the porch of the church, a quill in his hand. His bald pate and handlebar moustache (sadly damaged like the rest of the decaying monument) make him an unintentionally comical figure.

Kepler died in 1630. He had solved two of the greatest scientific problems of the Middle Ages – how the planets moved and how we can see. He did so driven by a relentless Christian faith and working in the medieval traditions of the universities. In putting Witelo's name in the title of his book on optics, Kepler was not afraid to admit to his sources. The same cannot be said for his contemporary Galileo Galilei. His achievements were just as great as Kepler's, but Galileo was a great deal more circumspect about where he was getting his ideas from.

CHAPTER 19

⧞

Galileo and Giordano Bruno

ccording to the traditional history of science, Galileo was a
man of unparalleled originality. He was, supposedly, the first
person to show that objects of different weights fall at the
same speed, the first to claim that vacuums could really exist and
the first to realise projectiles move in curves. He rejected Aristotle
when everyone else followed him slavishly. It is said that he proved
Copernicus was right and that the Inquisition cast him into prison as
a result. As it turns out, none of these things is exactly true. Galileo
never proved heliocentricism (as we have already seen, it was Kepler
who effectively did that) and his trial before the Inquisition was
based more on politics than science. Galileo's scientific achievement
was solidly based on the natural philosophy that came before him.
Appreciating that fact should not diminish our admiration of his genius. While almost all his theories can be traced back to earlier sources,
he was the first to mould them into a coherent whole and the first to
show how they could be experimentally demonstrated. In that sense,
the long road to modern science really does start with him.

The Early Career of Galileo
Galileo Galilei was born in the city of Pisa on 15 February 1564. His
father was a distinguished musician by profession who had acquired
some business interests in Pisa through his wife. Galileo entered the
university of Pisa to study medicine (although the family had moved
back to Florence by this stage), but left without a degree in 1585.[1]

He had no great interest in medicine but the preliminary courses in mathematics and natural philosophy encouraged him to devote himself to the study of numbers. The astronomy textbook that Galileo would most probably have used during his student days was Clavius's. This would have introduced him to the ideas of Copernicus but also demonstrated why they were wrong. At the same time, Galileo would have covered Aristotle's natural philosophy. The aspect that interested him most was the theory of motion, which he was convinced Aristotle had got badly wrong. His lifetime of opposition to Aristotle's physics certainly arose during his days as a Pisan student.

Despite lacking a degree, young Galileo's talent for mathematics was obvious. His family also had friends in high places and this was enough to secure him a mathematics professorship at Pisa in 1589.

The greatest monument of Pisa, its leaning tower, was begun in 1173 to function as a bell tower for the local cathedral. According to one of his early biographers, Galileo used the leaning tower to strike a mighty blow against Aristotle's ideas about free fall. The story goes that Galileo, together with a gaggle of his students, summoned the Aristotelian professors to the base of the tower. The followers of the Philosopher all insisted that heavy objects fall faster than light ones. In fact, they thought an object that is twice as heavy should fall twice as quickly. To prove this wrong, Galileo mounted the tower armed with two lead balls of very different sizes. They were the same size so that air resistance would not be a differentiating factor. He dropped the balls from the belfry at the top of the tower and, to the consternation of the Aristotelians, both hit the ground at the same time. In this way, Galileo refuted the doctrine of Aristotle and showed how his experimental method trumped the rational analysis of the philosophers.

Nowadays, many historians dismiss the story as a myth, but it is entirely possible that something similar did happen.[2] For example, we know that Galileo was dropping wooden and lead balls together at around this time, although his results were not that they both landed at the same time. Instead, he reported that the wooden ball began by falling faster and was then overtaken by the lead ball that reached the ground first. Modern physics cannot explain this result and it shows that even the great Galileo's experiments were fallible.[3] He was also not the only one at Pisa carrying out this sort of work. The Aristotelian professors were doing experiments of their own to prove their theories. They too achieved results that are clearly wrong.

The conclusion that objects of differing weights fall at the same speed was hardly new. John Philoponus had, we noted in chapter 11, made this observation in the sixth century and Thomas Bradwardine had later suggested that it also held true in a vacuum. Giovanni Battista Benedetti (1530–90), an Italian mathematician and student of Nicolò Tartaglia,[4] published the result as his own discovery in 1553 and his book was translated into English and German. Benedetti incorrectly said that the speed at which an object fell depended on its density – a point on which Galileo agreed with him.[5] Jean Taisnier, the priest who had plagiarised Peter the Pilgrim's treatise on magnetism, also claimed to have made Benedetti's breakthrough himself.[6] Taisnier's fraudulent work inspired Simon Stevin (1548–1620), a Dutch engineer, to carry out his own experiment on the matter. In a book of 1586, Stevin wrote:

> The experiment against Aristotle is this: let us take (as the very learned Mr Jan Cornets de Groot, most industrious investigator of the secrets of nature, and myself have done) two spheres of lead, one ten times larger and heavier than the other, and drop them together from the height of thirty feet on to a board or something on which they give a perceptible sound. Then it will be found the lighter will not be ten times longer on its way than the heavier but that they fall together on to the board so simultaneously that their two sounds seem to be one and same.[7]

He went on to refute Benedetti's conjectures on density, thinking they were Taisnier's. So, by the time Galileo came to address the question, just about every natural philosopher would have heard the evidence that Aristotle was wrong.

Clearly, Galileo's early doubts about Aristotle's account of motion were not the thoughts of a lone radical, but part of a scientific milieu where experimentation and criticism of Greek natural philosophy were becoming increasingly common. Of course, we know that this began with John Buridan way back in the fourteenth century and that the Jesuits had published well-regarded books on motion with which Galileo was familiar. His earliest attempt to explain projectile motion (that is, what happens to a cannon ball after it is fired, or when you throw something), drew directly on the concept of impetus that Buridan had found so useful.[8] Galileo's ideas are similar to those of Arabic thinkers who had in turn found inspiration from John Philoponus.[9] We also know from Galileo's own writings that he had not figured out the truth about free-falling bodies at this early stage of his career. That would come later.

Domingo de Soto and Falling Objects

Although Galileo did not yet understand how falling bodies actually move, someone else did. He was a Spanish Dominican called Domingo de Soto (1494–1560). After his early education was completed in his home country, Domingo moved to the university of Paris to study theology.[10] He had first to complete the course of Aristotle's philosophy and basic mathematics, which were both still viewed as an essential introduction to the 'queen of the sciences'. His course would have been dominated by the works of John Buridan and his contemporary followers such as Thomas Bricot. We know that Juan de Celaya (d.1558), Soto's teacher and compatriot at Paris, published a highly technical book on the science of physics in 1517 which makes full use of the work of the Merton Calculators and John Buridan.[11] Clearly, the discoveries of the masters of the fourteenth century were still found to be relevant in the sixteenth. Domingo himself returned to Spain before he completed the theological course but gained his degree shortly afterwards. In 1525, attracted by the order's academic reputation, he joined the Dominicans and spent the rest of his life teaching at their school in Salamanca. His interests included ethics – he criticised the treatment of Native Americans by Spanish colonists – and theology, as well as physics. The resulting textbooks sold so well that they financed a good deal of building work at his home priory.[12]

In 1545, the Dominicans sent Domingo to Italy to attend the Council of Trent, where the Catholic Church was laying the ground for the Counter Reformation, as one of their representatives. While he was there, he would have had the chance to learn about the work of Italian critics of Aristotle's theory of motion who had already been carrying out experiments to show that the ancient Greek was wrong. Domingo returned to Salamanca in 1551 and published his textbook on Aristotle's *Physics*. In it, he gives the first accurate description of how objects fall under gravity.[13]

We know today that the distance that a falling object travels is described by the mean speed theorem developed by the Merton Calculators and Nicole Oresme. What no one had managed to do up until the time of Domingo was relate the theorem to what actually happens to falling bodies. Domingo finally did so and reported the result in his textbook. It was widely distributed around southern Europe, especially in the schools run by the Dominican order.[14] His ideas were thus current during the period when Galileo was developing his own theories.

Galileo's Family Life

During his time at Pisa, Galileo had already marked himself out as a thinker of note and needed only suitable patronage to advance his career. In 1592, he left Pisa and used his connections to gain a job at the much more prestigious university of Padua, situated in the hinterland of Venice. He stayed there for eighteen years. At some point during this period, he met a local woman called Marina Gamba. She moved in with him and bore him three children, but they never married and parted when he left Padua in 1610. His son, Vincenzo, received a good education and was eventually decreed a legitimate child so that he could become Galileo's heir. The two girls were less lucky. Their father tried to have them packed off to a nunnery as early as possible. The law stated that a girl could not profess as a nun until she was sixteen and old enough to make an informed decision. Galileo was impatient and called on his influential friends to get both girls into a convent near Florence when they were barely thirteen. The elder one, Virginia, seemed quite content with this arrangement and remained close to her father until she died in 1634. The younger, Livia, was deeply unhappy at effectively being forced into a religious life that she was unsuited for.[15]

The main reason for Galileo's questionable treatment of his daughters was money. He never had enough of it. After his father had died in 1591, he had become head of the family and was responsible for finding dowries for his two sisters. He was not keen on taking on similar obligations for his offspring. His salary as a mathematics professor was not substantial and he was expected to supplement it by taking on extra pupils privately, some of whom also rented their lodging from him.

Galileo's research during his time at Padua was constantly interrupted by these other calls on his time. He also had to deal with plagiarism of his work and the task of keeping his ultimate masters, the Venetian senate, on side. Despite all these distractions, he made considerable progress in trying to formulate a realistic law of motion and what we would today call a scientific method. For while Galileo thought that Aristotle's conclusions were almost always wrong, he was a great admirer of his system of analysis. Italian philosophers had been arguing about and fine-tuning Aristotle's methodology since 1400 in an effort to explain exactly how to relate theory to physical reality.[16] Galileo's work comes at the end of this tradition. Essentially, Galileo believed that a scientific proposition was proved if it could be derived from properly grounded causes and then demonstrated by experience.

Today we have a much less demanding test for a valid scientific theory. Nothing is ever proved absolutely in science because there is always the possibility that some new evidence will come along and show that the old theory is inadequate. We now know that even some of the most famous scientific theories are not exactly correct, including Newton's laws and Kepler's model of the solar system. Galileo wanted to believe that science can discover things with certainty and this, as we will see, eventually got him into serious trouble. Another, more direct reason for his later problems was that while he was in Padua he became one of the few people to believe that Copernicus might be right in saying that the earth orbited the sun.

We first hear Galileo enthusiastically espousing Copernicanism in 1597 in a letter he wrote to Kepler acknowledging a copy of one of Kepler's books that he had received from a mutual acquaintance. Kepler was thrilled to find an ally, even one he had never heard of, and immediately wrote back, enclosing another couple of copies of the book for Galileo to distribute. By that time, though, Galileo had decided that he really had little in common with Kepler. The latter's religious mysticism was not to the Galilean taste and the text was probably just too opaque to bother with.[17] Galileo was a masterful writer and valued limpid prose in others. This had the unfortunate consequence that Galileo ignored Kepler's great discoveries and never made much use of them in his own work.

The Church Discovers the Theory of Copernicus

Religious reaction to Copernicus before 1600 had been subdued largely because so few people thought that his theory was correct. A few writers had pointed out that the idea of a moving earth clearly contradicted scripture, but the point was moot because no one believed it anyway.

In 1584, Didacus à Stunica (1536–97) a Spanish friar, wrote a biblical commentary in which he explained that all the passages in the Bible that said the earth did not move could easily be interpreted figuratively. He explained that the Bible was written from the point of view of an observer on earth rather than from a 'god's-eye' view of the heavens as a whole. This is, of course, exactly the same argument as Oresme had made 250 years earlier. Then, in 1597, Stunica wrote another book, this time actually about natural philosophy rather than a biblical commentary. By then, he had decided that even though the issue could not be settled from the Bible, Copernicus's ideas were physically absurd.[18] In other words, he thought a moving earth was religiously unobjectionable but scientifically untenable.

A Platonic philosopher by the name of Francisco Patrizi (1529–97) suggested that the earth rotated for very similar reasons to those suggested by Nicole Oresme. This was not even the most radical aspect of his work, which sought to replace Aristotelian materialism with a mystical Platonic alternative. The Catholic Church's censorship body, the Congregation of the Index, ordered him to amend his work but allowed the rotation of the earth to stand, despite initially questioning it. A revised version of Patrizi's book appeared two years later and he kept his job as professor of Platonic philosophy, having been appointed by the Pope himself.[19]

Patrizi's most fruitful suggestion was that vacuums might actually be real.[20] Since 1277, Christians had admitted that they were possible, if only through God's absolute power to do what he liked, but few natural philosophers had postulated that they really existed. This aspect of Patrizi's thinking may well have influenced Galileo, who was also becoming receptive to the possibility of vacuums.

Giordano Bruno: Martyr for Magic

No early proponent of Copernicus's hypothesis is more famous than a renegade Dominican and magician by the name of Giordano Bruno (1548–1600). He was born near Naples in 1548 and joined the local Dominican monastery in 1563. We know very little about his early career except that in 1576 he clashed with his superiors over his strange doctrines. Bruno cast off his friar's cloak and embarked upon a life on the road. His travels took him through France and then on to England where he arrived with a letter of recommendation addressed to the French ambassador in March 1583.[21] Within a few months, he was in Oxford arguing with the local professors about his magical philosophy.

Bruno went further than many of the magicians of his time by trying to add an entirely new religion of his own creation to the existing magical doctrines. His beliefs were loosely based on the same neo-Platonic writings that had inspired Ficino, but Bruno made no effort to conform his ideas to Christianity. He even referred to such controversial writers as Cecco D'Ascoli whose previously banned books were now in print.[22] He was also monumentally conceited, with a habit of writing about himself in the third person as some kind of genius. The combination of newfangled and absurd theology with an unerring ability to rub people up the wrong way meant that he could rarely stay put for long. People had a habit of running him out of town.

Among his various ideas, Bruno was right about at least one thing – the earth goes around the sun. He lauded Copernicus for realising this, even though he certainly could not handle the mathematics upon which Copernicus built his system. Instead, Bruno shared a veneration of the sun with other neo-Platonists. By way of explanation, he wrote:

> The cause of such motion [of the earth] is the renewal and the rebirth of this body, which cannot last forever under the same disposition. Just as things which cannot last forever through the species (speaking in common terms) endure through species, substances which cannot perpetuate themselves under the same countenance do so by changing their configuration.[23]

Don't worry if this doesn't make any sense. It is certainly not scientific in the way that we might understand today. What Bruno seems to be saying here is that the earth moves so that it will benefit from the seasons. But before we write it off as complete lunacy, it did have some influence on William Gilbert.[24] As we have seen, Gilbert equated the earth's magnetic field with its soul. That was the kind of thinking that Bruno inspired and it is entirely possible that Gilbert's mystical speculation was built on Bruno's.

Needless to say, none of this went down well in Oxford. Bruno decided to give some lectures and harangued his audience about how ignorant and backward-looking they were. This was how he treated everyone who disagreed with him. The trouble was that the Oxford masters were far from ignorant. They had been discussing

Copernicus for at least the previous ten years and had come to the same conclusion as everyone else – he was wrong.[25] The denouement of his Oxford visit went badly for Bruno. One of his audience realised that he was plagiarising Ficino and brought his own copy of Ficino's book with him to the next lecture. He caught Bruno misrepresenting his source and, after some procrastination, the magician slunk off back to the continent.[26] He wandered around Europe hawking his thoughts for almost another decade before, in 1591, he made the fateful decision to return to Italy into the arms of the Inquisition. A Venetian patrician had invited him to the city and after a few months, denounced him to the local inquisitor as a heretic.[27] It has been suggested that the invitation to Italy was a trap, but perhaps the experience of having Bruno in his house for a few months was quite sufficient to cause any sensible Catholic to hand him over to the authorities.

Initially, the Venetian Inquisition knew little about Bruno. They had arrested him on the say-so of an aristocrat who also made a series of hair-raising allegations. Bruno denied almost everything and without further evidence, the case looked like it would go nowhere. If he recanted the minor errors he admitted to, he would be given a penance and they would have to release him. Then things started to go wrong for Bruno. First, it turned out he had previous. His exit from the Dominican order in 1576 had generated a file in Rome and the inquisitors there wanted him extradited. Usually, the authorities in Venice refused to hand over prisoners, but Bruno was not a citizen and not their problem, so he was sent to Rome.

Bruno's difficulties started to mount. He had discussed his ideas with his fellow prisoners who started making depositions that needed to be investigated. Worse, the inquisitors got hold of his books which were full of alarming ideas. In the end the file ran to 600 pages and it took nine years for the case to be concluded.[28] Throughout, Bruno insisted on his innocence.

Eventually the file was handed to a Jesuit professor by the name of Robert Bellarmine (1542–1621). He drew up a list of eight heretical statements that no one could doubt Bruno held. The list has not survived and this leaves us free to speculate about what the eight statements were. The choice from Bruno's extant works alone is very considerable. Those who imagine that Bruno was a martyr for science assume that his support for a moving earth and an infinite universe featured on the list. This is impossible. As we will see in the next chapter, Copernicanism was not declared a heresy until 1616 and as for an infinite universe, he was simply echoing Cardinal Nicholas of Cusa. Both these beliefs were discussed in the Inquisition's files, but that in no way proves that they were deemed formally heretical.

In any case, Bruno agreed to recant the statements on Bellarmine's list and do penance. At the same time he wrote a sealed letter to the Pope claiming that the statements were not heresies at all. In doing so, he undermined his own confession. Now, finally, the inquisitors lost patience. He was given 40 days to repent or face the stake as an impudent and recalcitrant heretic. With incredible bravery, Bruno stuck to his guns and was burned alive in Rome on 17 February 1600. Nobody deserves this terrible fate and the Inquisition should not even have taken him seriously.[29]

As for Robert Bellarmine, historians agree on three things. He was in possession of a brilliant intellect, he was a religious fundamentalist and you could not hope to meet a nicer bloke.[30] His manners and kindness were legendary. If sentencing a man to be burnt alive sounds less than nice, we may be sure that he felt the same. Instead, like so many other people in history, he had a misguided sense that he had to do his duty, however unpleasant.

Although physically a very small man, Bellarmine was a big fish in the Vatican pool. He was Italian by birth and the nephew of a pope, but he had no need for nepotism to advance up the ecclesiastical ladder. He completed his education at the university of Louvain in the Netherlands where he lectured on natural philosophy. Bellarmine was no unquestioning follower of Aristotle but instead thought that science should be based on the Bible. As there is very little science in the Bible, many scientific questions were, for Bellarmine, unanswerable. The certainties of Aristotle were unacceptable, especially his claim that the heavens never changed and were made of the fifth element, 'ether'. The Bible hinted nothing of this and so, in Bellarmine's eyes, the theory was unfounded. As for the motion of the planets, Ptolemy's theory was no good because it could not exactly describe the phenomena we observe. Bellarmine himself seemed to think that the movements of the planets were simply too complicated to be accurately modelled.[31] The reality of their motion was beyond the ken of man.

In 1576, Bellarmine was recalled to Rome and put in charge of the Jesuits' offensive against Protestantism. His analysis of the various threats to the Catholic Church, published as the book *Controversies*, made him the number one target of Protestant propaganda. He was a worthy opponent because he had no time for the superstitions popular with the Catholic laity, such as ringing church bells to ward off thunderstorms, which also enraged Protestant theologians.[32] Bellarmine upset the Pope too by downgrading his political power to merely 'indirect'.[33] But he was able to shrug off papal disapproval and the next pontiff promoted him. In 1599, he became a cardinal and in 1606 narrowly escaped becoming Pope himself, something he did not want at all. However, he remained in Rome as the most important thinker in the Vatican and the first man to whom the Pope turned when he needed intellectual advice. Thus, when Galileo's new ideas about the planets became an issue, Bellarmine was closely involved in formulating the official response.

CHAPTER 20

⪎⪏

Galileo and
the New Astronomy

On 13 March 1610, Galileo went from obscure professor of mathematics to international sensation overnight. The book that brought about this transformation is called the *Sidereal Messenger*. On the very day of its publication, the English ambassador to Venice was writing home about it. Galileo, the ambassador concluded, 'runs the risk of either being extremely famous or exceedingly ridiculous.'[1]

The *Sidereal Messenger* publicised the discoveries that Galileo had made with a new piece of scientific equipment – the telescope. It had been invented in Holland a couple of years earlier and one Hans Lipperhey (d.1619) had unsuccessfully tried to obtain a patent for it.[2] By 1609, Galileo had built his own improved version. He quickly found that there were many more fixed stars than could be seen with the naked eye and that the planet Jupiter had four small companions in orbit around it. Most controversially, he revealed that the moon was not a perfect sphere, as Aristotle said all heavenly bodies must be, but disfigured by craters and mountains. He was not the only person who examined the heavens with the new instrument but he was the first to publish his observations, and he kept up the flow of new discoveries. Within a year he announced that Saturn had 'ears', which later turned out to be rings, and that Venus had phases like the moon.

This last observation was the most important. Galileo noted that sometimes, when he looked at Venus through his telescope, he saw almost a complete disk. At other times, only a semicircle or even less of the surface was visible. The periods of these phases were such that Ptolemy's model of the heavens had to be wrong. Venus did not orbit the earth; it orbited the sun.

Galileo's Discoveries

The controversy caused by these results was immediate. The first problem was that Galileo's telescope, for one reason or another, was better than anyone else's. He could see things that were not apparent to other observers, especially those who had decided in advance that he must be mistaken. One such critic was his colleague at the university of Padua, Cesare Cremonini (1550–1631). He was the university's professor of Aristotelian philosophy and not very receptive to ideas that conflicted with his master's teaching. However, he was no shirker of debate. His consistent refusal to contradict Aristotle's denial of the soul's immortality meant that the Inquisition were very keen to talk to him. Luckily, his residence in Venice, which only handed people over to the Vatican if it wanted to, kept him out of danger. However, Cremonini quickly became exasperated by the whole question of the telescope. 'I do not wish to approve of claims about which I do not have any knowledge, and about things which I have not seen', he said, 'and then to observe through those glasses gives me a headache. Enough! I do not want to hear anything more about this.'[3]

While Cremonini was soothing his sore head, Giulio Libri (c.1550–1610), professor of Aristotelian philosophy at Pisa, was also having trouble seeing the moons of Jupiter. When he died shortly afterwards, Galileo sarcastically said of Libri, 'never having wanted to see [the moons of Jupiter] on earth, perhaps he'll see them on the way to heaven?'[4] These remarks seem to be the source of the persistent legend that certain individuals refused even to look through the telescope. In fact, we know of no one who definitely declined to do so. The argument was over what they would see once they had peered through it.

As far as the Church was concerned, only one person's opinion mattered – Christopher Clavius's. He was now an old man, revered as the Jesuits' most senior astronomer and famous throughout Europe. Initially, he was suspicious of Galileo's discoveries. In 1610, no one had any idea how the telescope worked and it seemed very possible that it was distorting as well as magnifying the images. When one of Clavius's students built his own telescope it was of insufficient power to show all the things Galileo claimed to see. The student set to work on an improved version and this time it worked very well.[5]

Cardinal Bellarmine wrote to Clavius to ask whether he could confirm Galileo's discoveries. The Jesuits replied that they could, although they were not sure about how the results should be interpreted.[6] Nevertheless, when Galileo visited Rome in 1611, he received a hero's welcome and the Jesuits lauded him openly for his wonderful discoveries. In the same year, Johann Kepler drew on his experience of mathematical optics to explain how the telescope worked and suggest some improvements. Clavius added a section to his astronomy textbook outlining Galileo's findings and ending with a note that 'astronomers ought to consider how the celestial orbs should be rearranged to model these phenomena.'[7] Even Clavius, a supporter of Ptolemy, could see that reform was now necessary. Unfortunately, he died shortly afterwards and never let on what he thought the reforms should be.

However, it is possible to make a guess. When Ptolemy's system was rendered obsolete by Galileo's discoveries, there were two candidates to replace it – the ideas of Copernicus and those of Tycho Brahe. Galileo and Kepler supported Copernicus while just about everyone else thought Tycho's model the best. There were still overwhelming problems, both physical and religious, with the idea that the earth moved. This alone made Tycho's theory the much more attractive one. Galileo realised the question was wider and more important than simply which scientific theory should be adopted. After all, if Tycho turned out to be wrong, then his ideas could be dropped just like Ptolemy's had been. But the religious issues surrounding the heliocentric model were potentially toxic.

In 1610, cashing in his new-found celebrity, Galileo took up the post of official mathematician to the Grand Duke of Tuscany, Cosimo I Medici (1590–1621), a descendant of the Cosimo who had ordered Ficino to translate the Hermetic corpus. This meant that he no longer needed to take in students to make ends meet. He was being paid to produce books and research that would redound to the glory of the Medici family. Galileo had started well, naming the moons of Jupiter the 'Medicean Stars' in honour of his prospective employer. As a job application, this ploy of flattering the Grand Duke was completely successful. His old colleague, Cesare Cremonini, knew that Galileo was taking a risk leaving Venice for the Medicis' home in Florence. Once there, Galileo would no longer enjoy the protection from the Inquisition that Venice afforded Cremonini. Of course, at the time Galileo had nothing to fear and ignored the warnings. Besides, he was going home.

The mother of Grand Duke Cosimo was called Christina (1565–1637). She was an intelligent woman, interested in science and philosophy, who asked one of Galileo's friends about the compatibility of Copernicanism and scripture. He passed the query to Galileo, who as scientific adviser to the Medicis, was well placed to answer her question. This he did, in great detail.

There is no denying that a completely literal reading of the Bible strongly supports the stability of the earth and the movement of the sun. For example, Ecclesiastes 1:5 reads: 'The sun also arises, and the sun goes down, and hastens to the place where it arose', while Psalm 103:5 (104:5 to Protestants) reads 'He [God] laid the foundations of the earth, that it should not be moved forever.' Back in the fourth century, St Augustine of Hippo had wrestled with the matter of what to do when the Bible and science said different things in his commentary on the book of Genesis. Genesis clearly conflicted with the best available Greek science of the time. Augustine was worried that Christians who read their Bible too literally risked making their religion look ridiculous. In his commentary he wrote:

Usually, even a non-Christian knows something about the earth, the heavens and the other elements of this world, about the motions and orbits of the stars and even their sizes and relative positions ... Now it is a disgraceful and dangerous thing for an infidel to hear a Christian, presumably giving the meaning of the Holy Scriptures, talking nonsense on these topics, and we should take all means to prevent such an embarrassing situation, in which people show up vast ignorance in a Christian and laugh it to scorn.[8]

Augustine's solution was to set out the circumstances when the Bible should be read in a figurative rather than literal sense.

Carefully developing these earlier ideas, Galileo showed that the Bible should not be read as a scientific document, but one written in the language of the common people. Thus, when it referred to the sun rising and setting, it was simply using the everyday idiom that everyone is familiar with. The same principle applies to the two references in the Bible to the 'four corners of the earth'.[9] This is simply a phrase that is used in common parlance, but not even the most one-eyed biblical literalist thinks the earth is flat. Galileo summed up his argument with the famous words: 'The intention of the Holy Spirit is to teach us how one goes to heaven, not how the heavens go.'[10] While this aphorism is pithy, the idea was not original. We have seen how William of Conches expressed exactly the same sentiment four centuries earlier.

Galileo made one further argument that in retrospect seems the most astute of them all. If, he said, the issue of whether or not Copernicus was correct was remotely uncertain, the Church should leave the matter well alone. If it made a ruling that later turned out to be wrong, the damage to its authority would be incalculable.[11] Unfortunately, no one in the Vatican was listening.

The Church takes action against Copernicus

Matters came to a head in 1616. Galileo had been quite open about his support for Copernicus but, as it turned out, it was not his activities that pushed the Church into taking official action. Instead, a book appeared by an Italian friar called Paolo Foscarini (1565–1616), which argued along similar lines to Galileo that the earth's motion was consistent with a proper reading of the Bible. The Church could not ignore Foscarini's advocacy of Copernicus because he was a professional theologian.[12] His opinions on Biblical interpretation carried far more weight than Galileo's. At the centre of the Church's response to Foscarini's challenge was none other than Cardinal Bellarmine.

Bellarmine's opinion about the Bible was completely different from Galileo's and quite unusual among Catholics at the time. He believed that the Bible should always be treated as literally true unless there were ironclad reasons not to do so. Official Catholic doctrine laid down by the recent Council of Trent said that the Bible was only without error in matters of faith or morals. The clear implication was that incidental details might occasionally be wrong. Bellarmine disagreed. As the whole Bible was the word of God, then for him it was a matter of faith that it was completely without error.[13] This was why he had rejected Aristotle's natural philosophy – it conflicted with the Bible. He was also relaxed about the collapse of Ptolemy's astronomy because it had no biblical support either. Crucially, though, the Bible did say that the earth is immobile and the sun moves. On this point, Bellarmine was himself immovable. Well, almost. He did concede that if Copernicus was ever proven to be correct, then he would just have to accept that the Bible was being figurative on this point.[14] This attitude might sound commendably moderate, but this would mistake Bellarmine's meaning. As far as he was concerned, there was absolutely no chance that Copernicanism would be shown to be demonstrably true. Irrefutable proof in science is a rare enough thing and Bellarmine was insisting on complete certainty before he would reconsider what the Bible seemed to say. Clearly, science cannot operate in such an environment.

In March 1616, the Congregation of the Index of Forbidden Books made its decision on Copernicus: that to assert that the earth orbited the sun was scientifically 'foolish and absurd' as well as being contrary to scripture. Foscarini's book was banned outright and Copernicus's *Revolutions of the Heavenly Spheres* suspended until corrected.[15] As for

Galileo, he was officially warned by Bellarmine that he could not defend, hold or teach the views of Copernicus. At the same time, he was assured that the Inquisition was not formally accusing him of doing anything wrong, still less making out that he was a heretic. Despite the fact that he was personally safe, the Church's condemnation of Copernicus had realised Galileo's worst fears. From that moment on, he seemed to be on a personal mission to have the decision reversed. It was a hopeless quest but Galileo's self-belief was such that he was willing to try.

Prelude to Confrontation

On his return to Florence, Galileo could no longer publicly talk about Copernicus. Luckily, he had other controversies to take up his time. He had already been involved in a dispute with a Jesuit astronomer over who first discovered sunspots. Then, in 1618, three comets crossed the sky and these gave rise to another fierce argument. A Jesuit astronomer, Horatio Grassi (1583–1654), had measured the parallax of one of the comets and found that it was above the moon rather than being an atmospheric phenomenon. On this, Grassi agreed with the earlier findings of Tycho Brahe and others. Galileo, oddly enough, was not convinced and briefed a student to write a rebuttal that attacked the basis of saying anything definite about comets. From then on, the argument raged in an undignified fashion.[16] In 1623, Galileo wrote a short book called the *Assayer* which was intended to scotch Grassi's arguments. In this, he was resoundingly successful. The *Assayer* is a statement of what Galileo takes to be best practice in science; that is, to follow not the authority of other writers, but the authority of nature itself. He wrote:

> Philosophy is written in this grand book, the universe, which stands continually open to our gaze. But the book cannot be understood unless one first learns to comprehend the language and read the letters in which it is composed. It is written in the language of mathematics, and its characters are triangles, circles, and other geometric

figures without which it is humanly impossible to understand a single word of it; without these, one wanders about in a dark labyrinth.[17]

Of course, there was nothing new about the metaphor of nature as a book. Nor was the idea that it was mathematical particularly novel – we have already encountered it in the work of Nicholas of Cusa and Kepler. However, mathematics had long been a poor relation to natural philosophy and Galileo was adding his weight to Clavius's in order to ensure that it was now properly regarded. Unfortunately, the *Assayer* was primarily an attack on the Jesuits in Rome who took the whole affair quite badly. It is also worth noting that modern opinion agrees far more with Grassi than Galileo on the question of comets. Regardless, the *Assayer* was a literary *tour de force* and no one doubted that Galileo had won the debate.[18]

Alienating the Jesuits who had supported him in 1611 was probably not a good idea, but other things were going Galileo's way. In 1620, The Congregation of the Index issued its corrections to *Revolutions of the Heavenly Spheres*. They were surprisingly modest – just ten amendments to a book that was several hundred pages long. The purpose of the corrections was to demote Copernicus's ideas from a picture of reality to a mathematical hypothesis. Thus, when Copernicus wrote that he had no shame in 'admitting' that the earth moves around the sun, the Congregation of the Index changed this to 'assuming' that the earth so moves.[19] Rather than order that existing copies of Copernicus's work be destroyed, the Congregation ruled that their corrections be added as a special insert.

In 1621, Cardinal Bellarmine died and in 1623 there was a new Pope, Maffeo Barberini (1568–1644) who took the name Urban VIII. Barberini was already an admirer of Galileo. They were both from Tuscany and had known each other for many years. The new Pope had little time for the Jesuits and enormously enjoyed the *Assayer*, which was dedicated to him. This helped Galileo more than he ever knew because the *Assayer* supported the doctrine of atomism. This idea had landed Nicholas of Autrecourt in trouble back in the mid-fourteenth century and its close link to the Eucharist meant that it remained a touchy subject. A Jesuit assessor at the Congregation of the Index suggested that Galileo's book should be censored, but Pope Urban put one of his nephews – who was newly created a Cardinal together with several other members of the family – in charge of the inquiry.[20] Predictably, nothing came of it and Galileo remained of good standing.

The election of the new Pope presented an opportunity to see how far the ban on Copernicanism could be stretched. As Urban was sympathetic towards Galileo, the Pope was unlikely to throw the book at him if he tiptoed around the edges of the prohibition. Treating Copernicanism as a mathematical construction was quite acceptable as long as he did not assert that it represented reality. This meant that arguing about its virtues *as a model* would be legitimate, especially if he was comparing it to the other models available. Galileo planned a book that would do just that. In 1624, he came to Rome to meet Urban VIII. He was granted no less than six audiences with the Pope during his stay, as well as a papal pension for his son. While they wandered around the gardens of the Lateran Palace, Urban explained his opinion on the great cosmological debate. As far as he was concerned, it was beyond man's ability to work out how the heavens really worked. Whatever mathematical model was used to describe the movement of the planets, God could easily have arranged things differently but so as to give the same result. Thus, no matter how exactly any model predicts the movement of the planets, we cannot be sure that it is true. Urban's argument (technically known as the problem of underdetermination) is philosophically correct but falls down in practical terms. If we have a scientific theory that accurately explains the results of all our experiments, we are well justified in believing that the theory is true.

Galileo left Rome with enough encouragement to start writing his book. Urban had let on to a cardinal who was a friend of Galileo that he did not consider Copernicanism to be formally heretical anyway, merely rash.[21] Besides, no one would ever prove it conclusively, so the Church could afford to be relaxed.

The Astrological Pope

Aside from his relationship with Galileo, Urban VIII is most famous as a patron of the arts. During his pontificate, Rome was rebuilt in the new Baroque style that signalled the Catholic Church's resurgence after the Reformation.

A less well-known aspect of his character was his fervent belief in astrology. Despite a strong papal condemnation of horoscopes by one of Urban's predecessors in 1589, astrologers were still very widely consulted at the start of the seventeenth century. The Pope's enemies thought they could take advantage of his credulity by publicising their own version of his horoscope. They confidently predicted that a series of eclipses in 1628 and 1630 surely heralded the Pope's demise. Almost everyone, including Urban, believed this and preparations discreetly got under way for the election of a new Pope.[22]

The aim of the conspiracy must have been to kill or at least incapacitate Urban purely through the power of suggestion. Luckily for the Pope, he had a defender against this magical attack languishing in the papal prison. Tommaso Campanella (1568–1639) was, like Giordano Bruno, a renegade Dominican who had invented an entire system of esoteric philosophy. Since 1599, he had been in jail for his part in a rebellion centred on Naples. Horribly tortured, he had only escaped execution by convincing his captors that he was insane.[23] Like Bruno, Campanella believed that the earth orbited the sun for mystical reasons. In 1622 he wrote a *Defence of Galileo* from prison that echoed the *Letter to the Grand Duchess Christina* in saying that the Bible was the wrong place to look for information about how nature worked. For Campanella, though, the right place was in the neo-Platonic writings of Ficino and his followers.[24]

Campanella had convinced himself that Urban VIII was the universal monarch whom his astrological predictions had foreordained. The Pope, Campanella thought, would allow him to convert the world to his new religion of quasi-paganism. Whether or not Urban agreed, he badly needed Campanella's magical skills to protect him from the dangers of the upcoming eclipses. The prisoner was summoned to the papal apartments and ordered to protect the Pope from the baleful influence of the stars. The ritual that Campanella used was something he had made up especially for the occasion. He and the Pope shut themselves in a sealed room, blocking up any air holes. They sprinkled perfume, burnt rosemary and draped the walls with white sheets. In the middle of the room, five torches and two candles were lit to represent the seven planets.[25] As they believed that the eclipse had rendered the heavens defective, Campanella arranged that a miniature version of them should be created in the papal apartments to shield the Pope from the dangers outside. To Urban's great relief, Campanella's magic worked and the Pope survived. In return, he allowed the magician to set up a school in Rome to preach his ideas to whoever would listen, while ignoring his blatant heresies.

That Urban would ignore any unorthodoxy when it suited him boded well for Galileo. Not well enough, as it turned out.

CHAPTER 21

~~

The Trial and Triumph
of Galileo

G alileo called his defence of the heliocentric hypothesis *Dialogue concerning the Two Chief World Systems: the Ptolemaic and the Copernican* (1632). Its author intended it for lay people and so he wrote in Italian rather than in Latin. The two world systems of the title exclude Tycho Brahe's and so, to some extent, Galileo was not putting up the strongest competitor to Copernicus. Likewise, he made no use of Kepler's discovery that planets move in ellipses rather than circles, nor the extraordinary accuracy of Kepler's *Rudulphine Tables*. Thus the *Dialogue* did not represent the cutting edge of science at the time. Rather it was a popularisation of the issues intended to bring the advantages of Copernicus to as wide an audience as possible.

This means that Galileo's *Dialogue concerning the Two Chief World Systems* is not a masterwork of science. Instead, it is a first-class piece of rhetoric aimed squarely at non-experts. The modern genre it most resembles is popular science of the sort that tries to convince lay readers that they can understand relativity or string theory while glossing over all the difficult points. There are three characters in the *Dialogue*, Salviati (who represents Galileo), Simplicio (a naïve Aristotelian) and Sagredo (a not-terribly-neutral chairman) who engage in discussion over four days. The book as a whole is curiously unstructured but always reasonably entertaining. Simplicio's job is to put up the arguments that Galileo wants to refute, which inevitably makes him appear foolish. Sagredo is supposed to represent the reader who initially has no opinion, but always ends up agreeing with Galileo.

Galileo's arguments for Copernicus

The first part of the book refutes the Aristotelian doctrine that the region of the heavens is unchanging and perfect whereas the earth is subject to corruption. The contrary view was becoming widely accepted after the novas of 1572 and 1604 as well as occasional comets, most recently in 1618. Galileo seems now to accept the position of the Jesuit Grassi that comets are a heavenly and not an atmospheric event.[1] Galileo also brings forth his own telescopic discoveries that the moon is not a perfect sphere and the sun has blemishes.

The second day of discussion in the *Dialogue* is devoted to the question of the earth's rotation. Again, there is nothing much new here. Galileo trots out the old objection that the rotation of the earth should give rise to a great wind as the atmosphere is left behind and the old explanation about relative motion on board a ship. This had all been around since the fourteenth century but there is no reason why his lay readers should have been familiar with any of it. After all, Galileo's discussion is a great deal more detailed and he expands the number of explanatory thought experiments. In one scenario, he suggests dropping a stone from the mast of a ship while it is moving and again when it is stationary. In both cases the stone will land at the base of the mast.[2] There is no evidence that Galileo ever carried out any of these experiments, as he admits in the text, but he was correct in his deductions about their result.

In the third part, the conversation finally turns to Copernicus. Galileo runs through the reasons for favouring the sun being the centre of the planets' orbits. He covers the way that planets' movements all seem linked to the sun, as well as his discovery of the phases of Venus and sunspots (which show that the sun rotates on its axis).

The final part of the dialogue is devoted to the cause of the tides and represents Galileo's most serious scientific mistake. He thought that it was possible to prove a scientific theory – to *demonstrate* it, in the jargon of the time. For him, the tides were proof that Copernicus was right and that the earth was rotating.[3] He imagined that as the earth turned and moved through space, the oceans got slightly left behind and bunched up on one side of their basin before overcompensating and sloshing over to the other side.[4] This is an extremely bad argument that directly contradicts Galileo's own explanations about how the atmosphere is carried along as the earth rotates. He seemed to think that whereas air in the atmosphere moves with the earth, water in the seas is slightly retarded. In fact, the tides are caused by the gravitational pull of the moon dragging them around as it orbits the earth. They are perfectly explicable in the Ptolemaic universe and offer no evidence at all that Copernicus is correct. The link between the moon and the tides had been noted in antiquity and we have seen how it had been suspected, if not always admitted, throughout the early and late Middle Ages.

Still, Galileo believed that the tides argument was the demonstration that he needed to prove Copernicus correct. This left him with a problem. He was not allowed to defend the belief that the earth orbited the sun, only to treat it as a hypothesis. If he wrote that he had proved it, he would be stepping outside his remit. His solution was extremely foolish. What he should have done was to explain why the tides argument might be false (which would not have been difficult, because it is). Instead, he put the Pope's argument about how God could engineer circumstances to produce any end result into the mouth of Simplicio.[5] The Pope's argument was weak and inimical to natural philosophy. By slotting it in at the end, Galileo showed what he thought of it and thereby ensured that His Holiness would take umbrage.

The Trial of Galileo

Before the *Dialogue* could be printed, Galileo needed to obtain official sanction from the Congregation of the Index. After a certain amount of negotiation, this was forthcoming and the book duly came out in Florence in February 1632. When the Pope read it, he was furious. An eyewitness reported: 'His Holiness exploded into great anger and suddenly told me that even our Galileo had dared to enter where he should not have, into the most serious and dangerous subjects which could be stirred up at this time.'[6] The Pope ordered the book withdrawn and appointed a special commission to examine it. Galileo was summoned to Rome in February 1633 to stand trial for heresy. His efforts to persuade the Church to abandon its opposition to Copernicus had failed. He had only succeeded in reinforcing it.

Scholars still argue about exactly why Galileo's former friend Pope Urban put him on trial. The reason was probably that he felt betrayed by Galileo mocking his own argument and felt he had to either react or lose credibility. He had enough problems hanging on to his authority in Rome without some jumped-up mathematician making him look silly. The Church is culpable for banning heliocentricism back in 1616, but Galileo himself shares some of the blame for escalating the crisis in 1632, even if he was acting for the purest of motives.

Everyone knew he was guilty of defending and holding Copernicus's views. The problem for the Inquisition was that the Congregation of the Index had actually permitted the publication of the *Dialogue*. How could they convict Galileo of heresy when he had their prior approval? The solution presented itself during the run-up to the trial. It turned out that Galileo had not merely been told not to hold or defend the views of Copernicus back in 1616. He had actually been enjoined not to teach them in any way at all. This was a stronger prohibition than Galileo had revealed to the Congregation of the Index. He denied that he remembered exactly those words, and the memorandum on the Inquisition's files that contained them was not signed. This has led a few scholars to suggest that it was faked in order to secure a conviction. For some historians, its discovery just when it was needed is almost too convenient. That said, there is no evidence of forgery and Galileo would certainly have been convicted even without the suspect memorandum.[7]

His trial began on 12 April 1633. During the hearing, he was given comfortable quarters but this did not lessen the seriousness of his position. He started off by claiming that he did not agree with Copernicanism and that his book refuted the position. No one believed this but he stuck to his guns. Lying to the Inquisition was an extremely dangerous thing to do. There was adequate proof of his guilt and he could not expect mercy if he refused to own up. One of the inquisitors decided to meet Galileo informally to talk him around. As a result, at his next interrogation Galileo admitted that an uninformed reader of his book would get the impression that he thought Copernicanism was true. He also admitted to arrogance and vainglory in making his arguments appear stronger than they were. On 21 June, the Inquisition gave Galileo a final chance to admit that he did hold to Copernicanism. When he refused, he was reminded that the evidence against him was sufficiently strong for torture to be justified in obtaining a confession.[8] Still he refused but stated that he would submit to whatever the Church decided to do with him. The Inquisition decided that this was a sufficient reply and Galileo was not to be tortured.

The next day, a committee of cardinals chaired by Pope Urban met to consider the verdict. The committee found Galileo seriously suspected of heresy and ordered him to admit his errors. Furthermore, he was sentenced to life imprisonment for failing to obey the Inquisition's order not to hold, defend or teach Copernicanism. This sentence was immediately commuted to house arrest and Galileo returned home. In an unprecedented step, the Pope ordered that copies of Galileo's sentence should be dispatched throughout the Catholic world.

Galileo must have felt wretched. His plan to help the Church correct its mistake had failed and his friendship with the Pope was ruined. Soon after he returned home, his beloved daughter, now Sister Marie Celeste, died and Galileo's own health was rapidly declining. A lesser man might have given up the hard labour of natural philosophy to sink into endless melancholy. Not Galileo. He began to write again and produced the book that sealed his reputation as a titan among mankind's intellectual champions. Because of his disgrace, the new book, *Discourses on Two New Sciences*, had to be smuggled out and published by a printer in the Netherlands.[9] Despite this, it was never banned by the Congregation of the Index and there is no evidence that it provoked further trouble for its author from the authorities. Any action would have been purely vindictive in any case. By the time Galileo received his own copy of the book, he was completely blind. He was no longer a threat. When the English Protestant poet John Milton (1608–74) visited him in the autumn of 1638, no one objected.[10] Galileo's last years were peaceful and as comfortable as extreme old age could be in those days. He died on the night of 8 January 1642. Today, he lies in a grand tomb in the Franciscan Church of Santa Croce in Florence, outside of which Cecco D'Ascoli had burnt three centuries earlier. The machinations of Urban VIII denied Galileo the fitting memorial that we see today for almost a century after his death.

Two New Sciences: The Legacy of Medieval Science

Discourses on Two New Sciences (1638) represents the culmination of four centuries of work by medieval mathematicians and natural philosophers. For this book, Galileo brought back the three characters that he had created in *Dialogue concerning Two Chief World Systems* – Salviati, Sagredo and Simplico – and again provided a record of their fictitious discussions over four days. Some of the conclusions on local motion and projectiles had been previewed on the second day of dialogue in the earlier book, but now Galileo provides much more detail. According to the conceit of *Discourses on Two New Sciences*, Salviati has been reading a treatise by Galileo from which he quotes lengthy excerpts.

Galileo admits that not all of the material in the book originates from his own work and has rightly been criticised for not fully acknowledging his sources. Modern scholars have found that a great deal more of the *Discourses* owe a debt to medieval and sixteenth-century predecessors of Galileo than previously realised. In the book, Salviati admits, 'some of the conclusions have been reached by others, first of all by Aristotle',[11] but this is something of an understatement. There is more truth in his next remark: that Galileo had demonstrated the conclusions much more rigorously than hitherto.

The first day of the *Discourses* begins with Salviati musing on the arsenal of Venice which had, in 1571, produced many of the ships that had defeated the Turks in the sea battle of Lepanto. Nicolò Tartaglia's own work on projectiles, also called *New Science* and written in 1537, had been dedicated to Venice's efforts to defeat the Turks. Perhaps this is how Galileo chose to acknowledge his intellectual debts.

The conversation on the first day ranges widely. Galileo states unambiguously that a vacuum can really exist, at least instantaneously as two flat plates are parted.[12] There must briefly be a vacuum between the plates before the surrounding air can rush in to fill the gap. He goes on to explain that vacuums 'suck' because nature resists their formation and attempts to test the strength of this force. (In fact, Galileo is wrong about this. Vacuums do not 'suck', rather air pressure pushes.)

Later in the day, the conversation moves to falling objects. Sagredo points out to Simplicio, who is defending Aristotle's doctrine that heavy weights fall faster than light ones: 'I, who have made the test, can assure you that a cannon ball weighing one or two hundred pounds, or even more, will not reach the ground by as much as a span ahead of a musket ball weighing only half a pound.'[13] This repeats the observation that had been made a millennium earlier by John Philoponus and much more recently by Simon Stevin in Holland. In one of his early manuscripts dated around 1590 Galileo had already noted and criticised Philoponus's ideas on falling bodies, but appears now to essentially agree with his observation.[14]

Galileo goes on to provide some intriguing thought experiments. He considers the case where two objects of differing weights are connected together and let fall. If Aristotle were right, when a small stone was attached to a large stone, the former would have to retard the fall of the latter. Hence, the combination would fall more slowly than the heavy stone on its own.[15] This is nonsense, and Galileo thereby proves that objects fall at the same rate both experimentally and from first principles. His results may not be original, but his treatment is certainly definitive.

Galileo and Free Fall

The second day of discussion deals with the science of materials. We will leave that to one side and move straight on to day three. Here, Galileo tackles free fall – how objects move under gravity. Again, he is not quite honest about the extent to which he draws on prior work on the subject:

> I have discovered by experiment some properties of motion that are worth knowing and which have not hitherto been observed or demonstrated. Some superficial observations have been made, as for instance, that the free motion of a heavy falling body is continuously accelerated. But to just what extent this acceleration occurs has not yet been announced.[16]

In fact it had been announced, as we have seen, by Domingo de Soto. It is one of the great mysteries of the history of science how much Domingo's widely circulating textbook influenced Galileo, if only at second hand. We will probably never know. However, Galileo must have been familiar with the mean speed theorem of the Merton Calculators.[17] It had appeared in at least seventeen printed books before 1600, not to mention countless manuscripts.[18]

Galileo also claims that 'so far as I know, no one has yet pointed out that the distances traversed, during equal intervals of time, by a body falling from rest, stand to one another in the same relation as the odd numbers beginning with unity.'[19] With the aid of a diagram from the *Discourses*, it is easy to see what he means by this. The graph of time plotted against speed for an object subject to a constant rate of acceleration will be a straight but sloping line. It is straight because the acceleration is constant, and the slope corresponds to the acceleration. The steeper the slope, the higher the acceleration represented.

As Nicole Oresme had realised, the distance the object travels is represented by the area under the graph. Point A represents the object at rest. At time C, it has accelerated to speed B. Thus, the distance it has travelled is the area of the triangle, ABC. Recall

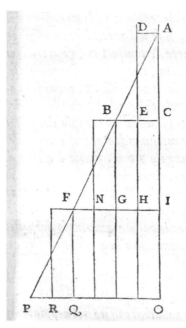

15. A reproduction of the mean speed theorem's proof in Galileo's
Discourses on Two New Sciences **(1638) from the 1914 English translation**

from the diagram in chapter 12 that the mean speed theorem of the
Merton Calculators implies that the area of triangle ABC, representing motion at a constant acceleration, equals the area of rectangle
ADEC, representing motion at the average speed of the accelerating
object. Clearly, Galileo is using this result here and demonstrating it
in the same way as Nicole Oresme had done.

If we define the distance travelled in time AC as one unit, represented by the triangle ABC, it is clear from the diagram that the
distance travelled in the next increment of equal time CI is the area
BCIF. This is, as you can see by rearranging the pieces, three times
the area of ABC. And the distance travelled in the following increment of time IO is the area FIOP which is five times the area of
ABC. This is what Galileo meant by 'the distances traversed stand to
one another in the same relation as the odd numbers beginning with
unity'. The distances covered increase in the series 1, 3, 5, 7 and so
on.

Compare this to the words of William Heytesbury in the early fourteenth century. He wrote: 'When the acceleration of a motion takes place uniformly from zero to some amount, the distance it will traverse in the first half of the time will be exactly one third of that which it will traverse in the second half of the time.'[20] And the diagram from the 1494 edition of William Heytesbury's *Rules for Solving Logical Puzzles* (following the original drafted by Nicole Oresme) really does look uncannily similar to the one used by Galileo to make the same point in the *Discourses*. William did not quite generalise the distance travelled beyond one unit and three units in the first two time intervals, but shortly afterwards Oresme did state that the distances follow the series 1, 3, 5, 7 et cetera as Galileo's would.[21]

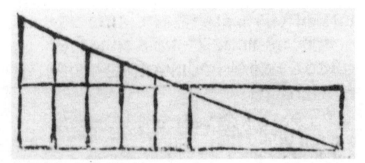

16. An illustration of the mean speed theorem from a 1494 printed edition of William Heytesbury's *Rules for Solving Logical Puzzles*

There is good evidence that Galileo knew of the very book by William Heytesbury quoted above, because in unpublished notes that he took as a student at Pisa he refers directly to it.[22] What is more, the same set of notes also mentions the Calculator Richard Swineshead and an Italian scholar, Gaetano di Thiene (1387–1465), who wrote the commentary printed in the same volume as William's text. It is true that humanists had obscured the achievements of medieval scholarship, but the relevant books were still easily available and familiar to specialists. Even if the natural philosophers of Galileo's time were beholden to Aristotle, mathematicians still used the conclusions of the Merton Calculators. One such mathematician, Niccolò Cabeo (1586–1650), was dismissive of Galileo's claims to priority. 'Another notable thing in Galileo is an intolerable boasting',

Cabeo says. 'He wants absolutely everyone to have been in shameful ignorance from the time of Adam to our own era such that they should not know by what proportions the speed of falling weights increase.'[23]

Galileo's contribution to the theory of falling objects was to realise that the distance fallen was proportional to the square of the time elapsed.[24] This result follows from the diagrams above quite easily but does not appear to have been specifically stated by the Merton Calculators. One axis of the graphs corresponds to time and the other to speed. The distance travelled, that is the area under the graph, is speed multiplied by distance divided by two. But where an object is accelerating at a constant rate, speed equals time multiplied by acceleration. That means that the other axis is proportional to time elapsed as well. So the distance must be acceleration ('a') multiplied by time ('t') squared, divided by two or $\frac{1}{2}at^2$. Anyone who remembers the mechanics they did at school will recognise this formula.

The triumph of Galileo was not his connection between the mean speed theorem and gravity. Nor was it his description of how an accelerated body moves. Rather it was his experimental proof that the relationships derived back in the fourteenth century held true in nature. He realised that in order to carry out accurate experiments, he had to slow the action down. To do this, he decided to replace dropping balls with rolling them down a slope (the slope is called an 'inclined plane' in scientific parlance). For timing, he used miniature water clocks – mechanical clocks were not yet sufficiently accurate for precision work. To time a short interval, he let water flow through a narrow spout and weighed the amount that escaped in the moment the spout was open.[25]

The idea of rolling balls down inclined planes was not new. John Marliani (d.1483), a physician in Pavia, had published a number of books on natural philosophy in the late fifteenth century. In his Question on Proportion, printed in 1482, Marliani discussed experiments he had carried out with pendulums and balls on slopes. Unlike Galileo, he was not testing the right hypothesis and consequently did not manage to make any breakthroughs.[26]

Before Galileo could rely on his experiments he had to deal with a complication: could he be sure that rolling a ball down an inclined plane was exactly equivalent to dropping it? For his results to be valid, there had to be a precise parallel between the two situations. The inclined plane was a problem that had exercised mathematicians since antiquity. How was it that a man could push up a slope a weight that he could not lift? Pushing it up the slope involved raising it just as high as picking it up. The ancient Greeks failed to provide a definitive answer to the question, but a medieval scholar, Jordanus de Nemore (c.1225–60), managed to crack it. He, or one of his early anonymous followers, proved that the force exerted by an object on a slope was proportional to the steepness of the slope.[27] This means that rolling balls down an inclined plane is analogous to dropping them. The speed of the balls will be slower but directly proportional to their speed when dropped.

In typical sixteenth-century style, Nicolò Tartaglia had reproduced and printed this important medieval result as his own work.[28] At the same time, the failed proofs attempted by ancient Greek writers came to light and mathematicians, Jerome Cardan among them, tried to understand where they had gone wrong.[29] Galileo, who must have been familiar with Jordanus's proof, spent a lot of time on the question and eventually managed his own solution based on classical methods.

Projectile Motion

The fourth and final day of the dialogue in *Discourses on Two New Sciences* covers projectile motion. Galileo begins with a statement of the law of inertia that finally puts paid to Aristotle's maxim that no object can move unless another object is moving it. Galileo tells us: 'Imagine any particle moving along a surface without any friction. Then we know ... this particle will move along a plane with a motion that is constant and perpetual, provided the surface has no limits.'[30] John Buridan had explained how the planets keep moving forever in a frictionless environment. Galileo now generalises the law and realises that strictly it applies to a flat plane rather than circles in the heavens. He then goes on to show that a projectile, such as a cannon ball, will follow a curved trajectory.

In his own book *New Science*, Tartaglia had already shown that projectiles follow curved paths, and Jerome Cardan had suggested the curve in question 'imitates the form of a parabola'.[31] However, Galileo rightly pointed out that no one had demonstrated that this was correct. He now did so from first principles. He assumed that a cannon will push the ball in the direction in which it is fired at a constant speed. Under the law of inertia, if there were no other forces acting on the ball, it would keep going in a straight line in that direction forever. But two other forces act upon it – air resistance (which we can ignore for the purposes of this discussion) and gravity. During the previous day of the dialogue, Galileo had shown what happens when an object falls under gravity. The path of the cannon ball, he explained, is the combination of it moving in a straight line and falling under gravity at the same time. Aristotle's statement that the violent motion from the cannon and the natural motion from gravity are incompatible is false. So, to calculate the path of the cannon ball you just add the two kinds of motion together. If you do so, you will find, as Cardan predicted, that the resulting path is a special kind of curve called a parabola.[32]

Galileo's achievement lay in bringing together what had been done before, disposing of the vast amount that was irrelevant or simply wrong, and then proving the remainder with controlled experiments and brilliant arguments. A kind of new science did indeed begin with him, but there is no denying that he built on medieval foundations. Without them, he would never have been able to cover a fraction of the ground that he did, even in the long life he was granted.

When he solved the problems with which medieval natural philosophers had long struggled, Galileo eclipsed his predecessors. Like Kepler on optics, he rendered much of what had gone before obsolete. That did not mean that it was not important. To historians who want to learn where Galileo and Kepler found their ideas, medieval natural philosophy is indispensable. The achievements of their generation, outstanding though they were, should not obscure the breakthroughs made by Thomas Bradwardine, John Buridan, Nicole Oresme and others.

However, the most significant contribution of the natural philosophers of the Middle Ages was to make modern science even conceivable. They made science safe in a Christian context, showed how it could be useful and constructed a worldview where it made sense. Their central belief that nature was created by God and so worthy of their attention was one that Galileo wholeheartedly endorsed. Without that awareness, modern science would simply not have happened.

CONCLUSION

❦

A Scientific Revolution?

In June 1833, the British Association for the Advancement of Science met in Cambridge. Among the matters discussed was the curious fact that there was no overarching name for the men (not many women in those days) who studied the workings of nature. Science was coming on in leaps and bounds but its practitioners had no official title. At the meeting was the poet, author of *Kublai Khan* and *Rime of the Ancient Mariner*, Samuel Taylor Coleridge (1772–1834). He posed the question of what would be a suitable name for the student of nature. The ensuing discussion was reported shortly afterwards:

> *Philosophers* was felt to be too wide and too lofty a term, and this was very properly forbidden them by Mr Coleridge, both in his capacity of philologer and metaphysician; *savans* was rather assuming, besides being French instead of English; some ingenious gentleman proposed that, by analogy with *artist*, they might form *scientist*, and added that there could be no scruple in making free with this termination when we have such words as economist and atheist – but this was not generally palatable.[1]

The ingenious person in question was William Whewell (1794–1866), president of the British Association and a notable author. He had helped feed the fashion for science in the Victorian era and could be called one of the first popular science writers. Although his suggestion

337

of 'scientist' did not go down well at the meeting, Whewell promptly started using it in his books and the term soon caught on.

The word 'scientist' was not coined until 1833 because only then did people realise it was needed. Science had become an autonomous subject that was completely separate from philosophy and theology. Although assumptions about God and creation had been necessary for science to get this far, it was now so successful that it no longer needed them. In this book, no one from the Middle Ages is called a scientist. Indeed the word isn't used at all, and not just because the term had not been invented. Rather, the kind of person we call a scientist did not exist in the Middle Ages. This book is the story of the gestation of science and obviously, before modern science was born, there could be no scientists.

The phrase 'before modern science was born' is not quite apposite. There was no moment of birth, no crisis from which science burst forth. All we can say is that by the time the British Association met in 1833, the process was complete and modern science undeniably existed as a younger version of what we have today.

At that point, people started asking where science came from. The general consensus rapidly emerged that it began with the ancient Greeks. Scientific progress was thought to have stalled when medieval Europe dropped the baton. Then, in the sixteenth century, Copernicus and Galileo picked it up again. But, as this book has explained, that picture is inaccurate and unfair. So, let us now briefly recap how the Middle Ages laid down the foundations of modern science.

There were four cornerstones, which can be described as institutional, technological, metaphysical and theoretical. Since this book tells the story in roughly chronological order, we have come across many instances of these cornerstones as they arose. For this summary, we can group the examples together to present a clearer picture of the medieval achievement.

The main scientific *institution* of the Middle Ages was the university. Although these were primarily intended to educate prospective members of the higher clergy, they also provided a home for natural philosophers. As it turned out, many theologians also wrote important works on natural philosophy and they considered the subject an essential part of their training. Protected by both Church and state, universities gave students and their professors unprecedented levels of security and intellectual freedom. Students enjoyed the same legal status as clerics and it was difficult for local secular authorities to restrict what they studied. For the Church's part, it never offered unqualified support for all branches of science and set limits beyond which natural philosophers were not allowed to go. However, we have seen that most of the stories about how the Church held back science are myths that arose after the Middle Ages had ended. Overall, the relationship between Christianity and natural philosophy that we have seen played out in this book might best be summed up with the words 'creative tension'.

Today, western universities remain the premier institutions for scientific research and training. The model pioneered in the Middle Ages, whereby the universities were self-governing corporations, has spread around the world, even to those places that were never colonised by Europeans. Almost all the earliest universities are still with us and some, like Oxford and Cambridge, are going from strength to strength.

Technological advances during the Middles Ages led to enormous increases in agricultural productivity and improvements in living standards. Some also had a direct impact on science. For example, we have seen how a glassmaker in Venice or Pisa invented spectacles in the thirteenth century. The Venetians became expert at grinding lenses and it is no coincidence that Galileo was working near Venice when he built his telescope. The navigational compass inspired the investigation of magnetism by Peter the Pilgrim and William Gilbert. Other inventions had less direct, but still profound, effects on scientific thinking. The mechanical clock, with its 24-hour cycle, closely resembled the medieval world-picture. The heavens themselves supposedly revolved around the earth each day. So the metaphor of the world as a clock built by a divine clockmaker came naturally to

Thomas Bradwardine and Nicole Oresme. This encouraged thinking about nature as a mechanism that could be investigated by reason.

Much of the technology we take for granted today, like the computer upon which this book was written, would not exist but for the achievements of modern science. In the Middle Ages, technology and engineering did not owe anything to natural philosophy. The relationship was the other way around. Technical advances gave natural philosophers clues about how the world worked as well as providing the equipment that they needed to investigate it. They could not point to any practical benefits of their work, still less applications that proved their theories were correct. This meant that they needed other ways of justifying their activities.

The *metaphysical* cornerstone of modern science is often overlooked. We take it for granted and we do not worry about why people began studying nature in the first place. Today you can enhance the credentials of any outlandish theory you like by labelling it 'scientific', as advertisers and quacks well appreciate. But, back in the Middle Ages, science did not enjoy the automatic authority that it has today.

To understand why science was attractive even before it could demonstrate its remarkable success in explaining the universe, it is necessary to look at things from a medieval point of view. The starting point for all natural philosophy in the Middle Ages was that nature had been created by God. This made it a legitimate area of study because through nature, man could learn about its creator. Medieval scholars thought that nature followed the rules that God had ordained for it. Because God was consistent and not capricious, these natural laws were constant and worth scrutinising. However, these scholars rejected Aristotle's contention that the laws of nature were bound by necessity. God was not constrained by what Aristotle thought. The only way to find out which laws God had decided on was by the use of experience and observation. The motivations and justification of medieval natural philosophers were carried over almost unchanged by the pioneers of modern science. Sir Isaac Newton explicitly stated that he was investigating God's creation, which was a religious duty because nature reflects the creativity of its maker. In 1713, he inserted into the second edition of his greatest work, *The Mathematical Principles of Natural Philosophy*, the words:

> Blind metaphysical necessity, which is certainly the same always and everywhere, could produce no variety of things. All that diversity of organisms which we find suited to different times and places could arise from nothing but the ideas and will of a Being necessarily existing ... And that is enough concerning God, to discourse of whom from the appearances of things does certainly belong to natural philosophy.[2]

It would take Charles Darwin (1809–82) to prove Newton wrong.

The final cornerstone is the set of *theories* about the world that the Middle Ages bequeathed to the early modern period. Here are a few of them: John Buridan's impetus theory was a vital stepping stone towards modern mechanics. He also showed how a planet that does not suffer any resistance keeps going forever and he explained why we cannot feel the rotation of the earth. William Heytesbury described the motion of an object subject to constant acceleration. Nicole Oresme proved Heytesbury's theorem right by using a graph. Nicholas of Cusa went so far as to suggest a limitless universe and life on other planets. And Peter the Pilgrim analysed magnets and suggested the use of a spherical magnet to William Gilbert. Without these scientific advances, it is hard to see how Copernicus, Kepler and Galileo could have made the progress that they did.

The power of many medieval theories was derived from the way that they combined mathematics with natural philosophy. Aristotle kept these two subjects distinct but in the Middle Ages they were increasingly linked together. Thomas Bradwardine had shown that a successful mathematical description of nature should work all the time, while the Merton Calculators popularised the idea of using numbers to study physical problems. This meant that by the sixteenth century, the boundary between mathematics and natural philosophy could effectively be ignored.

Nowadays, it is a commonplace to refer to the period between Copernicus and Newton as 'the scientific revolution'. Although this phrase was only invented in the mid-twentieth century, it has already become an unquestioned part of the language.[3] Most historians do think that something revolutionary happened, even if they dislike the label. Some lonely voices, like Andrew Cunningham at Cambridge and Steven Shapin in California, challenge this hegemony.[4] Cunningham believes that modern science was born in the years around 1800, which were revolutionary in other respects as well.[5] That modern medicine did not exist prior to the mid-nineteenth century and that it bears no resemblance to earlier medical practice is beyond dispute.[6]

This book should lend some support to the sceptics claiming that the term 'the scientific revolution' is another one of those prejudicial historical labels that explain nothing. You could call any century from the twelfth to the twentieth a revolution in science, with our own century unlikely to end the sequence. The concept of the scientific revolution does nothing more than reinforce the error that before Copernicus nothing of any significance to science took place at all.

Life in the Middle Ages was often short and violent. The common people were assailed by diseases they didn't understand; exploited by a distant ruling class; and dependent on a Christian church that rarely lived up to the ideals of its founder. It would be wrong to romanticise the period and we should be very grateful that we do not have to live in it. But the hard life that people had to bear only makes their progress in science and many other fields all the more impressive. We should not write them off as superstitious primitives. They deserve our gratitude.

SUGGESTIONS FOR FURTHER READING

There is now a good deal of scholarship devoted to medieval science but much of it is written with the specialist in mind. The books recommended in this section are intended as possible next steps for non-academic readers wishing to explore some of the issues raised in this book.

The standard textbook on medieval science is David C. Lindberg, *The Beginnings of Western Science*, second edition (Chicago: University of Chicago Press, 2007). This is an excellent introduction covering both ancient and medieval natural philosophy. Unfortunately, it is very obviously a textbook for university students and as a result rather dry. The articles in *Science in the Middle Ages* (Chicago: University of Chicago Press, 1978) present much more detail on the individual subjects that made up science in the Middle Ages. Marcia L. Colish, *Medieval Foundations of the Western Intellectual Tradition, 400–1400* (New Haven, CT: Yale University Press, 1997) is a comprehensive and fascinating introduction to all aspects of medieval culture. On the history of medicine, Roy Porter, *The Greatest Benefit to Mankind: A Medical History of Humanity from Antiquity to the Present* (London: HarperCollins, 1998) covers the subject thoroughly. Arthur Koestler, *The Sleepwalkers: A History of Man's Changing Vision of the Universe* (Harmondsworth: Penguin, 1990) is a novelist's biography of Copernicus, Kepler and Galileo, but the short medieval section is woefully inaccurate.

A complete history of science from prehistory to the present is James McClellan, *Science and Technology in World History: An Introduction* (Baltimore: The Johns Hopkins University Press, 1999). For scientific enterprise in Islam, Ehsan Masood, *Science and Islam:*

A History (Icon Books, 2009) and Howard Turner, *Science in Medieval Islam: An Illustrated Introduction* (Austin: University of Texas Press, 1995) are both enjoyable and reliable. Toby Huff, *The Rise of Early Modern Science: Islam, China and the West*, second edition (Cambridge University Press, 2003) is controversial and not always dependable, but a brave attempt to find what made western science unique.

On the relationship between science and religion, good articles can be found in the two collections *God and Nature: Historical Essays on the Encounter Between Christianity and Science* (Berkeley: University of California Press, 1986) and *When Science & Christianity Meet* (Chicago: University of Chicago Press, 2003). The articles in *Galileo Goes to Jail and Other Myths about Science and Religion* (Cambridge, MA: Harvard University Press, 2009) attempt to dispel some of the misinformation on this subject. Edward Peters, *Inquisition* (Berkeley: University of California Press, 1989) deals with both the myth and the reality of the notorious institution.

Finally, some academic books are just so good and so well written that they deserve as wide an audience as possible. M.T. Clanchy, *Abelard: A Medieval Life* (Oxford: Blackwell, 1997) brings the twelfth century alive. Francis Yates, *Giordano Bruno and the Hermetic Tradition* (London: Routledge Classics, 2002) explains the alien world of the Renaissance magicians better than anything else. Lynn White, *Medieval Technology and Social Change* (Oxford: Oxford University Press, 1966) revolutionised the study of medieval technology. David Wootton, *Bad Medicine: Doctors Doing Harm Since Hippocrates* (Oxford: Oxford University Press, 2006) is probably the most important book ever written on the history of medicine and is fantastically readable.

TIMELINE

~~

Year	Event
406	Barbarians invade the Roman Empire across the frozen River Rhine
410	The Goths sack Rome
455	The Vandals sack Rome
476	The last Roman Emperor in the West, Romulus Augustulus, is deposed
526	Boethius executed; *Consolation of Philosophy* published
622	Mohammed flees from Mecca to Medina; start of the Muslim calendar
732	Charles Martel defeats Muslim invaders of France at the Battle of Poitiers
800	Charlemagne crowned Emperor in Rome
999	Gerbert of Aurillac becomes Pope
1066	William of Normandy invades England; Battle of Hastings
1085	Muslim city of Toledo in Spain falls to Christians with its library intact
1086	Compilation of the *Domesday Book*
1092	Walcher makes an astronomical observation with an astrolabe
1093	Anselm appointed archbishop of Canterbury
1121	The first trial of Peter Abelard
1140	The second trial of Peter Abelard
1158	The world's first university in Bologna granted imperial privileges
1204	Constantinople falls to the fourth crusade
1210	Aristotle's books on natural philosophy banned in Paris

1231 The Pope reinstates Aristotle's books on natural philosophy

1257 Thomas Aquinas receives his doctorate in theology in Paris

1268 Roger Bacon completes his *Opus Maior*, *Opus Minor* and *Opus Tertium* for the Pope

1272 Earliest mention of a mechanical clock

1277 The bishop of Paris condemns 219 propositions derived from the philosophy of Averröes

1284 Traditional date for the invention of spectacles

1309 The Pope leaves Rome to take up residence in Avignon in Southern France

1320 John Buridan begins to teach philosophy at Paris

1323 Thomas Aquinas canonised

1327 Cecco D'Ascoli burnt in Florence for subjecting God to astrology

1335 William Heytesbury's *Rules for Solving Logical Puzzles* published, containing the mean speed theorem

1339 Failed effort to ban the ideas of William of Ockham from Paris

1347 The Black Death first appears in western Europe

1347 Nicholas of Autrecourt recants his errors after appeal to the Pope fails

1377 The Pope returns to Rome but an anti-Pope remains at Avignon. Beginning of the Western Schism

1417 The schism ends with the election of an undisputed pope.

1434 The dome on Florence's cathedral is complete

1453 Constantinople falls to the Turks

1455 Johann Gutenberg prints his great Bible with moveable type

1488 Bartolomeu Dias doubles the Cape of Good Hope

1492 Columbus sails across the Atlantic

1492 Granada becomes the last Muslim enclave in southern Spain to fall to Christians

1513 The Catholic Church proclaims that the immortality of the soul can be proved philosophically

1517 Martin Luther nails his 95 theses to a church door in Wittenberg and sparks the Reformation

1535   The medieval syllabus abolished at the universities of Oxford and Cambridge

1537   *New Science* by Niccolò Tartaglia shows projectiles move in curves

1540   Ignatius Loyola founds the Jesuits

1543   Publications of Copernicus's *On the Revolutions of the Heavenly Spheres*, Vesalius's *On the Fabric of the Human Body* and Archimedes' *Works*

1553   Michael Servetus burnt at the stake in Geneva

1559   The Catholic Church launches its first *Index of Prohibited Books*

1563   The Council of Trent concludes and starts the Counter-Reformation

1572   A supernova appears in the heavens and proves not to be an atmospheric phenomenon

1575   Death of Jerome Cardan in Rome

1577   Appearance of a comet observed by Tycho Brahe

1600   Publication of William Gilbert's *On the Magnet*

1610   Galileo announces his new discoveries with the telescope

1616   The Catholic Church condemns the ideas of Copernicus

1618   Beginning of the Thirty Years War between Catholics and Protestants in Germany

1627   Kepler publishes the *Rudulphine Astronomical Tables* whose accuracy leads to the acceptance of heliocentricism

1628   William Harvey publishes *On the Motion of the Heart*

1632   Publication of Galileo's *Dialogue Concerning the Two Chief World Systems*

1633   The trial of Galileo in Rome

1638   Publication of Galileo's *Discourses on Two New Sciences*

1642   Death of Galileo

LIST OF KEY CHARACTERS

Abelard, Peter (1079–1142): Outspoken logician and theologian who had a controversial career. Writer of *Story of My Calamities* and lover of Héloïse.

Adelard of Bath (c.1080–c.1160): Travelled to the Middle East in search of Arabic texts. Translator of Euclid's *Elements* and author of *Natural Questions*.

Albert of Saxony (c.1316–1390): German pupil of John Buridan whose textbooks helped spread his master's ideas. Showed a projectile following a curved path.

St Albert the Great (c.1200–1280): German natural philosopher and theologian who taught Thomas Aquinas. Called the 'Universal Doctor' and patron saint of natural science.

Alcuin of York (c.735–804): Minister and educational reformer under the Emperor Charlemagne.

Alderotti, Taddeo (c.1223–1295): Pioneer of learned medicine in Bologna.

Alhazen (Al-Haytham) (96–c.1039): Muslim philosopher whose work on light was unsurpassed during the Middle Ages.

Al-Khwarizmi (c.780–c.850): Muslim mathematician who wrote an essential guide to calculation and algebra.

Al-Tusi, Nasir al-Din (1201–1274): Persian astronomer whose calculations found their way into Copernicus's *On the Revolutions of the Heavenly Spheres*.

Amaury of Bène (d.c.1207): Heretic who taught at the university of Paris and whose followers, the Amalricians, were burnt at the stake in 1210.

St Anselm of Canterbury (1033–1109): Saint and archbishop of Canterbury who combined faith and reason in his ontological argument that God exists.

Aquinas, St Thomas (1225–74): The 'Angelic Doctor' whose massive volumes of philosophy and theology made Greek philosophy safe for Christians.

Archimedes (287–212BC): Ancient Greek engineer and mathematician who discovered the law of buoyancy and much else besides. Killed during the sack of Syracuse by the Romans.

Aristotle (384–322BC): Greek philosopher who was a pupil of Plato and tutor to Alexander the Great. In the Middle Ages he was called simply 'The Philosopher'.

St Augustine of Hippo (AD354–430): Father of the Church and Bishop of Hippo in North Africa. He wrote extremely influential books on theology including *The City of God*.

Averröes (Ibn Rushd) (1126–98): Philosopher from Muslim Spain whose detailed analysis of the work of Aristotle earned him the title of 'The Commentator'.

Avicenna (Ibn Sina) (c.980–1037): Muslim doctor and philosopher whose writings, including the *Canon of Medicine*, were extremely influential in medieval Europe.

Bacon, Francis (1561–1626): Lord Chancellor to King James I of England who attempted to reform the science of his day.

Bacon, Roger (1214–92): English Franciscan who investigated light and wrote voluminous works on the reform of natural philosophy for the Pope.

Bartolomeu Dias (d.1500): Portuguese sea captain who was the first European to sail around the Cape of Good Hope.

Bellarmine, St Robert (1542–1621): Jesuit cardinal and theologian who oversaw the trial of Giordano Bruno and the condemnation of heliocentricism.

Berenger of Tours (c.1000–88): Theologian who was accused of heresy for doubting the doctrine of transubstantiation.

St Bernard of Clairvaux (1090–1153): Radical monk and reformer of Christianity. Successfully accused Peter Abelard of heresy.

Bernardo Gui (c.1261–1331): Dominican inquisitor in Toulouse and first biographer of Thomas Aquinas.

Bessarion, John (1403–72): Byzantine émigré who became a Catholic cardinal and noted patron of Greek studies.

Boethius, Anicius Manlius Severinus (AD480–525): Roman aristocrat and philosopher. Wrote textbooks and the *Consolation of Philosophy*. Executed for treason by King Theodoric.

Bonatti, Guido (c.1210–c.1290): Astrologer condemned to hell by Dante.

Bradwardine, Thomas (c.1290–1349): Mathematician who studied at Merton College, Oxford and later became archbishop of Canterbury.

Brahe, Tycho (1546–1601): Scandinavian astronomer who carried out observations of unprecedented accuracy from the island of Hven, Denmark.

Bricot, Thomas (d.1516): Theologian at the university of Paris whose textbooks were very popular in the early sixteenth century.

Brunelleschi, Filippo (1377–1446): Architect and artist who designed the dome of Florence's cathedral and pioneered the use of perspective in paintings.

Bruno, Giordano (1548–1600): Neo-pagan and enthusiast for the Hermetic corpus. Burnt at the stake in Rome by the Inquisition.

Buridan, John (c.1300–c.1358): Rector and philosopher at the university of Paris who developed impetus theory and speculated on the rotation of the earth.

Calvin, John (1509–64): French religious reformer who founded the Reformed branch of Protestantism from his base in Geneva.

Campanella, Tommaso (1568–1639): Magician and occult philosopher employed by Pope Urban VIII to protect him from malign stars.

Cardan, Jerome (1501–76): Italian doctor, mathematician, astrologer and inventor.

Cecco D'Ascoli (c.1269–1327): Astrologer and teacher at the university of Bologna, twice condemned by the Inquisition and burnt at the stake in Florence.

Chaucer, Geoffrey (c.1343–1400): English poet who wrote a scientific treatise on the use of the astrolabe as well as the *Canterbury Tales*.

Cicero, Marcus Tullius (106–43BC): Roman politician and orator whose Latin style was esteemed by humanists.

Clavius, Christopher (c.1538–1612): Jesuit astronomer who taught in Rome and supported the old cosmology of Ptolemy.

Copernicus, Nicolaus (1473–1543): Polish canon who remodelled the universe with the sun instead of the earth at its centre.

Cosimo de Medici (1389–64): Oligarch who ruled Florence and was patron of Marsilio Ficino's translations from Greek to Latin.

Cosimo I de Medici (1590–1621): Grand Duke of Tuscany and patron of Galileo.

Cremonini, Cesare (1550–1631): Aristotelian philosopher and colleague of Galileo at the university of Padua.

Cromwell, Thomas (c.1485–1540): Adviser to King Henry VIII who masterminded the replacement of the medieval syllabus with its humanist equivalent at the English universities.

D'Ailly, Pierre (1350–1420): A graduate of the University of Paris who was made a cardinal in 1411. A noted writer of geography who helped inspire Columbus.

Dante Alighieri (1265–1321): Florentine poet whose *Divine Comedy* featured many of the personalities of his time assigned to places in heaven or hell.

Dee, John (1527–1609): English astrologer and magician who unsuccessfully tried to reform astrology into a mathematical discipline.

Domingo de Soto (1494–1560): Spanish Dominican friar whose textbook on Aristotle's *Physics* was the first accurate statement of the law of free fall.

Duns Scotus, John (c.1265–1308): Franciscan who taught theology at Oxford and Paris and whose difficult philosophy carried forward the work of Thomas Aquinas.

Erasmus, Desiderius (c.1469–1536): Dutch humanist whose satires attacked medieval learning and who published a New Testament in its original Greek.

Euclid of Alexandria (c.325–c.265BC): Master of Greek geometry whose book, the *Elements*, dominated the subject for over 2,000 years.

Fallopio, Gabriele (1523–62): Anatomist and student of Vesalius who first identified the fallopian tubes.

Ficino, Marsilio (1433–99): Humanist and Greek scholar who translated the dialogues of Plato and the Hermetic corpus into Latin.

Fludd, Robert (1574–1637): Astrologer and alchemist from Kent who debated with Johann Kepler over the proper place of mathematics in science.

Frederick II (1194–1250): Holy Roman Emperor who patronised science and wrote a treatise on birds. Popularly called the 'Wonder of the World'.

Galen of Pergamon (c.AD131–c.201): Greek doctor who served the emperors of Rome. His many books formed the foundation of medieval Arabic and Christian medicine.

Galilei, Galileo (1564–1642): Florentine mathematician and natural philosopher who made numerous important discoveries.

Geoffrey of Rots (fl.1086): The Norman lord of Otham in Kent.

Gerbert of Aurillac (Sylvester II) (c.940–1003): Scholar and administrator who became Pope. Introduced Arabic numerals to Christian Europe.

Gilbert, William (1540–1603): London doctor whose book *On the Magnet* is celebrated as a seminal work of experimental science.

Grassi, Horatio (1583–1654): Jesuit astronomer who quarrelled with Galileo over comets.

Grosseteste, Robert (c.1170–1253): Bishop of Lincoln who wrote on natural philosophy and optics while at Oxford.

Hamilton, John (1511–71): Archbishop of St Andrews in Scotland whose lung condition improved under the ministrations of Jerome Cardan.

Harvey, William (1578–1657): English doctor who discovered the function of the heart and the circulation of the blood.

Héloïse (d.1164): Lover of Peter Abelard and abbess of the Paraclete.

Hermes Trismegistus: Mythical Egyptian sage who was reputed to live as a contemporary of Moses. His spurious works were revered by some in the sixteenth century.

Heytesbury, William (c.1313–73): Mathematician from Merton College, Oxford who first wrote down the mean speed theorem.

Ibn al-Shatir (d.1375): Syrian astronomer whose mathematical models found their way into the work of Copernicus.

Innocent III (1160–1216): Pope who launched the Albigensian crusade against the Cathars and increased the effectiveness of canon law.

St Jerome (c.AD340–420): Translated the Bible from Hebrew and Greek into colloquial Latin. This *Vulgate* translation was the official Bible of the Middle Ages.

John XXI (Peter of Spain) (c.1215–77): Pope who wrote textbooks on logic and medicine before his enthronement. Ordered the 1277 condemnations at the university of Paris.

John XXII (1249–1334): Pope who condemned William of Ockham and fraudulent alchemists.

Jordanus de Nemore (c.1225–60): Mathematician who studied the science of statics and solved the inclined plane problem.

Kilwardby, Robert (d.1279): Archbishop of Canterbury and categoriser of the sciences.

Lanfranc (c.1005–89): Master of Anselm at the Abbey of Bec in Normandy and his predecessor as archbishop of Canterbury.

Lombard, Peter (c.1095–c.1160): Theologian who taught in Paris and produced a synthesis of the sayings of the Christian fathers called the *Sentences*.

Maimonides, Moses (1135–1204): Jewish theologian who lived under Islamic rule. His *Guide for the Perplexed* attempted to reconcile natural philosophy with the Bible.

Melanchthon, Philipp (1497–1560): Martin Luther's intellectual protégé best known as an educational reformer.

Mondino dei Luzzi (d.c.1326): Pioneer of human dissection who worked at the university of Bologna.

de Montaigne, Michael (1533–92): French writer whose *Essays* combine classical scholarship with scepticism and mordant wit.

More, St Thomas (1477–1535): English humanist executed by King Henry VIII for his support of the Catholic Queen Catherine of Aragon.

Nicholas of Autrecourt (c.1300–69): Paris theologian convicted of heresy in 1347 for his ideas on atoms and the Eucharist. Forced to leave the University.

Nicholas of Cusa (1400–64): Theologian and mathematician who became a cardinal in 1449. He speculated on a limitless universe and life on other planets.

Oresme, Nicole (c.1325–82): Student of Buridan and celebrated mathematician who developed the use of graphs to model physical problems. Became bishop of Lisieux in 1377.

Otto III (980–1002): Holy Roman Emperor in the Saxon dynasty and patron of Gerbert.

Paracelsus (Theophrastus Bombastus von Hohenheim) (1493–1541): Renegade doctor who attempted to reform medicine along occult and alchemical lines.

Patrizi, Francisco (1529–97): Platonic philosopher who believed in the rotation of the earth and extra-terrestrial vacuums.

Peckham, John (d.1292): Franciscan and archbishop of Canterbury who wrote a popular textbook on optical theory.

Peter the Pilgrim (Pierre de Maricourt) (fl.1269): Scholar who took part in the siege of Lucera in Italy and wrote a treatise on the properties of magnets.

Peter the Venerable (1092–1156): Abbot of the great monastery of Cluny who sheltered Peter Abelard in his old age.

Petrarch, Francesco (1304–74): Italian humanist who invented the concept of the Dark Ages.

Peurbach, George (1423–61): German astronomer who worked as the Imperial Astrologer in Vienna. Master and colleague of Regiomontanus.

Philoponus, John (c.AD490–570): A Neo-Platonic Christian philosopher who worked at the school of Alexandria. Criticised the thought of Aristotle.

Plato (429–347BC): Greek philosopher whose dialogues are masterpieces of prose as well as the founding documents of western thought.

Pletho, Gemistus (c.1355–1452): Neo-pagan thinker who attended the Council of Florence and influenced Cosimo de Medici.

Pliny the Elder (AD23–79): Roman naturalist whose encyclopaedia was popular throughout the Middle Ages. Killed during the eruption of Vesuvius that buried Pompeii.

Pomponazzi, Pietro (1462–1525): Aristotelian philosopher at the university of Padua who doubted it was possible to philosophically prove the immortality of the soul.

Ptolemy of Alexandria (fl.AD140–170): Greek astronomer whose books the *Almagest* and the *Geography* were the cream of ancient Greek mathematical science.

Ragimbold of Cologne and Radolf of Liege (fl. eleventh century): Correspondents who wrote to each other on geometrical problems and the astrolabe.

Record, Robert (1510–58): London doctor who wrote books in English on mathematics, medicine and astronomy.

Regiomontanus, Johann (Johann Müller) (1437–76): German astronomer who condensed and corrected the *Almagest* of Ptolemy. Student of Peurbach.

Richard of Wallingford (1292–1336): Astronomer and abbot of the monastery of St Albans. He invented new astronomical instruments and built a renowned clock.

Roscelin of Compiègne (c.1050–c.1125): Teacher whose writings on the Trinity were condemned as heretical.

Sacrobosco, John (d.c.1256): Englishman who taught at the university of Paris and wrote popular basic textbooks on astronomy and arithmetic.

Scaliger, Julius Caesar (1484–1558): Humanist who criticised the work of Jerome Cardan.

Servetus, Michael (1511–53): Unitarian theologian who was burnt at the stake on the orders of John Calvin. Discovered the purpose of the pulmonary artery.

Siger of Brabant (d.c.1282): Christian follower of Averröes who was accused of denying the immortality of the soul and the creation of the world.

Stevin, Simon (1548–1620): A Dutch engineer who experimentally verified that heavy and light objects fall at the same speed.

Swineshead, Richard (fl.1340–55): Mathematician from Merton College, Oxford who was called 'The Calculator' due to his impressive command of the subject.

Tagliacozzi, Gaspare (1545–99): Plastic surgeon who specialised in repairing noses.

Taisnier, Jean (1508–62): French priest and plagiarist who passed off medieval natural philosophy as his own work.

Tartaglia, Niccolò (c.1499–1557): Stuttering Italian mathematician who solved cubic equations, published the works of Archimedes and quarrelled with Jerome Cardan.

Trapezuntius (George of Trebizond) (1395–1486): Greek émigré and translator who produced the first Latin editions of many of the Greek classics.

Urban VIII (Maffeo Barberini) (1568–1644): The Pope who, although he was initially supportive, ordered the trial of Galileo and was a patron of Campanella.

Velcurio, Johann (d.1534): Protestant professor whose textbooks excluded medieval learning from the sciences.

Vesalius, Andreas (1514–64): Anatomist from the Netherlands whose book *On the Fabric of the Human Body* tried to reform the work of Galen.

St Virgil of Salzburg (c.700–784): Irish missionary who became a bishop and was involved in a controversy over the antipodes.

William of Conches (1085–c.1154): Platonic philosopher who said the Bible should be interpreted non-literally.

William of Malmesbury (d.c.1143): English chronicler who wrote about Gerbert of Aurillac.

William of Ockham (or Occam) (c.1287–1347): Oxford Franciscan accused of heresy in Avignon who fled to serve the Holy Roman Emperor in Germany.

Witelo (fl.1250–75): Polish clergyman who wrote the largest of the medieval treatises on optics and inspired Kepler's study of light.

NOTES

Introduction – The Truth about Science in the Middle Ages

1   Issac Newton, letter to Robert Hooke, 5 February 1676, cited in Richard S.
     Westfall, *Never at Rest: A Biography of Isaac Newton* (Cambridge: Cambridge
     University Press, 1983), p. 274.

2   Richard William Southern, *The Making of the Middle Ages* (London:
     Hutchinson's University Library, 1953), p. 203.

3   Daniel J. Boorstin, *The Discoverers* (New York: Vintage, 1985), p. 102;
     William Manchester, *A World Lit Only by Fire: The Medieval Mind and the
     Renaissance: Portrait of an Age* (London: Macmillan, 1993), p. 3; Charles
     Freeman, *The Closing of the Western Mind: The Rise of Faith and the Fall of
     Reason* (London: William Heinemann, 2002), p. 328.

4   Thomas H. Huxley, *Darwiniana*, vol. 2; *Essays by Thomas H. Huxley*
     (London: Macmillan, 1894), p. 52.

5   Edward Grant, *God and Reason in the Middle Ages* (Cambridge: Cambridge
     University Press, 2001), p. 324.

6   Jean Le Rond D'Alembert (trans. Richard H. Schwarb), *The Preliminary
     Discourse to the Encyclopaedia of Diderot* (Chicago: Chicago University Press,
     1995), p. 62.

7   Colin A. Russell, 'The Conflict of Science and Religion', in Gary B.
     Ferngren, ed., *Science and Religion: A Historical Introduction* (Baltimore: The
     Johns Hopkins University Press, 2002), p. 10.

8   Girolamo Cardano, *De subtilitate* l. 3 in *Opera* vol. iii, p. 602 cited in Joseph
     Gies and Frances Gies, *Cathedral, Forge and Waterwheel: Technology and
     Invention in the Middle Ages* (London: Harper Perennial, 1995), p. 2.

9   Ibid., p. 246.

10   Grant, *God and Reason in the Middle Ages*, p. 30.

11   Michael White, *Leonardo: The First Scientist* (London: St. Martin's Griffin,
     2001).

12   Charles Nicholl, *Leonardo da Vinci: The Flights of the Mind* (London: Allen
     Lane, 2004), p. 96.

13   Many of Leonardo's inventions were not as original as often supposed. See
      Jean Gimpel, *The Medieval Machine: The Industrial Revolution of the Middle Ages*
      (London: Pimlico, 1992), p. 142.
14   Jeffrey Burton Russell, *Inventing the Flat Earth: Columbus and Modern
      Historians* (New York: Praeger Paperback, 1997), p. 65.

Chapter 1 – After the Fall of Rome: Progress in the Early Middle Ages

1   Angela Care Evans, *The Sutton Hoo Ship Burial* (London: British Museum
      Publications, 1986), p. 57.
2   Roger Collins, *Early Medieval Europe, 300–1000*, 2nd edn (Basingstoke:
      Palgrave Macmillan, 1999), p. xxiii.
3   Jared M. Diamond, *Collapse: How Societies Choose to Fail or Succeed* (London:
      Viking, 2005), p. 297.
4   Adapted from Ann Williams and G. H Martin (eds), *Domesday Book: A
      Complete Translation* (London: Penguin, 2003), p. 20.
5   Francis Pryor, *Britain in the Middle Ages: An Archaeological History* (London:
      HarperPress, 2006), p. 65. The heavy plough was known to the Romans but
      does not appear to have been widely adopted in northern Europe, where
      it is most suitable to be used, until the centuries after the Western Empire
      fell.
6   Lynn White, *Medieval Technology and Social Change* (Oxford: Oxford
      University Press, 1966), p. 43.
7   Ibid., p. 2.
8   Paul Gans, 'The Medieval Horse Harness: Revolution or Evolution: A
      Case Study in Technological Change', in *Villard's legacy : Studies in Medieval
      Technology, Science and Art in Memory of Jean Gimpel* (Aldershot: Ashgate,
      2004), p. 179.
9   Ibid., p. 184.
10   Jean Gimpel, *The Medieval Machine: The Industrial Revolution of the Middle Ages*
      (London: Pimlico, 1992), p. 40.
11   Ibid., p. 12.
12   Lynn White, *Medieval Religion and Technology: Collected Essays* (Berkeley:
      University of California Press, 1978), p. 21.
13   Josiah Cox Russell, *The Control of Late Ancient and Medieval Population*
      (Philadelphia: American Philosophical Society, 1985), p. 36.
14   White, *Medieval Religion and Technology*, p. 14. Although White's work has
      been developed by more recent scholars, his fundamental thesis still stands.
15   Henry Chadwick, *The Early Church* (Harmondsworth: Penguin, 1967),
      p. 225.
16   Ibid., p. 171.

17 Among works that exist only in the Arab version are Ptolemy's *Optics* and the later books of Apollonius of Perga's *Conics*.

18 Howard R. Turner, *Science in Medieval Islam: An Illustrated Introduction* (Austin: University of Texas Press, 1995), p. 47.

19 *Conquerors and Chroniclers of Early Medieval Spain*, 2nd edn (Liverpool: Liverpool University Press, 1999), p. 144.

20 Edward Gibbon, *The History of the Decline and Fall of the Roman Empire*, vol. 3 (Harmondsworth: Penguin, 1994), p. 336.

21 L.D. Reynolds and N.G. Wilson, *Scribes and Scholars: A Guide to the Transmission of Greek and Latin Literature*, 3rd edn (Oxford: Clarendon, 1991), p. 94.

22 Ibid., p. 93.

23 Collins, *Early Medieval Europe, 300-1000*, ch. 20.

Chapter 2 – The Mathematical Pope

1 Horace K. Mann, *The Lives of the Popes in the Early Middle Ages*, vol. 5 (London: Kegan Paul, 1910), p. 63. I have taken much of the story of Gerbert from Mann's volume and Anna Marie Flusche, *The Life and Legend of Gerbert of Aurillac: The Organbuilder who became Pope Sylvester II* (Lewiston, NY: The Edwin Mellen Press, 2005).

2 Alexander Murray, *Reason and Society in the Middle Ages* (Oxford: Clarendon Press, 1985), p. 290.

3 An outstanding example is the Basilica of Santa Maria Maggiore in Rome.

4 Maurice Hugh Keen, *The Penguin History of Medieval Europe* (Harmondsworth: Penguin, 1969), p. 42.

5 Richer of Saint Remi, *Historiae*, MGH Scriptores, vol. 38 (Hanover, 2000).

6 Gerbert of Aurillac (trans. Harriet Pratt Lattin), *The Letters of Gerbert: with his Papal Privileges as Sylvester II* (New York: Columbia University Press, 1961).

7 Gerbert, *The Letters of Gerbert*, p. 140.

8 D.J. Struik, 'Gerbert', in *Dictionary of Scientific Biography*, ed. Charles Coulston Gillispie, vol. 5 (New York: Scribner, 1970), 365.

9 Gerbert, *The Letters of Gerbert*, p. 184.

10 Stephen C. McCluskey, *Astronomies and Cultures in Early Medieval Europe* (Cambridge: Cambridge University Press, 1998), p. 177.

11 Ibid., p. 180.

12 William of Malmesbury (trans. R.A.B. Mynors), *Gesta Regum Anglorum: The History of the English Kings*, vol. 1 (Oxford: Clarendon Press, 1998), p. 295 [II:173].

13 Gerbert, *The Letters of Gerbert*, p. 45.

14 For a critical discussion of the term see Paolo Squatriti, 'Pornocracy', in Christopher Kleinhenz et al, eds, *Medieval Italy: An Encyclopaedia* (London: Routledge, 2004), vol. 2, p. 928.

15 Lorenzo Minio-Paluello, 'Boethius, Anicius Manlius Severinus', in *Dictionary of Scientific Biography*, ed. Charles Coulston Gillispie, vol. 2 (New York: Scribner, 1970), p. 228.

16 Werner Telesko, *The Wisdom of Nature: The Healing Powers and Symbolism of Plants and Animals in the Middle Ages* (Munich: Prestel, 2001), p. 90. Among other things, lions were believed to sleep with their eyes open and to be stillborn, only waking from the dead three days after birth – the parallel with Jesus's resurrection on the third day is obvious.

17 Anicius Boethius, *The Consolation of Philosophy*, revised edn (Harmondsworth: Penguin, 1999), p. 41.

18 Quoted in John Henry, *Knowledge Is Power: How Magic, the Government and an Apocalyptic Vision Inspired Francis Bacon to Create Modern Science* (Cambridge: Icon Books, 2003), p. 85.

19 Naomi Reed Kline, *Maps of Medieval Thought: The Hereford Paradigm* (Woodbridge, Suffolk: Boydell Press, 2001), p. 13.

20 Pliny the Elder, *Natural History: A Selection* (trans. John F. Healy), (London: Penguin, 1991), p. 41. The precise figure depends on the exact interpretation of the size of a *stade* (an ancient Greek measure of distance derived from the length of a stadium). See chapter 13 of the present book.

21 John Carey, 'Ireland and the Antipodes: The Heterodoxy of Virgil of Salzburg', *Speculum* 64, no. 1 (January 1989): p. 1. Sir Francis Bacon's remarks, referred to above, may be based on a garbled recollection of this case.

22 These arguments were rehearsed in many ancient sources. See, for example, Ptolemy (trans. G.J. Toomer), *Ptolemy's Almagest* (London: Duckworth, 1984), p. 45 [I:7].

23 Boethius referred to the 'music of the world', which we call the 'music of the spheres'. Anicius Boethius, ed. G. Friedlein, *De institutione arithmetica, libri duo: di institutione musica, libri quinque* (Leipzig: Teubner, 1867), p. 187. The concept of the 'music of the spheres' is Pythagorean and many ancient commentators, including Aristotle, denied its existence. See *On the Heavens* in Aristotle, ed. Jonathan Barnes, *The Complete Works of Aristotle: The Revised Oxford Translation*, vol. 1 (Princeton: Princeton University Press, 1984), p. 479 [290b14].

24 Alfred North Whitehead, *Process and Reality* (New York: Free Press, 1979), p. 39.

25   The phrase is Dante's from Dante Alighieri (trans. Robert M. Durling), *Inferno* (New York: Oxford University Press, 1996), p. 77 [IV: 131].

Chapter 3 – The Rise of Reason

1   Eadmer (trans. R.W. Southern), *The Life of St Anselm, Archbishop of Canterbury* (Oxford: Clarendon Press, 1972), p. 7.

2   Ibid., p. 37.

3   Adapted from St Anselm (trans. Benedicta Ward), *Prayers and Meditations of St. Anselm with the Proslogion* (Harmondsworth: Penguin, 1973), p. 245.

4   Ibid., p. 239.

5   Bertrand Russell, *A History of Western Philosophy* (London: Routledge, 1961), p. 568.

6   Adapted from Diarmaid MacCulloch, *The Reformation: Europe's House Divided 1490–1700* (London: Allen Lane, 2003), p. 25.

7   A.J. MacDonald, *Berenger and the Reform of Sacramental Doctrine* (London: Longmans, Green & Co, 1930), p. 304.

8   Michael T. Clanchy, *Abelard: A Medieval Life* (Oxford: Blackwell, 1999), p. 292.

9   Peter Abelard and Héloïse, *The Letters of Abelard and Héloïse*, revised (Harmondsworth: Penguin, 1974), p. 58.

10   Ibid., p. 78.

11   Clanchy, *Abelard*, p. 215.

12   Ibid., p. 91.

13   Ibid.

14   Abelard and Héloïse, *The Letters of Abelard and Héloïse*, p. 66.

15   Ibid.

16   Ibid., p. 67.

17   Ibid., p. 147.

18   Clanchy, *Abelard*, p. 199.

19   Ibid., p. 107.

20   Ibid., p. 322.

21   Ibid.,p. 318.

Chapter 4 – The Twelfth-Century Renaissance

1   The concept of the 'twelfth-century renaissance' originated with Charles Homer Haskins, *The Renaissance of the Twelfth Century* (Cambridge, MA: Harvard University Press, 1927).

2   Lynn Thorndike, *History of Magic and Experimental Science*, vol. 2 (New York: Columbia University Press, 1923), p. 60.

3   Plato (trans. Desmond Lee), *Timaeus and Critias* (Harmondsworth: Penguin, 1977), p. 42 [29].

4   M.-D. Chenu (trans. Jerome Taylor and Lester K. Little), *Nature, Man and Society in the Twelfth Century: Essays on New Theological Perspectives in the Latin West*, (Chicago: Chicago University Press, 1968), p. 12.

5   Genesis 1:16.

6   Edward Grant, *God and Reason in the Middle Ages* (Cambridge: Cambridge University Press, 2001), p. 24.

7   Gilbert of La Porrée quoted in Chenu, *Nature, Man and Society in the Twelfth Century*, p. 40.

8   Peter Dronke, 'Thierry of Chartres', in *A History of Twelfth-Century Western Philosophy*, ed. Peter Dronke (Cambridge: Cambridge University Press, 1988), p. 367.

9   Hugh of St Victor quoted in Peter Harrison, *The Bible, Protestantism, and the Rise of Natural Science* (Cambridge: Cambridge University Press, 1998), p. 1.

10   Louise Cochrane, *Adelard of Bath: The First English Scientist* (London: British Museum Press, 1994), p. 32.

11   Michael Mahoney, 'Mathematics', in *Science in the Middle Ages*, ed. David C. Lindberg (Chicago: University Of Chicago Press, 1978), p. 153.

12   Adelard of Bath (trans. Charles Burnett), *Conversations with his Nephew, On the Same and the Different, Questions on Natural Science, and On Birds* (Cambridge: Cambridge University Press, 1998), pp. 85–9.

13   Ibid., p. 169.

14   Anon, 'A List of Translations Made From Arabic Into Latin in the Twelfth Century', in *A Source Book in Medieval Science*, ed. Edward Grant (Cambridge, MA: Harvard University Press, 1974), p. 35.

15   David C. Lindberg, 'Transmission of Greek and Arabic Learning', in *Science in the Middle Ages*, ed. David C. Lindberg (Chicago: Chicago University Press, 1978), p. 62.

16   Roger D. Masters, 'The Case of Aristotle's Missing Dialogues: Who Wrote the Sophist, the Statesman, and the Politics?', in *Political Theory* vol. 5, no. 1 (February 1977): p. 32.

17   The phrase is Dante's from Dante Alighieri (trans. Robert M. Durling), *Inferno* (New York: Oxford University Press, 1996), p. 77 [IV: 131].

18   *Brill's New Pauly: Encyclopaedia of the Ancient World*, vol. 10 (Leiden, Boston: Brill, 2002), p. 798.

19   Robin Lane Fox, *The Classical World: An Epic History of Greece and Rome* (London: Allen Lane, 2005), p. 211.

20   Keith Dix, 'Aristotle's Peripatetic Library', in *Lost Libraries: the Destruction of the Great Book Collections Since Antiquity*, ed. James Raven (Basingstoke: Palgrave MacMillan, 2004), p. 59.

21   Ibid., p. 61.

22   Pearl Kibre and Nancy G. Siraisi, 'The Institutional Setting: The Universities', in *Science in the Middle Ages*, ed. David C. Lindberg (Chicago: Chicago University Press, 1978), p. 120.

23   Olaf Pedersen, *The First Universities: Studium Generale and the Origins of University Education in Europe* (Cambridge: Cambridge University Press, 1997), p. 211.

24   Ibid., p. 139.

25   Ibid., p. 161.

26   Kibre and Siraisi, 'The Institutional Setting: The Universities', p. 125.

27   Pedersen, *The First Universities*, p. 154.

28   Charles Talbot, 'Medicine', in *Science in the Middle Ages*, ed. David C. Lindberg (Chicago: Chicago University Press, 1978), p. 400.

29   Richard William Southern, *The Making of the Middle Ages* (London: Hutchinson's University Library, 1953), p. 206.

30   Pedersen, *The First Universities*, p. 111.

Chapter 5 – Heresy and Reason

1   Walter L. Wakefield and Austin P. Evans, eds, *Heresies of the High Middle Ages: Selected Sources* (New York: Columbia University Press, 1969), p. 258.

2   J.M.M.H. Thijssen, 'Master Amalric and the Amalricians: Inquisitorial Procedure and the Suppression of Heresy at the University of Paris', *Speculum* vol. 71, no. 1 (1996), p. 48.

3   Ibid., p. 54.

4   Ibid., p. 61.

5   Lynn Thorndike, *University Records and Life in the Middle Ages* (New York: Columbia University Press, 1944), p. 26.

6   Amos Funkenstein, *Theology and the Scientific Imagination: From the Middle Ages to the Seventeenth Century* (Princeton, NJ: Princeton University Press, 1986), p. 122.

7   Edward Grant, *The Foundations of Modern Science in the Middle Ages: Their Religious, Institutional and Intellectual Contexts* (Cambridge: Cambridge University Press, 1996), p. 78.

8   Anthony Levi, *Renaissance and Reformation: The Intellectual Genesis* (New Haven: Yale University Press, 2002), p. 43.

9   Thorndike, *University Records and Life in the Middle Ages*, p. 40.

10   Ibid., p. 64.

11 M.-D. Chenu (trans. Jerome Taylor and Lester K. Little), *Nature, Man and Society in the Twelfth Century: Essays on New Theological Perspectives in the Latin West* (Chicago: Chicago University Press, 1968), p. 219.

12 Richard William Southern, *Western Society and the Church in the Middle Ages* (Harmondsworth: Penguin, 1970), p. 17.

13 Steven Runciman, *A History of the Crusades*, vol. 1 (Harmondsworth: Penguin, 1978), p. 137.

14 Wakefield and Evans, *Heresies of the High Middle Ages*, p. 129.

15 Jonathan Sumption, *The Albigensian Crusade* (London: Faber, 1978), p. 48.

16 Edward Peters, *Inquisition* (Berkeley: University of California Press, 1989), p. 47ff.

17 Edward Peters, ed., *Heresy and Authority in Medieval Europe: Documents in Translation* (London: Scolar Press, 1980), p. 197.

18 Keith Thomas, *Religion and the Decline of Magic: Studies in Popular Beliefs in Sixteenth- and Seventeenth-Century England* (Harmondsworth: Penguin, 1973), p. 259.

19 Peters, *Inquisition*, p. 52. The investigating magistrates trying to combat corruption in Italy today are the direct descendants of the medieval inquisitors.

20 Ibid., pp. 58–67.

21 Ibid., ch. 5.

22 Olaf Pedersen, *The First Universities: Studium Generale and the Origins of University Education in Europe* (Cambridge: Cambridge University Press, 1997), p. 97.

23 Peters, *Inquisition*, p. 117.

24 For example, Bernardo Gui burned 42 out of 930 cases in Toulouse whereas Jacques Fournier burned five from 114 cases in Pamiers. See James Buchanan Given, *Inquisition and Medieval Society: Power, Discipline and Resistance in Languedoc* (Ithaca, NY: Cornell University Press, 1997), p. 69 and Emmanuel Le Roy Ladurie, *Montaillou: Cathars and Catholics in a French Village, 1294–1324* (Harmondsworth: Penguin, 1978), p. xiv.

25 Southern, *Western Society and the Church in the Middle Ages*, p. 280.

26 Ibid., p. 296.

27 Maurice Hugh Keen, *The Penguin History of Medieval Europe* (Harmondsworth: Penguin, 1969), p. 158.

28 Sumption, *The Albigensian Crusade*, p. 48.

29 Roger French and Andrew Cunningham, *Before Science: The Invention of the Friars' Natural Philosophy* (Aldershot: Scolar Press, 1996), p. 120.

30 Augustine of Hippo (trans. Roger Green), *De Doctrina Christiana* (Oxford: Clarendon Press, 1995), p. 125 [II:144].

31   The classic statement of this point of view is in Thomas Aquinas, ed.
     Thomas Gilby, *Summa Theologiae*, vol. 1 (London: Blackfriars, 1964), p. 17
     [Pt I, Q 1, Art 5].

Chapter 6 – How Pagan Science was Christianised

1   Ptolemy of Lucca quoted in Lynn Thorndike, *History of Magic and
     Experimental Science*, vol. 2 (New York: Columbia University Press, 1923), p.
     522.
2   Albertus Magnus, *Opera Omnia*, 38 vols (Paris: Ludovicus Vives, 1890).
3   Paola Zambelli, *The Speculum Astronomiae and Its Enigma: Astrology, Theology,
     and Science in Albertus Magnus and His Contemporaries* (Dordrecht: Kluwer
     Academic, 1992).
4   Olaf Pedersen, *The First Universities: Studium Generale and the Origins of
     University Education in Europe* (Cambridge: Cambridge University Press,
     1997), p. 281.
5   Catholic University of America, *New Catholic Encyclopedia*, vol. 1 (New York:
     McGraw-Hill, 1967), p. 257.
6   Albertus Magnus, *Book of Minerals* (Oxford: Clarendon Press, 1967), p. 69
     [2.2.1].
7   Bernardo Gui is notorious as the fanatical inquisitor played by F. Murray
     Abraham in the film adaptation of Umberto Eco's *The Name of the Rose*.
8   Kenelm Foster, ed., *The Life of Thomas Aquinas: Biographical Documents*
     (London: Longmans, Green & Co, 1959), p. 30.
9   Ibid., p. 33.
10   Ibid., p. 32.
11   Ibid., p. 44.
12   Strictly speaking, these ideas may not fairly represent the views of Averröes
     but rather his Latin followers. See Willam Wallace, 'The Philosophical
     Setting of Medieval Science', in *Science in the Middle Ages*, ed. David C.
     Lindberg (Chicago: Chicago University Press, 1978), p. 104. To some
     extent the reader is left to infer the heterodox opinions of Siger and his
     followers from the condemnations of their opponents. No one would be
     foolish enough to put down their most radical beliefs in writing. This
     lack of written evidence has led some scholars to claim that no one really
     believed the condemned opinions, but this neglects the central place of oral
     discussion and questioning at medieval universities. For a spirited defence
     of Siger as a philosophical moderate, see B. Carlos Bazan, 'Siger of Brabant',
     in *A Companion to Philosophy in the Middle Ages*, eds Jorge J.E. Gracia and
     Timothy B. Noone (Malden, MA: Blackwell Pub, 2003), pp. 632–40.

13　Thomas Aquinas, ed. Timothy McDermott, *Summa Theologiae*, vol. 2 (London: Blackfriars, 1964), p. 13 [Pt 1, Q 2, Art 3].

14　Adapted from the analogy used in Patterson Brown, 'Infinite Causal Regression', *Philosophical Review* 75 (1966), pp. 510–25 The first of Aquinas's five ways deals specifically with motion as understood by Aristotle, while the second argues from causes more generally.

15　Ralph Lerner and Muhsin Mahdi, *Medieval Political Philosophy: A Sourcebook* (Ithaca, NY: Cornell University Press, 1972), p. 337.

16　Moses Maimonides (trans. M. Friedlander), *Guide for the Perplexed*, 2nd edn (New York: Dover, 1956), p. 2.

17　Aquinas, *Summa Theologiae*, vol. 2, p. 11 [Pt 1, Q 2, Art 2].

18　Thomas Aquinas, ed. Thomas Gilby, *Summa Theologiae*, vol. 1 (London: Blackfriars, 1964), p. 7 [Pt 1, Q 1, Art 1].

19　Lynn Thorndike, *University Records and Life in the Middle Ages* (New York: Columbia University Press, 1944), p. 85.

20　J.M.M.H. Thijssen, *Censure and Heresy at the University of Paris, 1200–1400* (Philadelphia: University of Pennsylvania Press, 1998), p. 48.

21　Roy Porter, *The Greatest Benefit to Mankind: A Medical History of Humanity from Antiquity to the Present* (London: HarperCollins, 1998), p. 110.

22　Dante Alighieri (trans. Robert and Jean Hollander), *Paradiso* (New York: Doubleday, 2007), p. 239 [X:136–8].

23　Adapted from Lerner and Mahdi, *Medieval Political Philosophy*, pp. 337–54.

24　Ibid., p. 352.

25　Thijssen, *Censure and Heresy at the University of Paris, 1200–1400*, p. 55.

26　R. Howgrave-Graham, *The Cathedrals of France* (London: Batsford, 1959), p. 118.

27　Lynn White, *Medieval Religion and Technology: Collected Essays* (Berkeley: University of California Press, 1978), p. 233.

28　Judith Förstel and Aline Magnien (trans. Diana Fowles), *Beauvais Cathedral* (Amiens: Inventaire Général, 2005), p. 10.

29　Ibid., p. 2.

Chapter 7 – Bloody Failure: Magic and Medicine in the Middle Ages

1　A short but hard-hitting analysis of pre-modern medicine is found in David Wootton, *Bad Medicine: Doctors Doing Harm Since Hippocrates* (Oxford: Oxford University Press, 2006).

2　Roy Porter, *The Greatest Benefit to Mankind: A Medical History of Humanity from Antiquity to the Present* (London: HarperCollins, 1998), p. 113.

3　Richard Kieckhefer, *Magic in the Middle Ages*, Canto edn (Cambridge, UK: Cambridge University Press, 2000), p. 15. There is a lot of debate as to how strictly this demarcation was observed in the Middle Ages. However, even if only made explicit in the sixteenth century, it is implied by much medieval comment on magic.

4   Marie-Christine Pouchelle, *The Body and Surgery in the Middle Ages* (Oxford: Polity, 1989), p. 20.

5   Books of folk cures, such as the English leech books, still survive. For some of the oldest see Linda Voigts, 'Anglo-Saxon Plant Remedies and the Anglo-Saxons', in *The Scientific Enterprise in Antiquity and Middle Ages: Readings from Isis*, ed. Michael Shank (Chicago: Chicago University Press, 2000), p. 163.

6   Stanton J. Linden, ed., *The Alchemy Reader: From Hermes Trismegistus to Isaac Newton* (Cambridge: Cambridge University Press, 2003), p. 137.

7   Bert Hansen, 'Science and Magic', in *Science in the Middle Ages*, ed. David C. Lindberg (Chicago: Chicago University Press, 1978), p. 493.

8   P.G. Maxwell-Stuart, ed., *The Occult in Mediaeval Europe, 500-1500: A Documentary History* (Basingstoke: Palgrave Macmillan, 2005), p. 174.

9   John Henry, *Knowledge Is Power: How Magic, the Government and an Apocalyptic Vision Inspired Francis Bacon to Create Modern Science* (Cambridge: Icon Books, 2003), p. 58.

10 Nancy G. Siraisi, *Medieval and Early Renaissance Medicine: An Introduction to Knowledge and Practice* (Chicago: University Of Chicago Press, 1990), p. 125.

11 Wootton, *Bad Medicine*, p. 17.

12 Siraisi, *Medieval and Early Renaissance Medicine*, p. 115.

13 Charles Talbot, 'Medicine', in *Science in the Middle Ages*, ed. David C. Lindberg (Chicago: Chicago University Press, 1978), p. 400.

14 Ibid., p. 403.

15 Porter, *The Greatest Benefit to Mankind*, p. 73.

16 George Sarton, *Galen of Pergamon* (Lawrence, Kansas: University of Kansas Press, 1954), p. 23.

17 Archimatthaeus (fl. 12th century) quoted in Edward Grant, *A Source Book in Medieval Science* (Cambridge, MA: Harvard University Press, 1974), p. 743.

18 Wootton, *Bad Medicine*, p. 56.

19 Geoffrey Chaucer (trans. Nevill Coghill), *The Canterbury Tales* (Harmondsworth: Penguin, 1951), p. 36 [I(A) 414–21].

20 Lynn Thorndike, *History of Magic and Experimental Science*, vol. 2 (New York: Columbia University Press, 1923), p. 856.

Chapter 8 – The Secret Arts of Alchemy and Astrology

1   Cornelius Tacitus, *The Annals of Imperial Rome*, revised edn (Harmondsworth: Penguin Books, 1989), p. 210 [VI:21].

2   Acts 19:19.

3   Augustine (trans. R.W. Dyson), *The City of God Against the Pagans* (Cambridge: Cambridge University Press, 1998), p. 197 [5:8].

4   Ptolemy (trans. F.E. Robbins), *Tetrabiblos*, (Cambridge, MA: Harvard University Press, 1994).

5   Adelard of Bath (trans. Charles Burnett), *Conversations with his Nephew, On the Same and the Different, Questions on Natural Science, and On Birds* (Cambridge: Cambridge University Press, 1998), p. 189.

6   Lynn Thorndike, *History of Magic and Experimental Science*, vol. 2 (New York: Columbia University Press, 1923), p. 828.

7   Dante Alighieri (trans. Robert M. Durling), *Inferno* (New York: Oxford University Press, 1996), p. 311 [XX:118].

8   Nancy G. Siraisi, *Medieval and Early Renaissance Medicine: An Introduction to Knowledge and Practice* (Chicago: University Of Chicago Press, 1990), p. 111.

9   Edward Grant, *A Source Book in Medieval Science* (Cambridge, MA: Harvard University Press, 1974), p. 657.

10   Thomas Aquinas, eds Thomas O'Meara and Michael Duffy, *Summa Theologiae*, vol. 40 (London: Blackfriars, 1968), p. 55 [Pt 2.2, Q 95, Art 5].

11   Ibid.

12   Ralph Lerner and Muhsin Mahdi, *Medieval Political Philosophy: A Sourcebook* (Ithaca, N.Y.: Cornell University Press, 1972), p. 350.

13   For example, the Paris condemnations of 1270 and 1398. See Lynn Thorndike, *University Records and Life in the Middle Ages* (New York: Columbia University Press, 1944), pp. 80, 266.

14   See chapter eighteen.

15   S.J. Tester, *A History of Western Astrology* (Woodbridge: Boydell Press, 1987), p. 194.

16   A different but unsubstantiated source states that Cecco's heresy was to claim that the virgin birth was a natural rather than a miraculous event. See Lynn Thorndike, 'The Relations of the Inquisition to Peter Abano and Cecco d'Ascoli', *Speculum* vol. 1, no. 3 (1926), p. 342.

17   Gerolamo Biscaro, *Inquisitori ed eretici Lombardi, 1292–1318*, vol. 19, Miscellanea di Storia Italiana 3 (Turin: Fratelli Bocca, 1922), p. 539.

18   G. Boffio, 'Perchè fu condannato al fuoco l'astrologo Cecco d'Ascoli?', *Studi e Documenti di Storia e Diritto*, vol. 20 (1899), p. 14.

19   Tester, *A History of Western Astrology*, p. 161.

20   Georg Agricola (trans. Herbert Hoover and Lou Henry Hoover), *De Re Metallica* (London: The Mining Magazine, 1912), p. xxviii.

21   Geoffrey Chaucer (trans. Nevill Coghill), *The Canterbury Tales* (Harmondsworth: Penguin, 1951), p. 482 [VIII (G) 862–7].

22 Stanton J. Linden, ed., *The Alchemy Reader: From Hermes Trismegistus to Isaac Newton* (Cambridge: Cambridge University Press, 2003), p. 103.

23 P.G. Maxwell-Stuart, ed., *The Occult in Mediaeval Europe, 500–1500: A Documentary History* (Basingstoke: Palgrave Macmillan, 2005), p. 229.

24 Thomas Aquinas, ed. Marcus Lefébure, *Summa Theologiae*, vol. 38 (London: Blackfriars, 1975), p. 221 [Pt 2.1, Q 77, Art 3].

25 William Newman, 'Technology and Alchemical Debate in the Late Middle Ages', in *The Scientific Enterprise in Antiquity and Middle Ages: Readings from Isis*, ed. Michael Shank (Chicago: Chicago University Press, 2000), p. 276.

26 Maxwell-Stuart, *The Occult in Mediaeval Europe, 500–1500*, p. 226.

27 David Knight, *Ideas in Chemistry: A History of the Science* (London: The Athlone Press, 1995), p. 19.

28 Allen George Debus, *The Chemical Philosophy: Paracelsian Science and Medicine in the Sixteenth and Seventeenth Centuries*, vol. 1 (New York: Science History Publications, 1977), p. 12.

29 Unsurprisingly, some Muslim scholars fiercely contest the idea that Christian alchemists should take credit for discoveries previously believed to have been made by the Arabs. However, to date, they have been unable to produce the original Arabic versions of the relevant Latin texts. This may change as more Middle Eastern libraries are explored and catalogued. See William Newman, *The Summa Perfectionis of Pseudo-Geber: A Critical Edition, Translation and Study* (Leiden: E.J. Brill, 1991), p. 61.

30 Knight, *Ideas in Chemistry: A History of the Science*, p. 18.

31 Robert P. Multhauf, 'The Science of Matter', in *Science in the Middle Ages*, ed. David C. Lindberg (Chicago: Chicago University Press, 1978), p. 381.

Chapter 9 – Roger Bacon and the Science of Light

1 Alan B. Cobban, *English University Life in the Middle Ages* (London: UCL Press, 1999), p. 161.

2 John Peckham, ed. Charles Trice Martin, *Registrum Epistolarum Fratris Johannis Peckham, Archiepiscopi Cantuariensis, Rerum Britannicarum medii aevi scriptores* (Rolls Series), vol. 77 (London: Longman, 1882), p. 944.

3 Richard Fletcher, *The Barbarian Conversion: From Paganism to Christianity* (New York: Henry Holt, 1997), p. 507.

4 Jonathan Sumption, *The Albigensian Crusade* (London: Faber, 1978), p. 148.

5 Lynn White, *Medieval Religion and Technology: Collected Essays* (Berkeley: University of California Press, 1978), p. 268.

6 Joseph Gies and Frances Gies, *Cathedral, Forge and Waterwheel: Technology and Invention in the Middle Ages* (London: Harper Perennial, 1995), p. 147.

7   *Physics* in Aristotle, ed. Jonathan Barnes, *The Complete Works of Aristotle: The Revised Oxford Translation*, vol. 1 (Princeton: Princeton University Press, 1984), p. 425 [254b12].

8   Marshall Clagett, *The Science of Mechanics in the Middle Ages* (Madison: University of Wisconsin Press, 1959), p. 514.

9   Peter Peregrinus, 'The Letter of Peregrinus', in *A Source Book in Medieval Science*, ed. Edward Grant (Cambridge, MA: Harvard University Press, 1974), p. 376.

10   The Muslim garrison was disarmed but spared, whereas rebel Christians in the city were put to death. Steven Runciman, *The Sicilian Vespers: A History of the Mediterranean World in the Later Thirteenth Century* (Cambridge: Cambridge University Press, 1992), p. 124.

11   Peregrinus, 'The Letter of Peregrinus', p. 368.

12   Edward Grant, *A Source Book in Medieval Science* (Cambridge, MA: Harvard University Press, 1974), p. 824.

13   Robert Bartlett, *The Natural and the Supernatural in the Middle Ages* (Cambridge: Cambridge University Press, 2008), p. 130.

14   Richard William Southern, *Western Society and the Church in the Middle Ages* (Harmondsworth: Penguin, 1970), p. 294.

15   Bartlett, *The Natural and the Supernatural in the Middle Ages*, p. 129.

16   Roger Bacon (trans. Robert Belle Burke), *The Opus Majus of Roger Bacon* (Philadelphia: University of Pennsylvania Press, 1928), p. 415.

17   David C. Lindberg, 'Science as Handmaiden: Roger Bacon and the Patristic Tradition', in *The Scientific Enterprise in Antiquity and the Middle Ages: Readings from Isis*, ed. Michael Shank (Chicago: Chicago University Press, 2000), p. 304.

18   Richard William Southern, *Robert Grosseteste: The Growth of an English Mind in Medieval Europe* (Oxford: Clarendon, 1986), p. 4.

19   Ibid., p. 65.

20   Daniel Angelo Callus, 'Robert Grosseteste as Scholar', in *Robert Grosseteste, Scholar and Bishop: Essays in Commemoration of the Seventh Centenary of His Death*, ed. Daniel Angelo Callus (Oxford: Clarendon, 1955), p. 7.

21   James McEvoy, *The Philosophy of Robert Grosseteste* (Oxford: Clarendon, 1982), p. 208.

22   Jeremiah Hackett, 'Roger Bacon: His Life, Career and Works', in *Roger Bacon and the Sciences: Commemorative Essays*, ed. Jeremiah Hackett (Leiden: Brill, 1997), p. 19.

23   David C. Lindberg, 'Medieval Science and Its Religious Context', *Osiris* vol. 10, 2nd series (1995), p. 76.

24   Quoted in James Riddick Partington, *A History of Greek Fire and Gunpowder* (Cambridge: Heffer, 1960), p. 78.

25   Roger Bacon (trans. Tenney Lombard Davis), *Roger Bacon's Letter Concerning the Marvelous Power of Art and of Nature and Concerning the Nullity of Magic* (Easton: Chemical Publishing Company, 1923), p. 26.

26   Robert Record, *The Pathway to Knowledge*, facsimile (Amsterdam: Theatrum Orbis Terra, 1974), preface.

27   Bacon, *The Opus Majus of Roger Bacon*, p. 582.

28   Southern, *Robert Grosseteste*, p. 136.

29   David C. Lindberg, *Theories of Vision from Al-Kindi to Kepler* (Chicago: University Of Chicago Press, 1976), p. 58.

30   Ibid., p. 12.

31   Ibid., p. 109.

32   David C. Lindberg, 'Light, Vision and the Universal Emanation of Force', in *Roger Bacon and the Sciences: Commemorative Essays*, ed. Jeremiah Hackett (Leiden: Brill, 1997), p. 268.

33   Lindberg, *Theories of Vision from Al-Kindi to Kepler*, p. 121.

34   Ibid., p. 118.

35   The regulations are not wholly unambiguous but are convincing evidence of spectacles. See Vincent Ilardi, *Renaissance Vision from Spectacles to Telescopes* (Philadelphia: American Philosophical Society, 2007), p. 8.

36   White, *Medieval Religion and Technology*, p. 221.

37   On the spurious story of Salvino degli Armati see Ilardi, *Renaissance Vision from Spectacles to Telescopes*, p. 15.

38   David S. Landes, *The Wealth and Poverty of Nations: Why Some Are so Rich and Some so Poor* (London: Abacus, 1999), p. 47.

Chapter 10 – The Clockmaker: Richard of Wallingford

1   John North, *God's Clockmaker: Richard of Wallingford and the Invention of Time* (London: Hambledon & London, 2007), p. 22.

2   Lynn Thorndike, *University Records and Life in the Middle Ages* (New York: Columbia University Press, 1944), p. 78.

3   Richard William Southern, 'From Schools to University', in *The Early Schools*, ed. Jeremy Catto, *The History of the University of Oxford* vol. 1 (Oxford: Clarendon Press, 1984), p. 26.

4   Diarmaid MacCulloch, *The Reformation: Europe's House Divided 1490–1700* (London: Allen Lane, 2003), p. 459.

5   Southern, 'From Schools to University', p. 26.

6   Geoffrey Chaucer (trans. Nevill Coghill), *The Canterbury Tales* (Harmondsworth: Penguin, 1951), p. 33 [I (A) 285].

7   Thomas More, ed. Daniel Kinney, *The Complete Works of St. Thomas More*, vol. 15 (New Haven: Yale University Press, 1986), p. 29.

8   Howard R. Turner, *Science in Medieval Islam: An Illustrated Introduction* (Austin: University of Texas Press, 1995), p. 47.

9   Alexander Murray, *Reason and Society in the Middle Ages* (Oxford: Clarendon Press, 1985), p. 170.

10   Anicius Boethius, ed. Michael Masi, *Boethian Number Theory: A Translation of the De Institutione Arithmetica (with Introduction and Notes)* (Amsterdam: Rodopi, 1983), p. 74.

11   Olaf Pedersen, *The First Universities: Studium Generale and the Origins of University Education in Europe* (Cambridge: Cambridge University Press, 1997), p. 262.

12   Robert Bartlett, *The Natural and the Supernatural in the Middle Ages* (Cambridge: Cambridge University Press, 2008), p. 73.

13   North, *God's Clockmaker*, p. 51.

14   Edward Grant, *Planets, Stars, and Orbs: The Medieval Cosmos, 1200–1687* (Cambridge: Cambridge University Press, 1994), p. 32.

15   North, *God's Clockmaker*, p. 359.

16   Ibid., p. 361.

17   Ibid., p. 78.

18   Ibid., p. 153.

19   Joseph Gies and Frances Gies, *Cathedral, Forge and Waterwheel: Technology and Invention in the Middle Ages* (London: Harper Perennial, 1995), p. 211.

20   David S. Landes, *The Wealth and Poverty of Nations: Why Some Are so Rich and Some so Poor* (London: Abacus, 1999), p. 48.

21   Jean Gimpel, *The Medieval Machine: The Industrial Revolution of the Middle Ages* (London: Pimlico, 1992), p. 164.

22   Lewis Mumford, *Technics and Civilization* (London: George Routledge & Sons, 1934), p. 17.

23   North, *God's Clockmaker*, p. 100.

24   Ibid., p. 221.

Chapter 11 – The Merton Calculators

1   Frederick Charles Copleston, *A History of Medieval Philosophy* (London: Methuen, 1972), p. 214.

2   David Luscombe, *Medieval Thought* (Oxford: Oxford University Press, 1997), p. 127.

3   Copleston, *A History of Medieval Philosophy*, p. 227.

4   Ibid., p. 214.

5   William J. Courtenay, *Ockham and Ockhamism: Studies in the Dissemination and Impact of His Thought* (Leiden: Brill, 2008), p. 98.

6   J.M.M.H. Thijssen, *Censure and Heresy at the University of Paris, 1200–1400* (Philadelphia: University of Pennsylvania Press, 1998), p. 14 and Courtenay, *Ockham and Ockhamism*, p. 101.

7   Bertrand Russell, *A History of Western Philosophy* (London: Routledge, 1961), p. 432.

8   Copleston, *A History of Medieval Philosophy*, p. 243.

9   Anthony Levi, *Renaissance and Reformation: The Intellectual Genesis* (New Haven: Yale University Press, 2002), p. 57.

10   Thijssen, *Censure and Heresy at the University of Paris, 1200–1400*, p. 59.

11   Paul Vincent Spade, 'Ockham's Nomalist Metaphysics: Some Main Themes', in *The Cambridge Companion to Ockham*, ed. Paul Vincent Spade (Cambridge: Cambridge University Press, 1999), p. 101.

12   *Physics* in Aristotle, ed. Jonathan Barnes, *The Complete Works of Aristotle: The Revised Oxford Translation*, vol. 1 (Princeton: Princeton University Press, 1984), p. 367 [216a15].

13   Aristotle does not use the exact phrase 'nature abhors a vacuum', but his argument is clear. See Ibid., p. 369 [217b20].

14   Ibid., p. 407 [241b34].

15   Ibid., p. 445 [267a3].

16   Morris R. Cohen and I.E. Drabkin, *A Source Book in Greek Science* (Cambridge, Mass: Harvard University Press, 1958), p. 223.

17   John E. Murdoch and Edith D. Sylla, 'The Science of Motion', in *Science in the Middle Ages*, ed. David C. Lindberg (Chicago: Chicago University Press, 1978), p. 212 and note 14 on p. 251.

18   Geoffrey Chaucer, ed. F.N. Robinson, *The Works of Geoffrey Chaucer*, 2nd edn (London: Oxford University Press, 1957), pp. 544–63.

19   Copleston, *A History of Medieval Philosophy*, p. 259.

20   Geoffrey Chaucer (trans. Nevill Coghill), *The Canterbury Tales* (Harmondsworth: Penguin, 1951), p. 249 [VII:3242].

21   *Posterior Analytics* in Aristotle, *The Complete Works of Aristotle*, p. 122 [75a38].

22   A.G. Molland, 'The Geometrical Background to the Merton School', *The British Journal for the History of Science* vol. 4, no. 2 (1968), p. 110.

23   Anneliese Maier (trans. Steven D. Sargent), *On the Threshold of Exact Science: Selected Writings of Anneliese Maier on Late Medieval Natural Philosophy* (Philadelphia: University of Pennsylvania Press, 1982), p. 157.

24   Carl B. Boyer, ed. Uta C. Merzbach, *A History of Mathematics*, 2nd edn (New York: Wiley, 1991), p. 348.

25   Quoted in Cohen and Drabkin, *A Source Book in Greek Science*, p. 220.

26   John North, *God's Clockmaker: Richard of Wallingford and the Invention of Time* (London: Hambledon & London, 2007), p. 277.

27   M.A. Hoskin and A.G. Molland, 'Swineshead on Falling Bodies: An Example of Fourteenth-Century Physics', *The British Journal for the History of Science* vol. 3, no. 2 (1966), p. 154.

28   Marshall Clagett, *The Science of Mechanics in the Middle Ages* (Madison: University of Wisconsin Press, 1959), p. 271.

Chapter 12 – The Apogee of Medieval Science

1   Anon, 'A List of Translations Made From Arabic Into Latin in the Twelfth Century', in *A Source Book in Medieval Science*, ed. Edward Grant (Cambridge, MA: Harvard University Press, 1974), p. 35.

2   Ernest A. Moody, *Studies in Medieval Philosophy, Science, and Logic: Collected Papers, 1933–1969* (Berkeley: University of California Press, 1975), p. 111.

3   Alexander Haggerty Krappe, 'The Legend of Buridan and the Tour de Nesle', *The Modern Language Review* vol. 23, no. 2 (1928), p. 216.

4   Ernest A. Moody, 'Buridan, Jean', in *Dictionary of Scientific Biography*, ed. Charles Coulston Gillispie, vol. 2 (New York: Scribner, 1970), p. 603.

5   Joël Biard, 'The Natural Order in John Buridan', in *The Metaphysics and Natural Philosophy of John Buridan*, eds J.M.M.H. Thijssen and Jack Zupko (Leiden: Brill Academic Publishers, 2001), p. 93.

6   Ibid., p. 91.

7   The earliest medieval reference to impetus is in a theological commentary on the *Sentences* of Peter Lombard that was written by Francis de Marchia in 1320. The idea was fully developed by Buridan in the following centuries. See Anneliese Maier (trans. Steven D. Sargent), *On the Threshold of Exact Science: Selected Writings of Anneliese Maier on Late Medieval Natural Philosophy* (Philadelphia: University of Pennsylvania Press, 1982), pp. 80–81.

8   Edward Grant, *The Foundations of Modern Science in the Middle Ages: Their Religious, Institutional and Intellectual Contexts* (Cambridge: Cambridge University Press, 1996), p. 95.

9   Maier, *On the Threshold of Exact Science*, p. 89.

10   Plato (trans. Desmond Lee), *Timaeus and Critias* (Harmondsworth: Penguin, 1977), p. 53 [38].

11   Edward Grant, *Planets, Stars, and Orbs: The Medieval Cosmos, 1200–1687* (Cambridge: Cambridge University Press, 1994), p. 527.

12   John North, *God's Clockmaker: Richard of Wallingford and the Invention of Time* (London: Hambledon & London, 2007), p. 202.

13   Maier, *On the Threshold of Exact Science*, p. 89.

14   Grant, *The Foundations of Modern Science in the Middle Ages*, p. 113.

15   Ptolemy (trans. G.J. Toomer), *Ptolemy's Almagest* (London: Duckworth, 1984), p. 45 [I:7].

16   Marshall Clagett, 'Oresme, Nicole', in *Dictionary of Scientific Biography*, ed. Charles Coulston Gillispie, vol. 10 (New York: Scribner, 1970), p. 223.

17   Lynn Thorndike, *History of Magic and Experimental Science*, vol. 3 (New York: Columbia University Press, 1934), p. 398.

18   Quoted in Edward Grant, *A Source Book in Medieval Science* (Cambridge, MA: Harvard University Press, 1974), p. 67.

19   John E. Murdoch and Edith D. Sylla, 'The Science of Motion', in *Science in the Middle Ages*, ed. David C. Lindberg (Chicago: Chicago University Press, 1978), p. 240.

20   Ernest A. Moody, 'Albert of Saxony', in *Dictionary of Scientific Biography*, ed. Charles Coulston Gillispie, vol. 1 (New York: Scribner, 1970), p. 93.

21   Ibid., p. 94.

22   J.H. Randall Jr., 'The Development of the Scientific Method in the School of Padua', *Journal of the History of Ideas* vol. 1, no. 2 (1940), p. 181.

23   Charles Talbot, 'Medicine', in *Science in the Middle Ages*, ed. David C. Lindberg (Chicago: Chicago University Press, 1978), p. 417.

24   Marshall Clagett, *The Science of Mechanics in the Middle Ages* (Madison: University of Wisconsin Press, 1959), p. 651.

25   Moody, 'Buridan, Jean', p. 603.

26   Grant, *A Source Book in Medieval Science*, p. 314.

27   Lambertus Marie de Rijk, ed., *Nicholas of Autrecourt: His Correspondence with Master Giles and Bernhard of Arezzo: A Critical Edition from The Two Parisian Manuscripts* (Leiden: E.J. Brill, 1994), p. 3.

28   Edward Grant, *God and Reason in the Middle Ages* (Cambridge: Cambridge University Press, 2001), p. 224.

29   Rijk, *Nicholas of Autrecourt*, p. 181.

30   Moody, *Studies in Medieval Philosophy, Science, and Logic*, pp. 150–2.

31   Grant, *A Source Book in Medieval Science*, p. 50.

32   Jean Gerson (trans. Brian Patrick McGuire), *Jean Gerson: Early Works* (New York: Paulist Press, 1998), p. 172.

33   North, *God's Clockmaker*, p. 326.

Chapter 13 – New Horizons

1   J.E. Hofman, 'Cusa, Nicholas of', in *Dictionary of Scientific Biography*, ed. Charles Coulston Gillispie (New York: Scribner, 1970), p, 515.

2   Ibid., p. 513.

3   Nicholas of Cusa (trans. Jasper Hopkins), 'On Learned Ignorance', in
     Complete Philosophical and Theological Treatises of Nicholas of Cusa, vol. 1
     (Minneapolis: Arthur J. Banning Press, 2001), p. 90 [II:11].

4   Ibid., p. 96 [II:12].

5   Robert P. Multhauf, 'The Science of Matter', in *Science in the Middle Ages*, ed.
     David C. Lindberg (Chicago: Chicago University Press, 1978), p. 386.

6   Laura Ackerman Smoller, *History, Prophecy, and the Stars: The Christian
     Astrology of Pierre d'Ailly, 1350–1420* (Princeton: Princeton University Press,
     1994), p. 7.

7   He predicted that in 1789 'there will be many great and marvellous
     alterations and changes in the world, chiefly in respect to laws and sects.'
     After this, he expected the Antichrist. See Ibid., p. 105.

8   Donald Engels, 'The Length of Eratosthenes' Stade', *American Journal of
     Philology* vol. 106, no. 3 (1985), p. 310.

9   Lynn Thorndike, *The Sphere of Sacrobosco and Its Commentators* (Chicago:
     University of Chicago Press, 1949), p. 122.

10   Ptolemy (trans. J.L. Berggren and Alexander Jones), *Ptolemy's Geography:
     An Annotated Translation of the Theoretical Chapters* (Princeton: Princeton
     University Press, 2000), p. 52.

11   Ptolemy believed that one degree of longitude at the equator was equal to
     500 stades. Ibid., p. 71 [I:11].

12   John Mandeville (trans. C.W.R.D. Moseley), *The Travels of Sir John
     Mandeville* (Harmondsworth: Penguin, 1983), p. 129.

13   *Meteorology* in Aristotle, ed. Jonathan Barnes, *The Complete Works of Aristotle:
     The Revised Oxford Translation*, vol. 1 (Princeton: Princeton University Press,
     1984), p. 587 [362a32].

14   Pliny the Elder (trans. H. Rackham), *Natural History: With an English
     Translation in Ten Volumes*, vol. 1 (London: Heinemann, 1938), p. 307 [II:68].

15   J.H. Parry, *The Age of Reconaissance* (London: Weidenfeld and Nicolson,
     1963), p. 137.

16   Joseph Gies and Frances Gies, *Cathedral, Forge and Waterwheel: Technology
     and Invention in the Middle Ages* (London: Harper Perennial, 1995), p. 281.

17   Edward Grant, *A Source Book in Medieval Science* (Cambridge, MA: Harvard
     University Press, 1974), p. 635.

18   Ibid., p. 637.

19   Ibid., p. 638.

20   Parry, *The Age of Reconaissance*, p. 150.

21   Christopher Columbus (trans. John G. Cummins), *The Voyage of Christopher
     Columbus: Columbus' Own Journal of Discovery Newly Restored and Translated*
     (London: Weidenfeld and Nicolson, 1992), p. 26.

22   John Huxtable Elliott, *The Old World and the New 1492–1650* (Cambridge: Cambridge University Press, 1992), p. 29.

23   Robert Record, *The Castle of Knowledge*, Facsimile (Amsterdam: Theatrum Orbis Terra, 1975), p. 70.

24   Steven Runciman, *The Fall of Constantinople, 1453* (Cambridge: Cambridge University Press, 1990), p. 76.

25   Ibid., p. 78.

26   The Sultan's gun is generally assumed to have been bronze, but it is more likely to have been of the same iron construction as Mons Meg. As it happened, Mehmet's lighter artillery was more effective. See Charles William Chadwick Oman, *A History of the Art of War in the Middle Ages*, vol. 2, 2nd edn (London: Methuen, 1924), p. 357.

27   Robert D. Smith and Kelly DeVries, *The Artillery of the Dukes of Burgundy 1363–1477* (Woodbridge, Suffolk: Boydell Press, 2005), p. 262.

28   Runciman, *The Fall of Constantinople, 1453*, p. 144.

29   Joseph Gies and Frances Gies, *Cathedral, Forge and Waterwheel*, p. 247.

30   Ibid.

31   Alexander Murray, *Reason and Society in the Middle Ages* (Oxford: Clarendon Press, 1985), p. 301.

32   Joseph Gies and Frances Gies, *Cathedral, Forge and Waterwheel*, p. 244.

33   Colin Clair, *A History of European Printing* (London: Academic Press, 1976), p. 16.

34   Desiderius Erasmus, ed. Charles Garfield Nauert (trans. Alexander Dalzell), *The Correspondence of Erasmus*, vol. 11 (Toronto: University of Toronto Press, 1994), p. 27.

Chapter 14 – Humanism and the Reformation

1   David C. Lindberg, *Theories of Vision from Al-Kindi to Kepler* (Chicago: University Of Chicago Press, 1976), p. 148.

2   Charles G. Nauert, *Humanism and the Culture of Renaissance Europe* (Cambridge: Cambridge University Press, 1995), p. 1.

3   Jacob Burckhardt (trans. S.G.C. Middlemore), *The Civilisation of the Renaissance in Italy* (Harmondsworth: Penguin, 1990), p. 98.

4   Paul Oskar Kristeller, *Renaissance Thought and its Sources* (New York: Columbia University Press, 1979), p. 22.

5   Ibid., p. 29.

6   Elizabeth Eisenstein, *The Printing Revolution in Early Modern Europe* (Cambridge: Cambridge University Press, 1993), p. 120.

7   Michael Reeve, 'Classical Scholarship', in *Cambridge Companion to Renaissance Humanism*, ed. Jill Kraye (Cambridge: Cambridge University Press, 1996), p. 34.

8   Charles Trinkaus, 'Marsilio Ficino', in *Contemporaries of Erasmus: A Biographical Register of the Renaissance and Reformation*, ed. Peter G Bietenholz and Thomas B Deutscher, vol. 2 (Toronto: University of Toronto Press, 1985), p. 27.

9   Kristeller, *Renaissance Thought and its Sources*, p. 188.

10   Ernst Cassirer, Paul Oskar Kristeller and John Herman Randall, eds, *The Renaissance Philosophy of Man: Selections in Translation* (Chicago: University of Chicago Press, 1948), p. 377.

11   Ibid., p. 381.

12   Kristeller, *Renaissance Thought and its Sources*, p. 191.

13   Michel de Montaigne (trans. M.A. Screech), *The Essays of Michel De Montaigne* (Harmondsworth: Penguin, 1993), p. 170.

14   For instance, anticipating infinite set theory. See Edward Grant, *God and Reason in the Middle Ages* (Cambridge: Cambridge University Press, 2001), p. 248.

15   J.S. Brewer, ed., *Letters and Papers, Foreign and Domestic, of the Reign of Henry VIII: Preserved in the Public Record Office, the British Museum, and Elsewhere in England*, vol. 9 (London: Longman, Roberts & Green, 1862), no. 350.

16   John E. Murdoch and Edith D. Sylla, 'Swineshead, Richard', in *Dictionary of Scientific Biography*, ed. Charles Coulston Gillispie, vol. 13 (New York: Scribner, 1970), p. 209.

17   James Farge, 'Thomas Bricot', in *Contemporaries of Erasmus: A Biographical Register of the Renaissance and Reformation*, eds Peter G. Bietenholz and Thomas B. Deutscher, vol. 1 (Toronto: University of Toronto Press, 1985), p. 199.

18   Marshall Clagett, *The Science of Mechanics in the Middle Ages* (Madison: University of Wisconsin Press, 1959), p. 638.

19   Sachiko Kusukawa, *The Transformation of Natural Philosophy: The Case of Philip Melanchthon* (Cambridge: Cambridge University Press, 1995), p. 110.

20   Adapted from Galileo Galilei (trans. Stillman Drake), *Dialogue Concerning the Two Chief World Systems: the Ptolemaic and Copernican* (Berkeley: University of California Press, 1953), p. 107.

21   N.R. Ker, 'The Provision of Books', in *A History of Oxford University: The Collegiate University*, ed. James McConica (Oxford: Clarendon Press, 1986), p. 466.

22   G.H. Martin, *A History of Merton College, Oxford* (Oxford: Oxford University Press, 1997), p. 79.

23   Alastair Hamilton, 'Humanists and the Bible', in *The Cambridge Companion to Renaissance Humanism*, ed. Jill Kraye (Cambridge: Cambridge University Press, 1996), p. 107.

24   Nauert, *Humanism and the Culture of Renaissance Europe*, p. 156.

25   Desiderius Erasmus (trans. Betty Radice), *Praise of Folly* (Harmondsworth: Penguin Classics, 1993), p. 88.

26   Diarmaid MacCulloch, *The Reformation: Europe's House Divided 1490–1700* (London: Allen Lane, 2003), p. 123.

27   Ibid., p. 131.

28   Ibid., p. 245.

29   Ibid., p. 679.

30   Peter Harrison, *The Bible, Protestantism, and the Rise of Natural Science* (Cambridge: Cambridge University Press, 1998), p. 114.

31   Eisenstein, *The Printing Revolution in Early Modern Europe*, p. 230.

32   Max Weber (trans. Stephen Kalberg), *The Protestant Ethic and the Spirit of Capitalism*, 3rd edn (Oxford: Blackwell, 2002).

33   Robert K. Merton, 'Science, Technology and Society in Seventeenth Century England', *Osiris* vol. 4, no. 1 (1938), pp. 360–632.

34   Robert Record, *The Castle of Knowledge*, Facsimile (Amsterdam: Theatrum Orbis Terra, 1975), p. 284. Spelling modernised.

35   Henry Howard (Oxford, Bodleian Library, 'MS Bodley 616'), fol. 5r. Spelling modernised.

36   Thomas Worcester, 'Introduction', in *The Cambridge Companion to the Jesuits*, ed. Thomas Worcester (Cambridge: Cambridge University Press, 2008), p. 3.

37   Richard G. Olson, *Science and Religion, 1450–1900: From Copernicus to Darwin* (Baltimore: The Johns Hopkins University Press, 2006), p. 69.

38   James Hannam, 'Teaching Natural Philosophy and Mathematics at Oxford and Cambridge 1500–1570' (PhD thesis, University of Cambridge, 2008), p. 213.

39   Rodney Stark, *For the Glory of God: How Monotheism Led to Reformations, Science, Witch-Hunts, and the End of Slavery* (Princeton: Princeton University Press, 2003), p. 161.

Chapter 15 – The Polymaths of the Sixteenth Century

1   Anna Rita Fantoni, ed. Lorenzo Crinelli, *Treasures from the Italian Libraries* (London: Thames and Hudson, 1997), p. 16.

2   Thomas Deutscher, 'Gemitos Plethon', in *Contemporaries of Erasmus: A Biographical Register of the Renaissance and Reformation*, eds Peter G. Bietenholz and Thomas B. Deutscher, vol. 2 (Toronto: University of Toronto Press, 1985), p. 85.

3   D.P. Walker, *Spiritual and Demonic Magic from Ficino to Campanella*, new edn (Stroud: Sutton, 2000), p. 62.

4   Frances A. Yates, *Giordano Bruno and the Hermetic Tradition* (London: Routledge Classics, 2002), p. 13.

5   Ibid., p. 11.

6   Angela Voss, ed., *Marsilio Ficino* (Berkeley: North Atlantic Books, 2006), p. 192.

7   Walker, *Spiritual and Demonic Magic from Ficino to Campanella*, p. 37.

8   Brian P. Copenhaver, trans., *Hermetica: The Greek Corpus Hermeticum and the Latin Asclepius* (Cambridge: Cambridge University Press, 1992).

9   Yates, *Giordano Bruno and the Hermetic Tradition*, p. 434.

10   Ibid., p. 127.

11   John Dee, eds Wayne Shumaker and J.L. Heilbron, *John Dee on Astronomy: 'Propaedeumata Aphoristica' (1558 and 1568)* (Berkeley: University of California Press, 1978), p. 5.

12   Sachiko Kusukawa, *The Transformation of Natural Philosophy: The Case of Philip Melanchthon* (Cambridge: Cambridge University Press, 1995), p. 135.

13   Dee, *John Dee on Astronomy*, p. 64.

14   Mario Gliozzi, 'Cardano, Girolamo', in *Dictionary of Scientific Biography*, ed. Charles Coulston Gillispie, vol. 3 (New York: Scribner, 1970), p. 64.

15   Girolamo Cardano (trans. Jean Stoner), *The Book of My Life* (New York: New York Review Books, 2002), p. 172.

16   Anthony Grafton, *Cardano's Cosmos: The Worlds and Works of a Renaissance Astrologer* (Cambridge, MA: Harvard University Press, 1999), p. 57.

17   Cardano, *The Book of My Life*, p. 38.

18   Gliozzi, 'Cardano, Girolamo', p. 66.

19   Cardano, *The Book of My Life*, p. 15.

20   Ibid., p. 87.

21   W.G. Waters, *Jerome Cardan: A Biographical Study* (London: Lawrence & Bullen, 1898), p. 128.

22   Grafton, *Cardano's Cosmos*, p. 112.

23   Ibid., p. 121.

24   P.G. Maxwell-Stuart, ed., *The Occult in Early Modern Europe: A Documentary History* (Basingstoke: Macmillan, 1999), p. 83.

25   Grafton, *Cardano's Cosmos*, p. 134.

26   Dee, *John Dee on Astronomy*, p. 52.

27 Grafton, *Cardano's Cosmos*, p. 78.

28 Ibid., p. 151.

29 Stillman Drake and I.E. Drabkin, *Mechanics in Sixteenth-Century Italy: Selections from Tartaglia, Benedetti, Guido Ubaldo, and Galileo* (Madison: University of Wisconsin Press, 1969), p. 27.

30 Gliozzi, 'Cardano, Girolamo', p. 66.

31 W. R. Laird, 'Archimedes among the Humanists', *Isis* vol. 82, no. 4 (1 January 1991), p. 635.

32 See the 'Life of Marcellus' in Plutarch (trans. Ian Scott-Kilvert), *Makers of Rome: Nine Lives* (Harmondsworth: Penguin, 1965), p. 100 [15].

33 Laird, 'Archimedes among the Humanists', p. 628.

34 Drake and Drabkin, *Mechanics in Sixteenth-Century Italy*, p. 17.

35 Paul Lawrence Rose, *The Italian Renaissance of Mathematics: Studies on Humanists and Mathematicians from Petrarch to Galileo* (Geneva: Droz, 1975), p. 152.

36 Ibid., p. 153.

37 Drake and Drabkin, *Mechanics in Sixteenth-Century Italy*, p. 23.

38 Cardano, *The Book of My Life*, p. 83.

39 Ibid.

40 Ibid., p. 63.

41 Anthony Grafton in the Introduction to Ibid., p. xiii.

42 Laird, 'Archimedes among the Humanists', p. 635.

43 Waters, *Jerome Cardan*, p. 153.

44 Cardano, *The Book of My Life*, p. 38.

45 Waters, *Jerome Cardan*, p. 219.

Chapter 16 – The Workings of Man: Medicine and Anatomy

1 Charles Webster, *Paracelsus: Medicine, Magic and Mission at the End of Time* (New Haven: Yale University Press, 2008), p. 10.

2 P.G. Maxwell-Stuart, ed., *The Occult in Early Modern Europe: A Documentary History* (Basingstoke: Macmillan, 1999), p. 198.

3 Charles Webster, *Paracelsus: Medicine, Magic and Mission at the End of Time*, p. 13.

4 Philip Ball, *The Devil's Doctor: Paracelsus and the World of Renaissance Magic and Science* (London: William Heinemann, 2006), p. 205.

5 Allen George Debus, *The French Paracelsians: The Chemical Challenge to Medical and Scientific Tradition in Early Modern France* (Cambridge: Cambridge University Press, 1991), p. 65.

6 Marie-Christine Pouchelle, *The Body and Surgery in the Middle Ages* (Oxford: Polity, 1989), p. 16.

7   Nancy G. Siraisi, *Medieval and Early Renaissance Medicine: An Introduction to Knowledge and Practice* (Chicago: University Of Chicago Press, 1990), p. 169.

8   Martha Teach Gnudi and Jerome Pierce Webster, *The Life and Times of Gaspare Tagliacozzi: Surgeon of Bologna, 1545–1599. With a Documented Study of the Scientific and Cultural Life of Bologna in the Sixteenth Century* (New York: H. Reichner, 1950), p. 118.

9   Roy Porter, *The Greatest Benefit to Mankind: A Medical History of Humanity from Antiquity to the Present* (London: HarperCollins, 1998), p. 190.

10   Gnudi and Webster, *The Life and Times of Gaspare Tagliacozzi*, p. 114.

11   Ibid., p. 288.

12   James Longrigg, 'Anatomy in Alexandria in the Third Century B.C.', *British Journal for the History of Science* vol. 21, no. 4 (1988), p. 457.

13   Porter, *The Greatest Benefit to Mankind*, p. 75.

14   Emilie Savage-Smith, 'Attitudes Toward Dissection in Medieval Islam', *Journal of the History of Medicine and Allied Sciences* vol. 50, no. 1 (1 January 1995), p. 87.

15   R.K. French, *Dissection and Vivisection in the European Renaissance* (Aldershot: Ashgate, 1999), p. 11.

16   Charles Talbot, 'Medicine', in *Science in the Middle Ages*, ed. David C. Lindberg (Chicago: Chicago University Press, 1978), p. 409.

17   Ball, *The Devil's Doctor*, p. 56.

18   Edward Grant, *A Source Book in Medieval Science* (Cambridge, MA: Harvard University Press, 1974), p. 739.

19   David Wootton, *Bad Medicine: Doctors Doing Harm Since Hippocrates* (Oxford: Oxford University Press, 2006), p. 76.

20   J.B. de C.M. Saunders and Charles Donald O'Malley, *The Anatomical Drawings of Andreas Vesalius: With Annotations and Translations, a Discussion of the Plates and Their Background, Authorship, and Influence, and a Biographical Sketch of Vesalius* (New York: Bonanza Books, 1982), p. 28.

21   Porter, *The Greatest Benefit to Mankind*, p. 183.

22   Wootton, *Bad Medicine*, p. 83.

23   Andreas Vesalius (trans. William Frank Richardson and John Burd Carman), *On the Fabric of the Human Body*, vol. 1 (Novato: Norman Publishing, 1998), p. 383.

24   Wootton, *Bad Medicine*, p. 91.

25   Porter, *The Greatest Benefit to Mankind*, p. 171.

26   C. Donald O'Malley, 'Andreas Vesalius' Pilgrimage', *Isis* vol. 45, no. 2 (1 January 1954), p. 138.

27   C.D. O'Malley, *Andreas Vesalius of Brussels, 1514–1564* (Berkeley: University of California Press, 1964), p. 304.

28 Wootton, *Bad Medicine*, p. 102.

29 Michael Servetus (trans. C.D. O'Malley), *Michael Servetus: A Translation of His Geographical, Medical, and Astrological Writings* (Philadelphia: American Philosophical Society, 1953), p. 204.

30 Columbo is also held to have discovered the clitoris, which had presumably evaded detection by any man until this point. See Wootton, *Bad Medicine*, p. 117.

31 Savage-Smith, 'Attitudes Toward Dissection in Medieval Islam', p. 102.

32 Geoffrey Keynes, *The Life of William Harvey* (Oxford: Clarendon Press, 1966), p. 137.

33 Wootton, *Bad Medicine*, p. 98.

34 William Harvey (trans. Kenneth Franklin), *Movement of the Heart and Blood in Animals: An Anatomical Essay* (Oxford: Blackwell, 1957), p. 39.

35 Charles Webster, 'William Harvey's Conception of the Heart as a Pump', *Bulletin of the History of Medicine* vol. 39 (1965), p. 510.

36 Harvey, *Movement of the Heart and Blood in Animals*, p. 88.

37 Ibid., p. 59.

38 John Aubrey, ed. John Buchanan-Brown, *Brief Lives* (Harmondsworth: Penguin, 2000), p. 144.

39 The story is engagingly told in Pete Moore, *Blood and Justice: The Seventeenth-Century Parisian Doctor Who Made Blood Transfusion History* (Chichester: John Wiley, 2002).

40 Porter, *The Greatest Benefit to Mankind*, p. 303.

Chapter 17 – Humanist Astronomy and Nicolaus Copernicus

1 Judith Rice Henderson, 'George of Trebizond', in *Contemporaries of Erasmus: A Biographical Register of the Renaissance and Reformation*, eds Peter G. Bietenholz and Thomas B. Deutscher, vol. 3 (Toronto: University of Toronto Press, 1985), p. 340.

2 Thomas Deutscher, 'Bessarion', in *Contemporaries of Erasmus: A Biographical Register of the Renaissance and Reformation*, ed. Peter G. Bietenholz and Thomas B. Deutscher, vol. 1 (Toronto: University of Toronto Press, 1985), p. 142.

3 Anna Rita Fantoni, ed. Lorenzo Crinelli, *Treasures from the Italian Libraries* (London: Thames and Hudson, 1997), p. 20.

4 Paul Lawrence Rose, *The Italian Renaissance of Mathematics: Studies on Humanists and Mathematicians from Petrarch to Galileo* (Geneva: Droz, 1975), p. 91.

5 Ibid., p. 101.

6   C. Doris Hellman and John E. Murdoch, 'Peurbach, Georg', in *Dictionary of Scientific Biography*, ed. Charles Coulston Gillispie, vol. 15 (New York: Scribner, 1970), p. 474.

7   Ibid.

8   Rose, *The Italian Renaissance of Mathematics*, p. 101.

9   John North, *God's Clockmaker: Richard of Wallingford and the Invention of Time* (London: Hambledon & London, 2007), p. 334.

10 Ibid., p. 64.

11 C.M. Linton, *From Eudoxus to Einstein: A History of Mathematical Astronomy* (Cambridge: Cambridge University Press, 2004), p. 20.

12 For a thorough and lucid explanation see John Henry, *Moving Heaven and Earth: Copernicus and the Solar System* (Cambridge: Icon Books, 2001), p. 32.

13 Moses Maimonides (trans. M. Friedlander), *Guide for the Perplexed*, 2nd edn (New York: Dover, 1956), p. 198.

14 Edward Grant, *A Source Book in Medieval Science* (Cambridge, MA: Harvard University Press, 1974), p. 525.

15 James M. Lattis, *Between Copernicus and Galileo: Christoph Clavius and the Collapse of Ptolemaic Cosmology* (Chicago: University of Chicago Press, 1994), p. 78.

16 A popular English work put the figure at at least 117 million miles. See C.S. Lewis, *The Discarded Image: An Introduction to Medieval and Renaissance Literature* (Cambridge: Cambridge University Press, 1994), p. 98.

17 Nicolaus Copernicus, *On the Revolutions of the Heavenly Spheres* (Amherst: Prometheus Books, 1995), p. 3.

18 Robert S. Westman, 'Proof, Poetics and Patronage: Copernicus's Preface to *De Revolutionibus*', in *Reappraisals of the Scientific Revolution*, ed. David C. Lindberg and Robert S. Westman (Cambridge: Cambridge University Press, 1990), p. 186.

19 For a more detailed explanation see Henry, *Moving Heaven and Earth*, pp. 87–90.

20 Rose, *The Italian Renaissance of Mathematics*, p. 119.

21 Linton, *From Eudoxus to Einstein*, p. 121.

22 Rose, *The Italian Renaissance of Mathematics*, p. 99.

23 Copernicus, *On the Revolutions of the Heavenly Spheres*, p. 27.

24 Ibid., p. 6.

25 Linton, *From Eudoxus to Einstein*, p. 121.

26 Angela Voss, ed., *Marsilio Ficino* (Berkeley: North Atlantic Books, 2006), p. 192.

27 D.P. Walker, *Spiritual and Demonic Magic from Ficino to Campanella*, new edn (Stroud: Sutton, 2000), p. 113.

28   Copernicus, *On the Revolutions of the Heavenly Spheres*, p. 25.

29   Grant, *A Source Book in Medieval Science*, p. 67.

30   Copernicus, *On the Revolutions of the Heavenly Spheres*, p. 17.

31   Ernest A. Moody, *Studies in Medieval Philosophy, Science, and Logic: Collected Papers, 1933–1969* (Berkeley: University of California Press, 1975), p. 442.

32   Paul W. Knoll, 'The Arts Faculty at the University of Cracow', in *The Copernican Achievement*, ed. Robert S. Westman (Berkeley: University of California Press, 1975), p. 150.

33   Nicholas of Cusa (trans. Jasper Hopkins), 'On Learned Ignorance', in *Complete Philosophical and Theological Treatises of Nicholas of Cusa*, vol. 1 (Minneapolis: Arthur J. Banning Press, 2001), p. 93 [II:12].

34   George Saliba, *Islamic Science and the Making of the European Renaissance* (Cambridge, MA: The MIT Press, 2007), p. 200.

35   Ibid., p. 196.

36   For some intriguing suggestions see Ibid., pp. 217–21.

37   Robert S. Westman, 'The Copernicans and the Churches', in *God and Nature: Historical Essays on the Encounter Between Christianity and Science*, eds David C. Lindberg and Ronald L. Numbers (Berkeley: University of California Press, 1986), p. 82.

38   E.G. Richards, *Mapping Time: The Calendar and Its History* (Oxford: Oxford University Press, 1999), p. 249.

39   Ibid., p. 253.

40   It was previously thought that there had been ten true Copernicans before 1600, but this number has now been reduced to nine. See Stephen Gaukroger, *The Emergence of a Scientific Culture: Science and the Shaping of Modernity 1210–1685* (Oxford: Clarendon Press, 2006), p. 122 note 118.

Chapter 18 – Reforming the Heavens

1   James M. Lattis, *Between Copernicus and Galileo: Christoph Clavius and the Collapse of Ptolemaic Cosmology* (Chicago: University of Chicago Press, 1994), p. 44.

2   *Meteorology* in Aristotle, ed. Jonathan Barnes, *The Complete Works of Aristotle: The Revised Oxford Translation*, vol. 1 (Princeton: Princeton University Press, 1984), p. 563 [344a15].

3   Robert Bartlett, *The Natural and the Supernatural in the Middle Ages* (Cambridge: Cambridge University Press, 2008), p. 70.

4   Lattis, *Between Copernicus and Galileo*, p. 147.

5   Ibid., p. 151.

6   C. Doris Hellman, 'Brahe, Tycho', in *Dictionary of Scientific Biography*, ed. Charles Coulston Gillispie, vol. 2 (New York: Scribner, 1970), p. 401.

7   C.M. Linton, *From Eudoxus to Einstein: A History of Mathematical Astronomy* (Cambridge: Cambridge University Press, 2004), p. 154.

8   Hellman, 'Brahe, Tycho', p. 404.

9   Ibid., p. 402.

10   Linton, *From Eudoxus to Einstein*, pp. 156, 163.

11   Ibid., p. 168.

12   Ibid., p. 162.

13   Lattis, *Between Copernicus and Galileo*, p. 206.

14   Ibid., p. 163.

15   Ibid., pp. 107, 206.

16   Stephen Pumfrey, *Latitude & the Magnetic Earth* (Cambridge: Icon Books, 2001), p. 16.

17   Ibid., p. 38.

18   Edward Grant, 'Peter Peregrinus', in *Dictionary of Scientific Biography*, ed. Charles Coulston Gillispie, vol. 10 (New York: Scribner, 1970), p. 538.

19   William Gilbert (trans. P. Fleury Mottelay), *De Magnete* (New York: Dover, 1991), p. 5.

20   Ibid., p. 170.

21   Max Caspar (trans. C. Doris Hellman), *Kepler* (London: Abelard-Schuman, 1959), p. 38.

22   Ibid., p. 46.

23   Linton, *From Eudoxus to Einstein*, p. 172.

24   Caspar, *Kepler*, p. 121.

25   Ibid., p. 122.

26   Ibid., p. 311.

27   Linton, *From Eudoxus to Einstein*, p. 181. The difference was not eight minutes of time, as is sometimes supposed.

28   Caspar, *Kepler*, p. 128.

29   Linton, *From Eudoxus to Einstein*, p. 170.

30   Psalm 19:1

31   Ibid., pp. 190, 192 and 197.

32   Ibid., p. 220.

33   Caspar, *Kepler*, p. 138.

34   David C. Lindberg, *Theories of Vision from Al-Kindi to Kepler* (Chicago: University Of Chicago Press, 1976), p. 200.

35   Caspar, *Kepler*, p. 206.

36   Ibid., p. 152.

37   Ibid., p. 256.

38   Keith Thomas, *Religion and the Decline of Magic: Studies in Popular Beliefs in Sixteenth- and Seventeenth-Century England* (Harmondsworth: Penguin, 1973), p. 87.

39   Brian P. Levack, *The Witch-Hunt in Early Modern Europe*, 2nd edn (London: Longman, 1995), p. 25.

40   William H. Huffman, *Robert Fludd and the End of the Renaissance* (London: Routledge, 1988), p. 16.

41   Diarmaid MacCulloch, *The Reformation: Europe's House Divided 1490–1700* (London: Allen Lane, 2003), p. 491.

42   Allen George Debus, *Man and Nature in the Renaissance* (Cambridge: Cambridge University Press, 1978), p. 122.

43   Allen George Debus, *The Chemical Philosophy: Paracelsian Science and Medicine in the Sixteenth and Seventeenth Centuries*, vol. 1 (New York: Science History Publications, 1977), p. 257.

44   Caspar, *Kepler*, p. 136.

Chapter 19 – Galileo and Giordano Bruno

1   Michael Sharratt, *Galileo: Decisive Innovator* (Cambridge: Cambridge University Press, 1996), p. 27.

2   Stillman Drake, *Galileo at Work: His Scientific Biography* (Chicago: University of Chicago Press, 1978), pp. 20, 416.

3   Sharratt, *Galileo*, p. 50. Attempts to repeat the experiment have shown that people unconsciously release the light object earlier and thus give it a head start. This does not explain how the heavy object can subsequently overtake the lighter one.

4   Paul Lawrence Rose, *The Italian Renaissance of Mathematics: Studies on Humanists and Mathematicians from Petrarch to Galileo* (Geneva: Droz, 1975), p. 154.

5   Stillman Drake and I.E. Drabkin, *Mechanics in Sixteenth-Century Italy: Selections from Tartaglia, Benedetti, Guido Ubaldo, & Galileo* (Madison: University of Wisconsin Press, 1969), p. 152.

6   Ibid., p. 34.

7   Simon Stevin, 'Appendix to the Art of Weighing', in *The Principal Works of Simon Stevin*, ed. Ernst Crone (trans. C. Dikshoorn), vol. 1 (Amsterdam: C.V. Swets & Zeitlinger, 1955), p. 511.

8   Drake, *Galileo at Work*, p. 23.

9   Ernest A. Moody, 'Galileo and Avempace: The Dynamics of the Leaning Tower Experiment', *Journal of the History of Ideas* vol. 12, no. 2 (1951), p. 193. For a more recent criticism of Moody's interpretation of Avempace's

physics see Abel B. Franco, 'Avempace, Projectile Motion, and Impetus Theory', *Journal of the History of Ideas* vol. 64, no. 4 (2003), pp. 521–46.

10   Willam Wallace, 'Domingo de Soto and the Iberian Roots of Galileo's Science', in *Hispanic Philosophy in the Age of Discovery*, ed. Keven White (Washington DC: The Catholic University of America Press, 1997), p. 113.

11   Ibid., p. 118.

12   Ibid., p. 113.

13   Ibid., p. 121.

14   Ibid., p. 122.

15   Sharratt, *Galileo*, p. 73.

16   J.H. Randall Jr., 'The Development of the Scientific Method in the School of Padua', *Journal of the History of Ideas* vol. 1, no. 2 (1940), pp. 177–206.

17   Sharratt, *Galileo*, p. 182.

18   Irving A. Kelter, 'The Refusal to Accommodate: Jesuit Exegates and the Copernican System', in *The Church and Galileo*, ed. Ernan McMullin (Notre Dame: University of Notre Dame Press, 2005), p. 40.

19   Frances A. Yates, *Giordano Bruno and the Hermetic Tradition* (London: Routledge Classics, 2002), p. 205.

20   Edward Grant, *Much Ado About Nothing: Theories of Space and Vacuum from the Middle Ages to the Scientific Revolution* (Cambridge: Cambridge University Press, 1981), p. 203.

21   Giordano Bruno (trans. Edward A. Gosselin and Lawrence S. Lerner), *The Ash Wednesday Supper* (Hamden, Conn: Archon Books, 1977), p. 18.

22   Yates, *Giordano Bruno and the Hermetic Tradition*, p. 217.

23   Bruno, *The Ash Wednesday Supper*, p. 213.

24   Hilary Gatti, *Giordano Bruno and Renaissance Science* (Ithaca: Cornell University Press, 1999), p. 88.

25   See, for instance, the disputation question at Andrew Clark, ed., *Register of the University of Oxford: 1571–1622* (Oxford: Oxford Historical Society, 1887), p. 170.

26   Yates, *Giordano Bruno and the Hermetic Tradition*, p. 228.

27   Maurice A. Finocchiaro, 'Philosophy versus Religion and Science versus Religion: the Trials of Bruno and Galileo', in *Giordano Bruno: Philosopher of the Renaissance*, ed. Hilary Gatti (Aldershot: Ashgate, 2002), p. 55.

28   Ibid., p. 61.

29   In the 1960s, Frances Yates destroyed Bruno's scientific credentials by showing him to be a Hermetic magus. Since then, a few scholars, mainly from Italy, have laboured to salvage his reputation, at least as a philosopher of note. See especially some of the essays in Hilary Gatti, ed., *Giordano Bruno: Philosopher of the Renaissance*. To date, they have not succeeded.

30   Annibale Fantoli, 'The Disputed Injunction and its Role in Galileo's Trial', in *The Church and Galileo*, ed. Ernan McMullin (Notre Dame: University of Notre Dame Press, 2005), p. 125.

31   Ernan McMullin, 'The Church's Ban on Copernicanism, 1616', in *The Church and Galileo*, p. 175.

32   Keith Thomas, *Religion and the Decline of Magic: Studies in Popular Beliefs in Sixteenth- and Seventeenth-Century England* (Harmondsworth: Penguin, 1973), p. 85.

33   James Brodrick, *Robert Bellarmine: Saint and Scholar* (London: Burns & Oates, 1961), p. 105.

Chapter 20 – Galileo and the New Astronomy

1   Adapted from Michael Sharratt, *Galileo: Decisive Innovator* (Cambridge: Cambridge University Press, 1996), p. 2.

2   Albert Van Helden, 'The Invention of the Telescope', *Transactions of the American Philosophical Society* vol. 67, no. 4, New Series (1977), p. 20.

3   The author wishes to thank Paul Newall for finding and translating this quotation and the next. Galileo Galilei, *Le Opere di Galileo Galilei*, eds Antonio Giorgio Garbasso and Giorgio Abetti (Florence: G. Barbèra, 1929–39), vol. 11, p. 165.

4   Ibid., vol. 10, p. 484.

5   James M. Lattis, *Between Copernicus and Galileo: Christoph Clavius and the Collapse of Ptolemaic Cosmology* (Chicago: University of Chicago Press, 1994), p. 183.

6   Ibid., p. 190.

7   Ibid., p. 198.

8   Ernan McMullin, 'Galileo's Theological Venture', in *The Church and Galileo*, ed. Ernan McMullin (Notre Dame: University of Notre Dame Press, 2005), p. 98.

9   Isaiah 11:12 and Revelation 7:1.

10   Galileo Galilei, 'Letter to the Grand Duchess Christina (1615)', in *The Galileo Affair: A Documentary History*, ed. Maurice A. Finocchiaro (Berkeley: University of California Press, 1989), p. 96.

11   Ibid., p. 114.

12   Ernan McMullin, 'The Church's Ban on Copernicanism, 1616', in *The Church and Galileo*, p. 170.

13   Ibid., p. 179.

14   Robert Bellarmine, 'Cardinal Bellarmine to Foscarini (1615)', in *The Galileo Affair*, p. 68.

15   Finocchiaro, *The Galileo Affair*, pp. 146, 149.

16   Sharratt, *Galileo*, p. 135.

17   Galileo Galilei, 'The Assayer', in *Discoveries and Opinions of Galileo* (trans. Stillman Drake) (New York: Doubleday, 1957), p. 237.

18   Sharratt, *Galileo*, p. 137.

19   Finocchiaro, *The Galileo Affair*, p. 201.

20   Mariano Artigas, Rafael Martínez, and William R. Shea, 'New Light on the Galileo Affair?', in *The Church and Galileo*, p. 222. Pietro Redondi's thesis that the atomism espoused in *Assayer* was the real reason for Galileo's trial has not won support among scholars.

21   Sharratt, *Galileo*, p. 144.

22   Michael Shank, 'Setting the Stage: Galileo in Tuscany, the Veneto and Rome', in *The Church and Galileo*, p. 75.

23   John M. Headley, *Tommaso Campanella and the Transformation of the World* (Princeton: Princeton University Press, 1997), p. 48.

24   Frances A. Yates, *Giordano Bruno and the Hermetic Tradition* (London: Routledge Classics, 2002), p. 407.

25   D.P. Walker, *Spiritual and Demonic Magic from Ficino to Campanella*, new edn (Stroud: Sutton, 2000), p. 207.

Chapter 21 – The Trial and Triumph of Galileo

1   Galileo Galilei (trans. Stillman Drake), *Dialogue Concerning the Two Chief World Systems: the Ptolemaic and Copernican* (Berkeley: University of California Press, 1953), p. 52.

2   Ibid., p. 144.

3   Michael Sharratt, *Galileo: Decisive Innovator* (Cambridge: Cambridge University Press, 1996), p. 168.

4   The tides argument still arouses controversy today and scholars continue to debate exactly how it was supposed to work. For a recent attempt to explain it, see Ron Naylor, 'Galileo's Tidal Theory', *Isis* vol. 98, no. 1 (2007), pp. 1–22.

5   Galileo, *Dialogue Concerning the Two Chief World Systems*, p. 464.

6   Maurice A. Finocchiaro, *The Galileo Affair: A Documentary History* (Berkeley: University of California Press, 1989), p. 229.

7   Annibale Fantoli, 'The Disputed Injunction and its Role in Galileo's Trial', in *The Church and Galileo*, ed. Ernan McMullin (Notre Dame: University of Notre Dame Press, 2005), pp. 132, 140.

8   Finocchiaro, *The Galileo Affair*, p. 287. There is no evidence that Galileo was 'shown the instruments of torture' as is often alleged.

9   Sharratt, *Galileo*, p. 185.

10   Ibid., p. 205.

11   Galileo Galilei (trans. Henry Crew and Alfonso de Salvio), *Dialogues Concerning Two New Sciences*, (New York: Dover, 1954), p. 6.

12   Ibid., p. 12.

13   Ibid., p. 62.

14   Michael Wolff, 'Philoponus and the Rise of Preclassical Dynamics', in *Philoponus: And the Rejection of Aristotelian Science*, ed. Richard Sorabji (London: Duckwork, 1987), p. 92.

15   Galileo, *Dialogues Concerning Two New Sciences*, p. 64.

16   Ibid., p. 153.

17   Even Stillman Drake, generally hostile to theories about Galileo's medieval influences, admitted this. See Stillman Drake, *Galileo at Work: His Scientific Biography* (Chicago: University of Chicago Press, 1978), p. 370.

18   Stillman Drake and I.E. Drabkin, *Mechanics in Sixteenth-Century Italy: Selections from Tartaglia, Benedetti, Guido Ubaldo, and Galileo* (Madison: University of Wisconsin Press, 1969), p. 54.

19   Galileo, *Dialogues Concerning Two New Sciences*, p. 153.

20   Marshall Clagett, *The Science of Mechanics in the Middle Ages* (Madison: University of Wisconsin Press, 1959), p. 272.

21   Ibid., p. 344.

22   Christopher Lewis, *The Merton Tradition and Kinematics in Late Sixteenth and Early Seventeenth Century Italy* (Padua: Antenore, 1980), p. 124.

23   Translated from the Latin in Ibid., p. 294.

24   Galileo, *Dialogues Concerning Two New Sciences*, p. 174.

25   Sharratt, *Galileo*, p. 199.

26   Marshall Clagett, *Giovanni Marliani and Late Medieval Physics* (New York: Columbia University Press, 1941), p. 140.

27   Edward Grant, 'Jordanus de Nemore', in *Dictionary of Scientific Biography*, ed. Charles Coulston Gillispie, vol. 7 (New York: Scribner, 1970), p. 173.

28   Paul Lawrence Rose, *The Italian Renaissance of Mathematics: Studies on Humanists and Mathematicians from Petrarch to Galileo* (Geneva: Droz, 1975), p. 153.

29   Drake and Drabkin, *Mechanics in Sixteenth-Century Italy*, pp. 16, 24.

30   Galileo, *Dialogues Concerning Two New Sciences*, p. 244.

31   Mario Gliozzi, 'Cardano, Girolamo', in *Dictionary of Scientific Biography*, ed. Charles Coulston Gillispie, vol. 3 (New York: Scribner, 1970), p. 66.

32   Galileo, *Dialogues Concerning Two New Sciences*, p. 245.

Conclusion – A Scientific Revolution?

1   William Whewell, 'Mrs Somerville on the Connexion of the Sciences', *The Quarterly Review* vol. 51, no. 1 (1834), p. 59.

2   From the 'General Scholium' in Isaac Newton, ed. Florian Cajori (trans. Andrew Motte), *Mathematical Principles of Natural Philosophy and the System of the World* (Berkeley: University of California Press, 1934), p. 546.

3   Andrew Cunningham and Perry Williams, 'De-Centring the "Big Picture": The Origins of Modern Science and the Modern Origins of Science', *The British Journal for the History of Science* vol. 26, no. 4 (1993), p. 410.

4   Steven Shapin memorably begins his book: 'There was no such thing as the Scientific Revolution and this is a book about it.' Steven Shapin, *The Scientific Revolution* (Chicago: University of Chicago Press, 1998), p. 1.

5   Cunningham and Williams, 'De-Centring the "Big Picture"', p. 418.

6   David Wootton, *Bad Medicine: Doctors Doing Harm Since Hippocrates* (Oxford: Oxford University Press, 2006), p. 3.

BIBLIOGRAPHY OF WORKS CITED

Abelard, Peter, and Héloïse, *The Letters of Abelard and Heloise*, revised edition (Harmondsworth: Penguin, 1974)

Adelard of Bath, *Conversations with his Nephew, On the Same and the Different, Questions on Natural Science, and On Birds*, trans. Charles Burnett (Cambridge: Cambridge University Press, 1998)

Agricola, Georg, *De Re Metallica*, trans. Herbert Hoover and Lou Henry Hoover (London: The Mining Magazine, 1912)

Albertus Magnus, *Book of Minerals* (Oxford: Clarendon Press, 1967)

Albertus Magnus, *Opera Omnia*, 38 vols (Paris: Ludovicus Vives, 1890)

Anselm, St, *Anselm of Canterbury: The Major Works* (New York: Oxford University Press, USA, 2008)

Anselm, St (trans. Benedicta Ward) *Prayers and Meditations of St Anselm with the Proslogion* (Harmondsworth: Penguin, 1973)

Aquinas, Thomas, *Summa Theologiae*, ed. Thomas Gilby, 61 vols (London: Blackfriars, 1964)

Aristotle, *The Complete Works of Aristotle: The Revised Oxford Translation*, ed. Jonathan Barnes, two vols (Princeton: Princeton University Press, 1984)

Artigas, Mariano, Rafael Martínez, and William R. Shea, 'New Light on the Galileo Affair?' in *The Church and Galileo*, ed. Ernan McMullin (Notre Dame: University of Notre Dame Press, 2005)

Aubrey, John, *Brief Lives*, ed. John Buchanan-Brown (Harmondsworth: Penguin, 2000)

Augustine, St, *The City of God Against the Pagans*, trans. R.W. Dyson (Cambridge: Cambridge University Press, 1998)

Augustine, St, *De Doctrina Christiana*, trans. Roger Green (Oxford: Clarendon Press, 1995)

Bacon, Roger, *Roger Bacon's Letter Concerning the Marvellous Power of Art and of Nature and Concerning the Nullity of Magic*, trans. Tenney Lombard Davis (Easton: Chemical Publishing Company, 1923)

Bacon, Roger, *The Opus Majus of Roger Bacon*, trans. Robert Belle Burke, 2 vols (Philadelphia: University of Pennsylvania Press, 1928)

Ball, Philip, *The Devil's Doctor: Paracelsus and the World of Renaissance Magic and Science* (London: William Heinemann, 2006)

Bartlett, Robert, *The Making of Europe* (Princeton: Princeton University Press, 1994)

Bartlett, Robert, *The Natural and the Supernatural in the Middle Ages* (Cambridge: Cambridge University Press, 2008)

Bazan, B. Carlos, 'Siger of Brabant', in *A Companion to Philosophy in the Middle Ages*, eds Jorge J.E. Gracia and Timothy B. Noone (Malden, MA: Blackwell Pub, 2003)

Biard, Joël, 'The Natural Order in John Buridan', in *The Metaphysics and Natural Philosophy of John Buridan*, eds J.M.M.H. Thijssen and Jack Zupko (Leiden: Brill Academic Publishers, 2001)

Biscaro, Gerolamo, *Inquisitori ed eretici Lombardi, 1292–1318*, Miscellanea di Storia Italiana, series 3, volume 19 (Turin: Fratelli Bocca, 1922)

Boethius, Anicius, *Boethian Number Theory: A Translation of the De Institutione Arithmetica (with Introduction and Notes)*, ed. Michael Masi (Amsterdam: Rodopi, 1983)

Boethius, Anicius, *De institutione arithmetica, libri duo : di institutione musica, libri quinque*, ed. G Friedlein (Leipzig: Teubner, 1867)

Boethius, Anicius, *Fundamentals of Music* (New Haven: Yale University Press, 1989)

Boethius, Anicius, *The Consolation of Philosophy*, revised edition (Harmondsworth: Penguin, 1999)

Boffio, G, 'Perchè fu condannato al fuoco l'astrologo Cecco d'Ascoli?' *Studi e Documenti di Storia e Diritto* vol. 20 (1899)

Bonatti, Guido, *The Book of Astronomy*, trans. Benjamin Nykes (The Cazimi Press, 2007)

Boorstin, Daniel J., *The Discoverers* (New York: Vintage, 1985)

Boyer, Carl B., *A History of Mathematics*. ed. Uta C. Merzbach, second edition (New York: Wiley, 1991)

Brewer, J.S, ed., *Letters and Papers, Foreign and Domestic, of the Reign of Henry VIII: Preserved in the Public Record Office, the British Museum, and Elsewhere in England*, 23 vols (London: Longman, Roberts & Green, 1862)

Bricot, Thomas, *Tractatus Insolubilium*, trans. E. J. Ashworth (Nijmegen: Ingenium, 1986)

Brill's New Pauly: Encyclopaedia of the Ancient World (Leiden, Boston: Brill, 2002)

Brodrick, James, *Robert Bellarmine: Saint and Scholar* (London: Burns & Oates, 1961)

Brooke, John Hedley, *Science and Religion: Some Historical Perspectives*. (Cambridge: Cambridge University Press, 1991)

Brown, Patterson, 'Infinite Causal Regression', *Philosophical Review* vol. 75 (1966), pp. 510–25

Bruno, Giordano, *The Ash Wednesday Supper*, trans. Edward A. Gosselin and Lawrence S. Lerner (Hamden, Conn: Archon Books, 1977)

Burckhardt, Jacob, *The Civilization of the Renaissance in Italy*, trans. S.G.C. Middlemore (Harmondsworth: Penguin, 1990)

Callus, Daniel Angelo, 'Robert Grosseteste as Scholar', in *Robert Grosseteste, Scholar and Bishop: Essays in Commemoration of the Seventh Centenary of His Death*, ed. Daniel Angelo Callus (Oxford: Clarendon Press, 1955)

Cardano, Girolamo, *The Book of My Life*, trans. Jean Stoner (New York: New York Review Books, 2002)

Carey, John, 'Ireland and the Antipodes: The Heterodoxy of Virgil of Salzburg', *Speculum* vol. 64, no. 1 (January 1989), pp. 1–10

Caspar, Max, *Kepler* (trans. C. Doris Hellman) (London: Abelard-Schuman, 1959)

Cassirer, Ernst, Paul Oskar Kristeller, and John Herman Randall, eds, *The Renaissance Philosophy of Man: Selections in Translation* (Chicago: University of Chicago Press, 1948)

Catholic University of America, *New Catholic Encyclopedia* (New York: McGraw-Hill, 1967)

Chadwick, Henry, *The Early Church* (Harmondsworth: Penguin, 1967)

Chaucer, Geoffrey, *The Canterbury Tales*, trans. Nevill Coghill (Harmondsworth: Penguin, 1951)

Chaucer, Geoffrey, *The Works of Geoffrey Chaucer*, ed. F.N Robinson, second edition (London: Oxford University Press, 1957)

Chenu, M.-D., *Nature, Man and Society in the Twelfth Century: Essays on New Theological Perspectives in the Latin West*, trans. Lester K. Little and Jerome Taylor (Chicago: Chicago University Press, 1968)

Clagett, Marshall, *Giovanni Marliani and Late Medieval Physics* (New York: Columbia University Press, 1941)

Clagett, Marshall, 'Oresme, Nicole', in *Dictionary of Scientific Biography*, ed. Charles Coulston Gillispie (New York: Scribner, 1970)

Clagett, Marshall, *The Science of Mechanics in the Middle Ages* (Madison: University of Wisconsin Press, 1959)

Clair, Colin, *A History of European Printing* (London: Academic Press, 1976)

Clanchy, Michael T., *Abelard: A Medieval Life* (Oxford: Blackwell, 1999)

Clark, Andrew, ed., *Register of the University of Oxford: 1571–1622* (Oxford: Oxford Historical Society, 1887)

Cobban, Alan B., *English University Life in the Middle Ages* (London: UCL Press, 1999)

Cochrane, Louise, *Adelard of Bath: The First English Scientist* (London: British Museum Press, 1994)

Cohen, Morris R, and I.E. Drabkin, *A Source Book in Greek Science* (Cambridge, MA: Harvard University Press, 1958)

Collins, Roger, *Early Medieval Europe, 300–1000*, second edition (Basingstoke: Palgrave Macmillan, 1999)

Columbus, Christopher, *The Voyage of Christopher Columbus: Columbus' Own Journal of Discovery Newly Restored and Translated*, trans. John G. Cummins (London: Weidenfeld and Nicolson, 1992)

Conquerors and Chroniclers of Early Medieval Spain, second edition (Liverpool: Liverpool University Press, 1999)

Copenhaver, Brian P, ed., *Hermetica: The Greek Corpus Hermeticum and the Latin Asclepius* (Cambridge: Cambridge University Press, 1992)

Copenhaver, Brian P, ed., *Renaissance Philosophy* (New York: Oxford University Press, USA, 1992)

Copernicus, Nicolaus, *On the Revolutions of the Heavenly Spheres* (Amherst: Prometheus Books, 1995)

Copleston, Frederick Charles, *A History of Medieval Philosophy* (London: Methuen, 1972)

Courtenay, William J, *Ockham and Ockhamism: Studies in the Dissemination and Impact of His Thought* (Leiden: Brill, 2008)

Crombie, A.C., *Robert Grosseteste and the Origins of Experimental Science, 1100–1700* (Oxford: Clarendon Press, 1953)

Cunningham, Andrew, and Perry Williams, 'De-Centring the "Big Picture": The Origins of Modern Science and the Modern Origins of Science', *The British Journal for the History of Science* vol. 26, no. 4 (1993), pp. 407–32

Dante Alighieri, *Inferno*, trans. Robert M. Durling (New York: Oxford University Press, 1996)

Dante Alighieri, *Paradiso*, trans. Robert and Jean Hollander (New York: Doubleday, 2007)

Darnton, Robert, *The Great Cat Massacre: And Other Episodes in French Cultural History* (New York: Vintage, 1985)

Debus, Allen G, *The Chemical Philosophy: Paracelsian Science and Medicine in the Sixteenth and Seventeenth Centuries*, two vols (New York: Science History Publications, 1977)

Debus, Allen G, *Man and Nature in the Renaissance* (Cambridge: Cambridge University Press, 1978)

Debus, Allen G, *The French Paracelsians: The Chemical Challenge to Medical and Scientific Tradition in Early Modern France* (Cambridge: Cambridge University Press, 1991)

Dee, John, *John Dee on Astronomy: 'Propaedeumata Aphoristica' (1558 and 1568)*, eds Wayne Shumaker and J.L. Heilbron (Berkeley: University of California Press, 1978)

Deutscher, Thomas, 'Bessarion', in *Contemporaries of Erasmus: A Biographical Register of the Renaissance and Reformation*, eds Peter G. Bietenholz and Thomas B. Deutscher (Toronto: University of Toronto Press, 1985)

Deutscher, Thomas, 'Gemitos Plethon', in *Contemporaries of Erasmus: A Biographical Register of the Renaissance and Reformation*, eds Peter G. Bietenholz and Thomas B. Deutscher (Toronto: University of Toronto Press, 1985)

Diamond, Jared M., *Collapse: How Societies Choose to Fail or Succeed* (London: Viking, 2005)

Dix, Keith, 'Aristotle's Peripatetic Library', in *Lost Libraries: the Destruction of the Great Book Collections Since Antiquity*, ed. James Raven (Basingstoke: Palgrave MacMillan, 2004)

Domesday Book: A Complete Translation, ed. Ann Williams and G.H. Martin (London: Penguin, 2003)

Drake, Stillman, *Galileo at Work: His Scientific Biography* (Chicago: University of Chicago Press, 1978)

Drake, Stillman, and I.E. Drabkin, eds *Mechanics in Sixteenth-Century Italy: Selections from Tartaglia, Benedetti, Guido Ubaldo, & Galileo* (Madison: University of Wisconsin Press, 1969)

Dronke, Peter, 'Thierry of Chartres', in *A History of Twelfth-Century Western Philosophy*, ed. Peter Dronke (Cambridge: Cambridge University Press, 1988)

Eadmer, *The Life of St Anselm, Archbishop of Canterbury*, trans. R.W. Southern (Oxford: Clarendon Press, 1972)

Eisenstein, Elizabeth, *The Printing Revolution in Early Modern Europe* (Cambridge: Cambridge University Press, 1993)

Elliott, John Huxtable, *The Old World and the New 1492–1650*, Canto edition (Cambridge: Cambridge University Press, 1992)

Engels, Donald, 'The Length of Eratosthenes' Stade', *American Journal of Philology* vol. 106, no. 3 (1985), pp. 298–311

Erasmus, Desiderius, *Praise of Folly*, trans. Betty Radice (Harmondsworth: Penguin Classics, 1993)

Erasmus, Desiderius, *The Correspondence of Erasmus*, ed. Charles Garfield Nauert, trans. Alexander Dalzell, twelve vols (Toronto: University of Toronto Press, 1994)

Evans, Angela Care, *The Sutton Hoo Ship Burial* (London: British Museum Publications, 1986)

Fantoli, Annibale, 'The Disputed Injunction and its Role in Galileo's Trial', in *The Church and Galileo*, ed. Ernan McMullin (Notre Dame: University of Notre Dame Press, 2005)

Fantoni, Anna Rita, *Treasures from the Italian Libraries*, ed. Lorenzo Crinelli (London: Thames and Hudson, 1997)

Farge, James, 'Thomas Bricot', in *Contemporaries of Erasmus: A Biographical Register of the Renaissance and Reformation*, eds Peter G. Bietenholz and Thomas B. Deutscher (Toronto: University of Toronto Press, 1985)

Finocchiaro, Maurice A., 'Philosophy versus Religion and Science versus Religion: the Trials of Bruno and Galileo', in *Giordano Bruno: Philosopher of the Renaissance*, ed. Hilary Gatti (Ithaca, NY: Cornell University Press, 1999)

Finocchiaro, Maurice A., *The Galileo Affair: A Documentary History* (Berkeley: University of California Press, 1989)

Fletcher, Richard, *The Barbarian Conversion: From Paganism to Christianity* (New York: Henry Holt, 1997)

Flusche, Anna Marie, *The Life and Legend of Gerbert of Aurillac: The Organbuilder who became Pope Sylvester II* (Lewiston, NY: The Edwin Mellen Press, 2005)

Förstel, Judith, and Aline Magnien, *Beauvais Cathedral*, trans. Diana Fowles (Amiens: Inventaire général, 2005)

Foster, Kenelm, ed., *The Life of Thomas Aquinas: Biographical Documents* (London: Longmans, Green & Co, 1959)

Franco, Abel B., 'Avempace, Projectile Motion, and Impetus Theory', *Journal of the History of Ideas* vol. 64, no. 4 (2003), pp. 521–46.

Freeman, Charles, *The Closing of the Western Mind: The Rise of Faith and the Fall of Reason* (London: William Heinemann, 2002)

French, R.K., *Dissection and Vivisection in the European Renaissance* (Aldershot: Ashgate, 1999)

French, Roger, and Andrew Cunningham, *Before Science: The Invention of the Friars' Natural Philosophy* (Aldershot: Scolar Press, 1996)

Funkenstein, Amos, *Theology and the Scientific Imagination: From the Middle Ages to the Seventeenth Century* (Princeton, NJ: Princeton University Press, 1986).

Galilei, Galileo, *Dialogue concerning the Two Chief World Systems: the Ptolemaic and Copernican*, trans. Stillman Drake (Berkeley: University of California Press, 1953)

Galilei, Galileo, *Dialogues concerning Two New Sciences*, trans. Henry Crew and Alfonso de Salvio (New York: Dover, 1954)

Galilei, Galileo, *Discoveries and Opinions of Galileo*, trans. Stillman Drake (Anchor, 1957)

Galilei, Galileo, *Le Opere di Galileo Galilei*, eds Antonio Giorgio Garbasso and Giorgio Abetti, 20 vols (Florence: G. Barbera, 1929–39)

Gans, Paul, 'The Medieval Horse Harness: Revolution or Evolution: A Case Study in Technological Change', in *Villard's Legacy: Studies in Medieval Technology, Science and Art in Memory of Jean Gimpel* (Aldershot: Ashgate, 2004)

Gatti, Hilary, *Giordano Bruno and Renaissance Science* (Ithaca, NY: Cornell University Press, 1999)

Gatti, Hilary, ed., *Giordano Bruno: Philosopher of the Renaissance* (Aldershot: Ashgate, 2002)

Gaukroger, Stephen, *The Emergence of a Scientific Culture: Science and the Shaping of Modernity 1210–1685* (Oxford: Clarendon Press, 2006)

Gerbert of Aurillac, *The Letters of Gerbert: with his Papal Privileges as Sylvester II*, trans. Harriet Pratt Lattin (New York: Columbia University Press, 1961)

Gerson, Jean, *Jean Gerson: Early Works*, trans. Brian Patrick McGuire (New York: Paulist Press, 1998)

Gibbon, Edward, *The History of the Decline and Fall of the Roman Empire*, three vols (Harmondsworth: Penguin, 1994)

Gies, Joseph, and Frances Gies, *Cathedral, Forge and Waterwheel: Technology and Invention in the Middle Ages* (London: Harper Perennial, 1995)

Gilbert, William, *De Magnete*, trans. P. Fleury Mottelay (New York: Dover, 1991)

Gillispie, Charles Coulston, ed., *Dictionary of Scientific Biography* (New York: Scribner, 1970)

Gimpel, Jean, *The Medieval Machine: The Industrial Revolution of the Middle Ages* (London: Pimlico, 1992)

Given, James Buchanan, *Inquisition and Medieval Society: Power, Discipline and Resistance in Languedoc* (Ithaca, NY: Cornell University Press, 1997)

Gliozzi, Mario, 'Cardano, Girolamo', in *Dictionary of Scientific Biography*, ed. Charles Coulston Gillispie (New York: Scribner, 1970)

Gnudi, Martha Teach, and Jerome Pierce Webster, *The Life and Times of Gaspare Tagliacozzi: Surgeon of Bologna, 1545–1599. With a Documented Study of the Scientific and Cultural Life of Bologna in the Sixteenth Century* (New York: H. Reichner, 1950)

Grafton, Anthony, *Cardano's Cosmos: The Worlds and Works of a Renaissance Astrologer* (Cambridge, MA: Harvard University Press, 1999)

Grant, Edward, *A Source Book in Medieval Science* (Cambridge, MA: Harvard University Press, 1974)

Grant, Edward, *God and Reason in the Middle Ages*. (Cambridge: Cambridge University Press, 2001)

Grant, Edward 'Jordanus de Nemore', in *Dictionary of Scientific Biography*, ed. Charles Coulston Gillispie (New York: Scribner, 1970)

Grant, Edward, *Much Ado About Nothing: Theories of Space and Vacuum from the Middle Ages to the Scientific Revolution* (Cambridge: Cambridge University Press, 1981)

Grant, Edward, 'Peter Peregrinus', in *Dictionary of Scientific Biography*, ed. Charles Coulston Gillispie (New York: Scribner, 1970)

Grant, Edward, *Planets, Stars, and Orbs: The Medieval Cosmos, 1200– 1687* (Cambridge: Cambridge University Press, 1994)

Grant, Edward, *The Foundations of Modern Science in the Middle Ages: Their Religious, Institutional and Intellectual Contexts* (Cambridge: Cambridge University Press, 1996).

Grendler, Paul, 'Printing and Censorship', in *The Cambridge History of Renaissance Philosophy*, eds Charles Schmitt and Quentin Skinner (Cambridge: Cambridge University Press, 1988)

Hackett, Jeremiah, 'Roger Bacon: His Life, Career and Works', in *Roger Bacon and the Sciences: Commemorative Essays*, ed. Jeremiah Hackett (Leiden: Brill, 1997)

Hamilton, Alastair, 'Humanists and the Bible', in *Cambridge Companion to Renaissance Humanism*, ed. Jill Kraye (Cambridge: Cambridge University Press, 1996)

Hannam, James, 'Teaching Natural Philosophy and Mathematics at Oxford and Cambridge 1500–70', PhD thesis, University of Cambridge, 2008

Hansen, Bert, 'Science and Magic', in *Science in the Middle Ages*, ed. David C. Lindberg (Chicago: Chicago University Press, 1978)

Harrison, Peter, *The Bible, Protestantism, and the Rise of Natural Science* (Cambridge: Cambridge University Press, 1998)

Harvey, William, *Movement of the Heart and Blood in Animals: An Anatomical Essay*, trans. Kenneth Franklin (Oxford: Blackwell, 1957)

Haskins, Charles Homer, *The Renaissance of the Twelfth Century* (Cambridge, MA: Harvard University Press, 1927)

Headley, John M., *Tommaso Campanella and the Transformation of the World* (Princeton: Princeton University Press, 1997)

Hellman, C. Doris, 'Brahe, Tycho', in *Dictionary of Scientific Biography*, ed. Charles Coulston Gillispie (New York: Scribner, 1970)

Hellman, C. Doris, and John E. Murdoch, 'Peurbach, Georg', in *Dictionary of Scientific Biography*, ed. Charles Coulston Gillispie (New York: Scribner, 1970)

Henderson, Judith Rice, 'George of Trebizond', in *Contemporaries of Erasmus: A Biographical Register of the Renaissance and Reformation*, eds Peter G. Bietenholz and Thomas B. Deutscher (Toronto: University of Toronto Press, 1985)

Henry, John, *Knowledge Is Power: How Magic, the Government and an Apocalyptic Vision Inspired Francis Bacon to Create Modern Science* (Cambridge: Icon Books, 2003)

Henry, John, *Moving Heaven and Earth: Copernicus and the Solar System* (Cambridge: Icon Books, 2001)

Hofman, J.E., 'Cusa, Nicholas of', in *Dictionary of Scientific Biography*, ed. Charles Coulston Gillispie (New York: Scribner, 1970)

Hoskin, M.A., and A.G. Molland, 'Swineshead on Falling Bodies: An Example of Fourteenth-Century Physics', *The British Journal for the History of Science* vol. 3, no. 2 (1966), pp. 150–82

Howard, Henry, Oxford, Bodleian Library, MS Bodley 616

Howgrave-Graham, R, *The Cathedrals of France* (London: Batsford, 1959)

Huff, Toby E., *The Rise of Early Modern Science: Islam, China and the West* (Cambridge: Cambridge University Press, 1993)

Huffman, William H, *Robert Fludd and the End of the Renaissance* (London: Routledge, 1988)

Huxley, Thomas H., 'The Origin of Species', *Westminster Review* 17 (1860), pp. 541–70

Ilardi, Vincent, *Renaissance Vision from Spectacles to Telescopes* (Philadelphia: American Philosophical Society, 2007)

Jones, Brian W., *The Emperor Domitian* (London: Bristol Classical Press, 1996)

Keen, Maurice Hugh, *The Penguin History of Medieval Europe* (Harmondsworth: Penguin, 1969)

Kelter, Irving A., 'The Refusal to Accommodate: Jesuit Exegates and the Copernican System', in *The Church and Galileo*, ed. Ernan McMullin (Notre Dame: University of Notre Dame Press, 2005)

Ker, N.R., 'The Provision of Books', in *A History of Oxford University: The Collegiate University*, ed. James McConica (Oxford: Clarendon Press, 1986)

Keynes, Geoffrey, *The Life of William Harvey* (Oxford: Clarendon Press, 1966)

Kibre, Pearl, and Nancy G. Siraisi, 'The Institutional Setting: The Universities', in *Science in the Middle Ages*, ed. David C. Lindberg (Chicago: Chicago University Press, 1978)

Kieckhefer, Richard, *Magic in the Middle Ages*, Canto edition (Cambridge, UK: Cambridge University Press, 2000)

Kline, Naomi Reed, *Maps of Medieval Thought: The Hereford Paradigm* (Woodbridge, Suffolk: Boydell Press, 2001)

Knight, David, *Ideas in Chemistry: A History of the Science* (London: The Athlone Press, 1995)

Knoll, Paul W., 'The Arts Faculty at the University of Cracow', in *The Copernican Achievement*, ed. Robert S. Westman (Berkeley: University of California Press, 1975)

Krappe, Alexander Haggerty, 'The Legend of Buridan and the Tour de Nesle', *The Modern Language Review* vol. 23, no. 2 (1928), pp. 216–22

Kraye, Jill, *The Cambridge Companion to Renaissance Humanism* (Cambridge: Cambridge University Press, 1996)

Kristeller, Paul Oskar, *Renaissance Thought and its Sources* (New York: Columbia University Press, 1979)

Kuhn, Thomas S, *The Copernican Revolution: Planetary Astronomy in the Development of Western Thought* (Cambridge, MA: Harvard University Press, 1957)

Kuhn, Thomas S., *The Structure of Scientific Revolutions* (Chicago: University Of Chicago Press, 1996)

Kusukawa, Sachiko, *The Transformation of Natural Philosophy: The Case of Philip Melanchthon* (Cambridge: Cambridge University Press, 1995)

Laird, W.R., 'Archimedes among the Humanists', *Isis* vol. 82, no. 4 (1 January 1991), pp. 628–37

Landes, David S., *The Wealth and Poverty of Nations: Why Some Are so Rich and Some so Poor* (London: Abacus, 1999)

Lane Fox, Robin, *The Classical World: An Epic History of Greece and Rome* (London: Allen Lane, 2005)

Lattis, James M., *Between Copernicus and Galileo: Christoph Clavius and the Collapse of Ptolemaic Cosmology* (Chicago: University of Chicago Press, 1994)

Le Rond D'Alembert, Jean, *The Preliminary Discourse to the Encyclopaedia of Diderot*, trans. Richard H. Schwarb (Chicago: Chicago University Press, 1995)

Le Roy Ladurie, Emmanuel, *Montaillou: Cathars and Catholics in a French Village, 1294–1324* (Harmondsworth: Penguin, 1978)

Leijenhorst, Cornelis Hendrik, Christoph Luthy, J.M.M.H. Thijssen, and Cees Leijenhorst, *The Dynamics of Aristotelian Natural Philosophy from Antiquity to the Seventeenth Century* (Leiden: Brill Academic Publishers, 2002)

Lerner, Ralph, and Muhsin Mahdi eds, *Medieval Political Philosophy: A Sourcebook* (Ithaca: Cornell University Press, 1972)

Levack, Brian P., *The Witch-Hunt in Early Modern Europe*, second edition (London: Longman, 1995)

Levi, Anthony, *Renaissance and Reformation: The Intellectual Genesis* (New Haven: Yale University Press, 2002)

Lewis, C.S., *The Discarded Image: An Introduction to Medieval and Renaissance Literature* (Cambridge: Cambridge University Press, 1994)

Lewis, Christopher, *The Merton Tradition and Kinematics in Late Sixteenth and Early Seventeenth Century Italy* (Padua: Antenore, 1980)

Lindberg, David C., 'Light, Vision and the Universal Emanation of Force', in *Roger Bacon and the Sciences: Commemorative Essays*, ed. Jeremiah Hackett (Leiden: Brill, 1997)

Lindberg, David C., 'Medieval Science and Its Religious Context', *Osiris* vol. 10 (1995), pp. 61–79

Lindberg, David C., *Roger Bacon's Philosophy of Nature, A Critical Edition, with English Translation* (Oxford: Clarendon Press, 1983)

Lindberg, David C., ed., 'Science as Handmaiden: Roger Bacon and the Patristic Tradition', in *The Scientific Enterprise in Antiquity and Middle Ages: Readings from Isis*, ed. Michael Shank (Chicago: Chicago University Press, 2000)

Lindberg, David C. ed., *Science in the Middle Ages* (Chicago: University of Chicago Press, 1978)

Lindberg, David C., *The Beginnings of Western Science: The European Scientific Tradition in Philosophical, Religious, and Institutional Context, Prehistory to AD 1450*, second edition (Chicago: University Of Chicago Press, 2007)

Lindberg, David C., 'The Science of Optics', in *Science in the Middle Ages*, ed. David C. Lindberg (Chicago: Chicago University Press, 1978)

Lindberg, David C., *Theories of Vision from Al-Kindi to Kepler* (Chicago: University Of Chicago Press, 1976)

Lindberg, David C., 'Transmission of Greek and Arabic Learning', in *Science in the Middle Ages*, ed. David C. Lindberg (Chicago: Chicago University Press, 1978)

Lindberg, David C., 'Witelo', in *Dictionary of Scientific Biography*, ed. Charles Coulston Gillispie (New York: Scribner, 1970)

Lindberg, David C., and Ronald L. Numbers, eds, *God and Nature: Historical Essays on the Encounter between Christianity and Science* (Berkeley: University of California Press, 1986)

Lindberg, David C. and Ronald L. Numbers, eds, *When Science and Christianity Meet* (Chicago: University Of Chicago Press, 2003)

Lindberg, David C., and Robert S. Westman, eds, *Reappraisals of the Scientific Revolution* (Cambridge: Cambridge University Press, 1990)

Linden, Stanton J, ed., *The Alchemy Reader: From Hermes Trismegistus to Isaac Newton* (Cambridge: Cambridge University Press, 2003)

Linton, C.M, *From Eudoxus to Einstein: A History of Mathematical Astronomy* (Cambridge: Cambridge University Press, 2004)

Longrigg, James, 'Anatomy in Alexandria in the Third Century B.C.', *British Journal for the History of Science* vol. 21, no. 4 (1988), pp. 455–88

Luscombe, David, *Medieval Thought* (Oxford: Oxford University Press, 1997)

MacCulloch, Diarmaid, *The Reformation: Europe's House Divided 1490–1700* (London: Allen Lane, 2003)

MacDonald, A.J., *Berenger and the Reform of Sacramental Doctrine* (London: Longmans, Green & Co, 1930)

Mahoney, Michael, 'Mathematics', in *Science in the Middle Ages*, ed. David C. Lindberg (Chicago: University of Chicago Press, 1978)

Maier, Anneliese, *On the Threshold of Exact Science: Selected Writings of Anneliese Maier on Late Medieval Natural Philosophy*, trans. Steven D. Sargent (Philadelphia: University of Pennsylvania Press, 1982)

Maimonides, Moses, *Guide for the Perplexed*, trans. M. Friedlander, second edition (New York: Dover, 1956)

Manchester, William, *A World Lit Only by Fire: The Medieval Mind and the Renaissance: Portrait of an Age* (London: Macmillan, 1993)

Mandeville, John, *The Travels of Sir John Mandeville*, trans. C.W.R.D Moseley (Harmondsworth: Penguin, 1983).

Mann, Horace K., *The Lives of the Popes in the Early Middle Ages*, eighteen vols (London: Kegan Paul, 1910).

Martin, G.H., *A History of Merton College, Oxford* (Oxford: Oxford University Press, 1997)

Masters, Roger D., 'The Case of Aristotle's Missing Dialogues: Who Wrote the Sophist, the Statesman, and the Politics?' *Political Theory* vol. 5, no. 1 (February 1977), pp. 31–60

Maxwell-Stuart, P.G., ed., *The Occult in Early Modern Europe: A Documentary History* (Basingstoke: Macmillan, 1999)

Maxwell-Stuart, P.G., ed., *The Occult in Mediaeval Europe, 500–1500: A Documentary History* (Basingstoke: Palgrave Macmillan, 2005)

McCluskey, Stephen C., *Astronomies and Cultures in Early Medieval Europe* (Cambridge: Cambridge University Press, 1998)

McEvoy, James, *The Philosophy of Robert Grosseteste* (Oxford: Clarendon, 1982)

McMullin, Ernan, 'Galileo's Theological Venture', in *The Church and Galileo*, ed. Ernan McMullin (Notre Dame: Univesity of Notre Dame Press, 2005)

Merton, Robert H., 'Science, Technology and Society in Seventeenth-Century England', *Osiris*, vol. 4, no. 1 (1938), pp. 360–632

Minio-Paluello, Lorenzo, 'Boethius, Anicius Manlius Severinus', in *Dictionary of Scientific Biography*, ed. Charles Coulston Gillispie (New York: Scribner, 1970)

Molland, A.G., 'The Geometrical Background to the Merton School', *The British Journal for the History of Science* vol. 4, no. 2 (1968), pp. 108–125

Montaigne, Michel de, *The Essays of Michel De Montaigne*. trans. M.A Screech (Harmondsworth: Penguin, 1993)

Moody, Ernest A., 'Albert of Saxony', in *Dictionary of Scientific Biography*, ed. Charles Coulston Gillispie (New York: Scribner, 1970)

Moody, Ernest A., 'Buridan, Jean', in *Dictionary of Scientific Biography*, ed. Charles Coulston Gillispie (New York: Scribner, 1970)

Moody, Ernest A., 'Galileo and Avempace: The Dynamics of the Leaning Tower Experiment', *Journal of the History of Ideas*, vol. 12, no. 2 (1951), pp. 163–193

Moody, Ernest A., *Studies in Medieval Philosophy, Science, and Logic: Collected Papers, 1933–1969* (Berkeley: University of California Press, 1975)

Moore, Pete, *Blood and Justice: the Seventeenth-Century Parisian Doctor who Made Blood Transfusion History* (Chichester: John Wiley, 2002)

More, Thomas, *The Complete Works of St Thomas More*, ed. Daniel Kinney, fifteen vols (New Haven: Yale University Press, 1986)

Multhauf, Robert P., 'The Science of Matter', in *Science in the Middle Ages*, ed. David C. Lindberg (Chicago: Chicago University Press, 1978)

Mumford, Lewis, *Technics and Civilization* (London: George Routledge & Sons, 1934)

Murdoch, John E., and Edith D. Sylla, 'Swineshead, Richard', in *Dictionary of Scientific Biography*, ed. Charles Coulston Gillispie (New York: Scribner, 1970)

Murdoch, John E., and Edith D. Sylla, 'The Science of Motion', in *Science in the Middle Ages*, ed. David C. Lindberg (Chicago: Chicago University Press, 1978)

Murray, Alexander, *Reason and Society in the Middle Ages* (Oxford: Clarendon Press, 1985)

Nauert, Charles G., *Humanism and the Culture of Renaissance Europe* (Cambridge: Cambridge University Press, 1995)

Naylor, Ron, 'Galileo's Tidal Theory', *Isis* vol. 98, no. 1 (2007), pp. 1–22

Newman, William, 'Technology and Alchemical Debate in the Late Middle Ages', in *The Scientific Enterprise in Antiquity and Middle Ages: Readings from Isis*, ed. Michael Shank (Chicago: Chicago University Press, 2000)

Newman, William, *The Summa Perfectionis of Pseudo-Geber: A Critical Edition, Translation and Study* (Leiden: E.J. Brill, 1991)

Newton, Isaac, *Mathematical Principles of Natural Philosophy and the System of the World*, ed. Florian Cajori, trans. Andrew Motte (Berkeley: University of California Press, 1934)

Nicholas of Cusa, 'On Learned Ignorance', in *Complete Philosophical and Theological Treatises of Nicholas of Cusa*, trans. Jasper Hopkins (Minneapolis: Arthur J. Banning Press, 2001)

Nicholl, Charles, *Leonardo da Vinci: The Flights of the Mind* (London: Allen Lane, 2004)

North, John, *God's Clockmaker: Richard of Wallingford and the Invention of Time* (London: Hambledon & London, 2007)

Olson, Richard G., *Science and Religion, 1450–1900: From Copernicus to Darwin* (Baltimore: The Johns Hopkins University Press, 2006)

O'Malley, C. Donald, *Andreas Vesalius of Brussels, 1514–1564* (Berkeley: University of California Press, 1964)

O'Malley, C. Donald, 'Andreas Vesalius' Pilgrimage', *Isis* vol. 45, no. 2 (1 January 1954), pp. 138–144

Oman, Charles William Chadwick, *A History of the Art of War in the Middle Ages*, second edition (London: Methuen, 1924)

Parry, J.H., *The Age of Reconaissance* (London: Weidenfeld and Nicolson, 1963)

Partington, James Riddick, *A History of Greek Fire and Gunpowder* (Cambridge: Heffer, 1960)

Peckham, John, *John Peckham and the Science of Optics: Perspectiva Communis*, trans. David C. Lindberg (Madison: University of Wisconsin Press, 1970)

Peckham, John, *Registrum Epistolarum Fratris Johannis Peckham, Archiepiscopi Cantuariensis*, ed. Charles Trice Martin, Rerum Britannicarum medii aevi scriptores (Rolls Series) (London: Longman, 1882)

Pedersen, Olaf, *The First Universities: Studium Generale and the Origins of University Education in Europe* (Cambridge: Cambridge University Press, 1997)

Peregrinus, Peter, 'The Letter of Peregrinus', in *A Source Book in Medieval Science*, ed. Edward Grant (Cambridge, MA: Harvard University Press, 1974)

Peters, Edward, *Inquisition* (Berkeley: University of California Press, 1989)

Peters, Edward, ed., *Heresy and Authority in Medieval Europe: Documents in Translation* (London: Scolar Press, 1980)

Plato, *Timaeus and Critias*, trans. Desmond Lee (Harmondsworth: Penguin, 1977)

Pliny the Elder, *Natural History: A Selection*, trans. John F. Healy (London: Penguin, 1991)

Pliny the Elder, *Natural History.*, trans. H Rackham, ten vols (London: Heinemann, 1938)

Plutarch, *Makers of Rome: Nine Lives*, trans. Ian Scott-Kilvert (Harmondsworth: Penguin, 1965)

Porter, Roy, *The Greatest Benefit to Mankind: A Medical History of Humanity from Antiquity to the Present* (London: HarperCollins, 1998)

Pouchelle, Marie-Christine, *The Body and Surgery in the Middle Ages* (Oxford: Polity, 1989)

Pryor, Francis, *Britain in the Middle Ages: An Archaeological History* (London: HarperPress, 2006)

Ptolemy, *Ptolemy's Almagest*, trans. G.J. Toomer (London: Duckworth, 1984)

Ptolemy, *Ptolemy's Geography: An Annotated Translation of the Theoretical Chapters*, trans. J.L. Berggren and Alexander Jones (Princeton: Princeton University Press, 2000)

Ptolemy, *Tetrabiblos*, trans. F.E. Robbins (Cambridge, Mass: Harvard University Press, 1994)

Pumfrey, Stephen, *Latitude & the Magnetic Earth* (Cambridge: Icon Books, 2001)

Randall Jr., J.H., 'The Development of the Scientific Method in the School of Padua', *Journal of the History of Ideas* vol. 1, no. 2 (1940), pp. 177–206

Record, Robert, *The Castle of Knowledge*, facsimile edition (Amsterdam: Theatrum Orbis Terrarum, 1975)

Record, Robert, *The Pathway to Knowledge*, facsimile edition (Amsterdam: Theatrum Orbis Terrarum, 1974)

Reeve, Michael, 'Classical Scholarship', in *Cambridge Companion to Renaissance Humanism*, ed. Jill Kraye (Cambridge: Cambridge University Press, 1996)

Reynolds, L.D, and N.G Wilson, *Scribes and Scholars: A Guide to the Transmission of Greek and Latin Literature*, third edition (Oxford: Clarendon Press, 1991)

Richards, E.G, *Mapping Time: The Calendar and Its History* (Oxford: Oxford University Press, 1999)

Richer of Saint Remi, *Historiae*, MGH Scriptores (Hanover, 2000)

Rijk, Lambertus Marie de, ed., *Nicholas of Autrecourt: His Correspondence with Master Giles and Bernhard of Arezzo: A Critical Edition from The Two Parisian Manuscripts* (Leiden: E.J. Brill, 1994)

Rose, Paul Lawrence, *The Italian Renaissance of Mathematics: Studies on Humanists and Mathematicians from Petrarch to Galileo* (Geneva: Droz, 1975)

Rosen, Edward, 'Copernicus, Nicholas', in *Dictionary of Scientific Biography*, ed. Charles Coulston Gillispie (New York: Scribner, 1970)

Rosen, Edward, 'Regiomontanus, Johannes', in *Dictionary of Scientific Biography*, ed. Charles Coulston Gillispie (New York: Scribner, 1970)

Runciman, Steven, *A History of the Crusades*, three vols (Harmondsworth: Penguin, 1978)

Runciman, Steven, *The Fall of Constantinople, 1453* (Cambridge: Cambridge University Press, 1990).

Runciman, Steven, *The Sicilian Vespers: A History of the Mediterranean World in the Later Thirteenth Century* (Cambridge: Cambridge University Press, 1992)

Russell, Bertrand, *A History of Western Philosophy* (London: Routledge, 1961)

Russell, Colin A., 'The Conflict of Science and Religion', in *Science and Religion: A Historical Introduction*, ed. Gary B. Ferngren (Baltimore: The Johns Hopkins University Press, 2002)

Russell, Jeffrey Burton, *Inventing the Flat Earth: Columbus and Modern Historians* (New York: Praeger Paperback, 1997).

Russell, Josiah Cox, *The Control of Late Ancient and Medieval Population* (Philadelphia: American Philosophical Society, 1985)

Saliba, George, *Islamic Science and the Making of the European Renaissance* (Cambridge, MA: The MIT Press, 2007)

Sarton, George, *Galen of Pergamon* (Lawrence, Kansas: University of Kansas Press, 1954)

Saunders, J.B. de C.M, and Charles Donald O'Malley, *The Anatomical Drawings of Andreas Vesalius: With Annotations and Translations, a Discussion of the Plates and Their Background, Authorship, and Influence, and a Biographical Sketch of Vesalius* (New York: Bonanza Books, 1982)

Savage-Smith, Emilie, 'Attitudes Toward Dissection in Medieval Islam', *Journal of the History of Medicine and Allied Sciences* vol. 50, no. 1 (1 January 1995), pp. 67–110

Schmitt, Charles B, and Quentin Skinner, eds, *The Cambridge History of Renaissance Philosophy* (Cambridge: Cambridge University Press, 1988)

Servetus, Michael, *Michael Servetus: A Translation of His Geographical, Medical, and Astrological Writings*, trans. C.D. O'Malley (Philadelphia: American Philosophical Society, 1953)

Shank, Michael, 'Setting the Stage: Galileo in Tuscany, the Veneto and Rome', in *The Church and Galileo*, ed. Ernan McMullin (Notre Dame: University of Notre Dame Press, 2005)

Shapin, Steven, *The Scientific Revolution* (Chicago: University of Chicago Press, 1998)

Sharratt, Michael, *Galileo: Decisive Innovator* (Cambridge: Cambridge University Press, 1996)

Siraisi, Nancy G., *Medieval and Early Renaissance Medicine: An Introduction to Knowledge and Practice* (Chicago: University Of Chicago Press, 1990)

Smith, Robert D., and Kelly DeVries, *The Artillery of the Dukes of Burgundy 1363–1477* (Woodbridge, Suffolk: Boydell Press, 2005)

Smoller, Laura Ackerman, *History, Prophecy, and the Stars: The Christian Astrology of Pierre d'Ailly, 1350–1420* (Princeton: Princeton University Press, 1994)

Southern, R.W, *Robert Grosseteste: The Growth of an English Mind in Medieval Europe* (Oxford: Clarendon Press, 1986)

Southern, R.W, *Western Society and the Church in the Middle Ages* (Harmondsworth: Penguin, 1970)

Southern, R.W, *The Making of the Middle Ages* (London: Hutchinson's University Library, 1953)

Southern, R.W, 'From Schools to University', in *The Early Schools: The History of the University of Oxford*, ed. Jeremy Catto (Oxford: Clarendon Press, 1984)

Spade, Paul Vincent, 'Ockham's Nominalist Metaphysics: Some Main Themes', in *The Cambridge Companion to Ockham*, ed. Paul Vincent Spade (Cambridge: Cambridge University Press, 1999)

Squatriti, Paolo, 'Pornocracy', in *Medieval Italy: An Encyclopaedia*, eds Christopher Kleinhenz et al (London: Routledge, 2004)

Stark, Rodney, *For the Glory of God: How Monotheism Led to Reformations, Science, Witch-Hunts, and the End of Slavery* (Princeton: Princeton University Press, 2003)

Stevin, Simon, 'Appendix to the Art of Weighing', in *The Principal Works of Simon Stevin*, ed. Ernst Crone, trans. C. Dikshoorn (Amsterdam: C.V. Swets & Zeitlinger, 1955)

Struik, D.J., 'Gerbert', in *Dictionary of Scientific Biography*, ed. Charles Coulston Gillispie (New York: Scribner, 1970)

Sumption, Jonathan, *The Albigensian Crusade* (London: Faber, 1978)

Tacitus, Cornelius, *The Annals of Imperial Rome*, revised edition (Harmondsworth: Penguin Books, 1989)

Talbot, Charles, 'Medicine', in *Science in the Middle Ages*, ed. David C. Lindberg (Chicago: Chicago University Press, 1978)

Telesko, Werner, *The Wisdom of Nature: The Healing Powers and Symbolism of Plants and Animals in the Middle Ages* (Munich: Prestel, 2001)

Tester, S.J., *A History of Western Astrology* (Woodbridge: Boydell Press, 1987)

Thijssen, J.M.M.H., *Censure and Heresy at the University of Paris, 1200–1400* (Philadelphia: University of Pennsylvania Press, 1998)

Thijssen, J.M.M.H., 'Master Amalric and the Amalricians: Inquisitorial Procedure and the Suppression of Heresy at the University of Paris', *Speculum* vol. 71, no. 1 (1996), pp. 43–65

Thijssen, J.M.M.H., and Jack Zupko, *The Metaphysics and Natural Philosophy of John Buridan* (Leiden: Brill Academic Publishers, 2000)

Thomas, Keith, *Religion and the Decline of Magic: Studies in Popular Beliefs in Sixteenth- and Seventeenth-Century England* (Harmondsworth: Penguin, 1973)

Thorndike, Lynn, *History of Magic and Experimental Science*, eight vols (New York: Columbia University Press, 1923)

Thorndike, Lynn, 'The Relations of the Inquisition to Peter Abano and Cecco d'Ascoli', *Speculum* vol. 1, no. 3 (1926), pp. 338–43

Thorndike, Lynn, *The Sphere of Sacrobosco and Its Commentators* (Chicago: University of Chicago Press, 1949)

Thorndike, Lynn, *University Records and Life in the Middle Ages* (New York: Columbia University Press, 1944)

Trinkaus, Charles, 'Marsilio Ficino', in *Contemporaries of Erasmus: A Biographical Register of the Renaissance and Reformation*, eds Peter G. Bietenholz and Thomas B. Deutscher (Toronto: University of Toronto Press, 1985)

Turner, Howard R., *Science in Medieval Islam: An Illustrated Introduction* (Austin: University of Texas Press, 1995)

Van Helden, Albert, 'The Invention of the Telescope', *Transactions of the American Philosophical Society* vol. 67, no. 4 (1977), pp. 5–67

Vesalius, Andreas, *On the Fabric of the Human Body*, trans. William Frank Richardson and John Burd Carman (Novato: Norman Publishing, 1998)

Voigts, Linda, 'Anglo-Saxon Plant Remedies and the Anglo-Saxons', in *The Scientific Enterprise in Antiquity and Middle Ages: Readings from Isis*, ed. Michael Shank (Chicago: Chicago University Press, 2000)

Voss, Angela, ed., *Marsilio Ficino* (Berkeley: North Atlantic Books, 2006)

Wakefield, Walter L., and Austin P. Evans, eds, *Heresies of the High Middle Ages: Selected Sources* (New York: Columbia University Press, 1969)

Walker, D.P., *Spiritual and Demonic Magic from Ficino to Campanella*, new edition (Stroud: Sutton, 2000)

Wallace, William, 'Domingo de Soto and the Iberian Roots of Galileo's Science', in *Hispanic Philosophy in the Age of Discovery*, ed. Keven White (Washington DC: The Catholic University of America Press, 1997)

Wallace, William, 'The Philosophical Setting of Medieval Science', in *Science in the Middle Ages*, ed. David C. Lindberg (Chicago: Chicago University Press, 1978)

Waters, W.G, *Jerome Cardan: A Biographical Study* (London: Lawrence & Bullen, 1898)

Weber, Max, *The Protestant Ethic and the Spirit of Capitalism*, trans. Stephen Kalberg, third edition (Oxford: Blackwell, 2002)

Webster, Charles, *Paracelsus: Medicine, Magic and Mission at the End of Time* (New Haven: Yale University Press, 2008)

Webster, Charles, 'William Harvey's Conception of the Heart as a Pump', *Bulletin of the History of Medicine* vol. 39 (1965), pp. 508–17

Westfall, Richard S., *Never at Rest: A Biography of Isaac Newton* (Cambridge: Cambridge University Press, 1983)

Westman, Robert S., 'Proof, Poetics and Patronage: Copernicus's Preface to De Revolutionibus', in *Reappraisals of the Scientific Revolution*, eds David C. Lindberg and Robert S. Westman (Cambridge: Cambridge University Press, 1990)

Westman, Robert S., 'The Copernicans and the Churches', in *God and Nature: Historical Essays on the Encounter Between Christianity and Science*, eds David C. Lindberg and Ronald L. Numbers (Berkeley: University of California Press, 1986)

Whewell, William, 'Mrs Somerville on the Connexion of the Sciences', *The Quarterly Review* vol. 51, no. 1 (1834), pp. 54–68

White, Andrew Dickson, *A History of the Warfare of Science With Theology in Christendom*, two vols (Amherst: Prometheus Books, 1993)

White, Lynn, *Medieval Religion and Technology: Collected Essays* (Berkeley: University of California Press, 1978).

White, Lynn, *Medieval Technology and Social Change* (Oxford: Oxford University Press, 1966)

White, Michael, *Leonardo: The First Scientist* (London: St. Martin's Griffin, 2001)

Whitehead, Alfred North, *Process and Reality* (New York: Free Press, 1979)

William of Malmesbury, *Gesta Regum Anglorum: The History of the English Kings*, trans. R.A.B. Mynors, two vols (Oxford: Clarendon Press, 1998)

Wolff, Michael, 'Philoponus and the Rise of Preclassical Dynamics', in *Philoponus: And the Rejection of Aristotelian Science*, ed. Richard Sorabji (London: Duckworth, 1987)

Wootton, David, *Bad Medicine: Doctors Doing Harm Since Hippocrates* (Oxford: Oxford University Press, 2006)

Worcester, Thomas, 'Introduction', in *The Cambridge Companion to the Jesuits*, ed. Thomas Worcester (Cambridge: Cambridge University Press, 2008)

Yates, Frances A., *Giordano Bruno and the Hermetic Tradition* (London: Routledge Classics, 2002)

Zambelli, Paola, *The Speculum Astronomiae and Its Enigma: Astrology, Theology, and Science in Albertus Magnus and His Contemporaries* (Dordrecht: Kluwer Academic, 1992)

ACKNOWLEDGEMENTS

The author would like to thank Christopher Barton, Bjørn-Are Davidsen, Paul Newall, Christopher Price, Brian Trafford, Richard Carrier and Nigel Brough for their valuable comments on earlier drafts. He would also like to thank his agent Andrew Lownie for believing in the book and Andrew's team of readers for their criticism and input. The staff of the London Library make working there a real pleasure. Finally, the team at Icon – Simon, Sarah, Andrew and Najma – have been fantastic throughout.

INDEX